Springer Series in Statistics

Springer Series in Statistics

D. F. Andrews and A. M. Herzberg, Data: A Collection of Problems from Many Fields for the Student and Research Worker. xx, 442 pages, 1985.

F. J. Anscombe, Computing in Statistical Science through APL. xvi, 426 pages, 1981.

J. O. Berger, Statistical Decision Theory: Foundations, Concepts, and Methods, 2nd edition. xiv, 425 pages, 1985.

P. Brémaud, Point Processes and Queues: Martingale Dynamics. xviii, 354 pages, 1981.

K. Dzhaparidze, Parameter Estimation and Hypothesis Testing in Spectral Analysis of Stationary Time Series. xii, 300 pages, 1985.

R. H. Farrell, Multivariate Calculation. xvi, 367 pages, 1985.

L. A. Goodman and W. H. Kruskal, Measures of Association for Cross Classifications. x, 146 pages, 1979.

J. A. Hartigan, Bayes Theory. xii, 145 pages, 1983.

H. Heyer, Theory of Statistical Experiments. x, 289 pages, 1982.

H. Kres, Statistical Tables for Multivariate Analysis. xxii, 504 pages, 1983.

M. R. Leadbetter, G. Lindgren and H. Rootzén, Extremes and Related Properties of Random Sequences and Processes. xii, 336 pages, 1983.

R. G. Miller, Jr., Simultaneous Statistical Inference, 2nd edition. xvi, 299 pages, 1981.

F. Mosteller, D. S. Wallace, Applied Bayesian and Classical Inference: The Case of The Federalist Papers. xxxv, 301 pages, 1984.

D. Pollard, Convergence of Stochastic Processes. xiv, 215 pages, 1984.

J. W. Pratt and J. D. Gibbons, Concepts of Nonparametric Theory. xvi, 462 pages, 1981.

L. Sachs, Applied Statistics: A Handbook of Techniques. xxviii, 706 pages, 1982.

E. Seneta, Non-Negative Matrices and Markov Chains. xv, 279 pages, 1981.

D. Siegmund, Sequential Analysis: Tests and Confidence Intervals. xii, 272 pages, 1985.

V. Vapnik, Estimation of Dependences based on Empirical Data. xvi, 399 pages, 1982.

K. M. Wolter, Introduction to Variance Estimation. xii, 428 pages, 1985.

Kirk M. Wolter

Introduction to Variance Estimation

With 16 Illustrations

Springer-Verlag
New York Berlin Heidelberg Tokyo

Kirk M. Wolter
U.S. Bureau of the Census
Washington, D.C. 20233
and The George Washington University
Washington, D.C. 20052
U.S.A.

AMS Classification: 62D05

Library of Congress Cataloging in Publication Data
Wolter, Kirk M.
   Introduction to variance estimation.
   (Springer series in statistics)
   Bibliography: p.
   Includes index.
   1. Analysis of variance. 2. Estimation theory.
I. Title. II. Series
QA279.W65 1985 519.5′352 85-4617

Typeset by J. W. Arrowsmith Ltd., Bristol, England.
Printed and bound by R. R. Donnelley and Sons, Harrisonburg, Virginia.
Printed in the United States of America.

9 8 7 6 5 4 3 2 1

ISBN 0-387-96119-4 Springer-Verlag New York Berlin Heidelberg Tokyo
ISBN 3-540-96119-4 Springer-Verlag Berlin Heidelberg New York Tokyo

Preface

I developed this book for statisticians who face the problem of variance estimation for large complex sample surveys. Many of the important variance estimating techniques have been relatively inaccessible to the general survey statistician. The existing literature on variance estimation has been available only in widely disparate places and usually in a highly theoretical form; heretofore there has been no single reference offering practical advice on the various variance estimating methodologies. By the late 1970s, when I first began working on the book, it was clear that a central reference text was needed in this area.

After preparing an early draft of the book, I gave a short course on variance estimation at the U.S. Bureau of the Census. This draft later formed the basis for another short course offered to statisticians in the Washington, D.C. area through the Washington Statistical Society. Beginning in the Fall of 1979 I used the emerging book in a one-semester, graduate level course on variance estimation at The George Washington University (GWU). The GWU classes were composed primarily of mathematical statisticians working at various agencies of the Federal Government and of graduate students in statistics. Prerequisites for the course were a first year graduate course in mathematical statistics and either a rigorous course in the theory and practice of sample surveys or the equivalent in terms of working experience. Although the background, interests, and needs of the students were varied, they shared a common interest in the application of the various variance estimating techniques to real survey data.

I improved the draft book considerably in the summer of 1980, and in August 1980 presented a short course based on this draft to a group of about 100 statisticians at the national meetings of the American Statistical Association in Houston, Texas. David W. Chapman and Joseph Sedransk

assisted me in presenting this course. By February 1983 I had made further
improvements and I presented a week-long course on variance estimation
at The Netherlands Central Bureau of Statistics in The Hague. I have
continued to offer the one-semester course at the GWU on an intermittent
basis.

The book is organized in a way that emphasizes both the theory and
applications of the various variance estimating techniques. Each technique
is presented in a separate chapter, and each chapter divided into several
main sections. The opening sections deal with the theory and motivation
for the particular method of variance estimation. Results are often presented
in the form of theorems; proofs are deleted when trivial or when a reference
is readily available. The latter sections of each chapter present numerical
examples where the particular technique was applied (and perhaps
modified) to a real survey. The objectives of this organizational format are
to provide the student with a firm technical understanding of the methods
of variance estimation; to stimulate further research on the various techniques, particularly as they apply to large, complex surveys; and to provide
an easy reference for the survey researcher who is faced with the problem
of estimating variances for real survey data.

The topics, in order of presentation, are the following:

1. Introduction
2. The Method of Random Groups
3. Variance Estimation Based on Balanced Half-Samples
4. The Jackknife Method
5. Generalized Variance Functions
6. Taylor Series Methods
7. Variance Estimation for Systematic Sampling
8. Summary of Methods for Complex Surveys
A. Hadamard Matrices
B. Asymptotic Theory of Variance Estimators
C. Transformations
D. The Effect of Measurement Errors on Variance Estimation
E. Computer Software for Variance Estimation

Chapters 2, 3, and 4 are closely related, each discussing a different member
of the general class of techniques that produce an estimator from each of
several "replicates," and estimate the variance by computing the variability
among the replicate estimates. Appendix A presents the orthogonal matrices,
known in mathematics as Hadamard matrices, that are useful in implementing the balanced half-sample method (Chapter 2). In many cases it is
important to use a transformation with the replicate-type methods, and this
is discussed in Appendix C.

Sometimes it is possible to model the variance as a function of certain
simple population parameters. Such models, which we shall call Generalized
Variance Functions (GVFs), are discussed in Chapter 5. Chapter 6 intro-

duces a method of variance estimation based on local linear approximation. The important topic of variance estimation for systematic samples is discussed in Chapter 7.

Appendix B provides the asymptotic underpinning for the replication methods and for the Taylor series method. The effects of measurement or response errors on variance calculations are discussed in Appendix D. And software for variance calculations is discussed in the closing portion of the book, Appendix E.

Since I began work on the book, the bootstrap method of variance estimation has emerged and garnered considerable attention, particularly among theoretical statisticians. This is a new and attractive method that may hold considerable promise for the future. Its utility for survey sampling problems is questionable, however, and as a consequence I have not included the bootstrap in the book at this stage. Work is now ongoing by a number of researchers to modify the basic bootstrap principles so that it can accommodate problems of finite population sampling. At this time I know of no successful applications of the bootstrap to complex survey data. But I intend to watch developments in this area carefully, and if the theory and applications are solved successfully, I'll plan to add a chapter on bootstrap methods to the next edition.

The inferential approach taken in the book is that of the randomization theory of survey sampling. Inferences derive mainly from the sampling distribution created by the survey design. I do not discuss variance estimation from the prediction-theory point of view nor from a Bayesian viewpoint. At times I employ superpopulation models, but only as a guide in choosing among alternative sampling strategies, never as a basis for the inference.

It is a pleasure to acknowledge Barbara Bailar for initial encouragement to develop the book and the subsequent courses based on the book. I thank Cary Isaki for contributing to Sections 7.6 to 7.9 and David W. Chapman for contributing to Sections 5.6 and 7.6–7.9. I am indebted to many people for providing data for the numerical examples, including W. Edwards Deming, Ben Tepping, Cathy Dippo, and Dwight Brock. I am grateful to Larry Cahoon for collaborating on the Current Population Survey (CPS) example in Chapter 5; to Dan Krewski for reading and commenting on Appendix B; to Phil Smith and Joe Sedransk for assistance in preparing Appendix E; to Colm O'Muircheartaigh and Paul Biemer for reading and commenting on Appendix D; and especially to Mary Mulry-Liggan for collaborating in the development of Appendix C and for general review of the manuscript. Lillian Principe typed the entire manuscript, with some assistance from Jeanne Ostenso, and I thank them for careful and diligent work.

Kirk M. Wolter

Contents

CHAPTER 1

Introduction

1.1. The Subject of Variance Estimation

The theory and applications of survey sampling have grown dramatically in the last 40 years. Hundreds of surveys are now carried out each year in the private sector, the academic community, and various governmental agencies, both in this country and abroad. Examples include market research and public opinion surveys; surveys associated with academic research studies; and large nationwide surveys about labor force participation, health care, energy usage, and economic activity. Survey samples now impinge upon almost every field of scientific study, including agriculture, demography, energy, transportation, health care, economics, politics, sociology, geology, forestry, and so on. Indeed, it is not an overstatement to say that much of the data undergoing any form of statistical analysis is collected in surveys.

As the number and uses of sample surveys have increased, so has the need for methods of analyzing and interpreting the resulting data. A basic requirement of nearly all forms of analysis, and indeed a principal requirement of good survey practice, is that a measure of precision be provided for each estimate derived from the survey data. The most commonly used measure of precision is the variance of the survey estimator. In general, variances are not known but must be estimated from the survey data itself. The problem of constructing such estimates of variance is the main problem treated in this book.

As a preliminary to any further discussion, it is important to recognize that the variance of a survey statistic is a function both of the form of the statistic and of the nature of the sampling design (i.e., the procedure used in selecting the sample). A common error of the survey practitioner or of the beginning student is the belief that simple random sampling formulae may be used to estimate variances, regardless of the design or estimator actually employed. This belief is false. An estimator of variance must take account of both the estimator and the sampling design.

Subsequent chapters in this book focus specifically on variance estimation methodologies for modern complex sample surveys. Although the terminology "modern complex sample survey" has never been rigorously defined, the following discussion may provide an adequate meaning for present purposes.

Important dimensions of a modern complex sample survey include

(i)   the degree of complexity of the sampling design;
(ii) the degree of complexity of the survey estimator(s);
(iii) multiple characteristics or variables of interest;
(iv) descriptive and analytical uses of the survey data;
(v)   the scale or size of the survey.

It is useful to discuss dimensions (i) and (ii) in the following terms:

	Simple design	Complex design
Linear estimators	a	b
Nonlinear estimators	c	d

Much of the basic theory of sample surveys deals with case a, while the modern complex survey often involves case b, c, or d. The complex survey often involves design features such as stratification, multiple stage sampling, unequal selection probabilities, double sampling, and multiple frames, and estimation features such as adjustments for nonresponse and undercoverage, large observation or outlier procedures, poststratification, and ratio or regression estimators. This situation may be distinguished from the basic survey which may involve only one or two of these estimation and design features. Regarding dimension (iii), most modern complex sample surveys involve tens or hundreds of characteristics of interest. This may be contrasted with the basic survey discussed in most existing textbooks, where only one characteristic or variable of interest is considered. Dimension (iv) captures the idea that many such surveys include both descriptive and analytical uses. In a simple survey the objective may amount to little more than describing various characteristics of the target population, such as the number of men or women that would vote for a certain candidate in a political election. The complex survey usually includes some descriptive objectives, but may also include analytical objectives where it is desired to build models and test hypotheses about the forces and relationships in the population. For example, instead of merely describing how many would vote for a certain political candidate, the survey goals may include study of how voter preference is related to income, years of education, race, religion, geographic region and other exogenous variables. Finally, the scale of the survey effort (dimension (v)) is important in classifying a survey as simple or complex. The complex survey usually involves hundreds, if not thousands, of individual respondents and a large field organization.

Of course, the distinction between simple and complex surveys is not clear-cut on any of the five dimensions and some surveys may be considered complex along certain dimensions but not on others.

In the context of these dimensions of the modern complex sample survey, how is one to choose an appropriate variance estimator? The choice is typically a difficult one, involving the *accuracy* of the variance estimator, *timeliness, cost, simplicity*, and other *administrative considerations*.

The *accuracy* of a variance estimator may be assessed in a number of different ways. One important measure is the mean square error (MSE) of the variance estimator. Given this criterion the estimator with minimum MSE is preferred. Since it is often the case that the variance estimates will be used to construct interval estimates for the main survey parameters, a second criterion of accuracy has to do with the quality of the resulting intervals. The variance estimator that provides the best interval estimates is preferred. Unfortunately, there may be a conflict between these criteria; it is frequently the case that the estimator of variance with minimum MSE provides poorer interval estimates than some other variance estimators with larger MSE. Finally, the survey specifications may include certain multivariate, time series, or other statistical analyses of the survey data. It would then be appropriate to prefer the variance estimator that has the best statistical properties for the proposed analysis. Of course, compromises will have to be made because different analyses of the same data may suggest different variance estimators.

In summary, the accuracy of alternative variance estimators may be assessed by any of the above criteria, and the planned uses and analyses of the survey data should guide the assessment.

Although accuracy issues should dominate decisions about variance estimators, *administrative considerations* such as *cost* and *timing* must also play an important role, particularly in the complex surveys with which this book is primarily concerned. The publication schedule for such surveys may include tens of tables, each with a hundred or more cells, or it may include estimates of regression coefficients, correlation coefficients, and the like. The cost of computing a highly accurate estimate of variance for each survey statistic may be very formidable indeed, far exceeding the cost of the basic survey tabulations, and quite possibly, the total survey budget. In such circumstances, methods of variance estimation that are cost effective may be highly desirable, even though they may involve a certain loss of accuracy. Timing is another important practical consideration, because modern complex surveys often have rather strict closeout dates and publication deadlines. The methods of variance estimation must be evaluated in light of such deadlines and the efficiency of the computer environment to be used in preparing the survey estimates.

A final issue, though perhaps subordinate to the accuracy, cost, and timing considerations, is the *simplicity* of the variance estimating methodology. Although this issue is closely interrelated with the previous considerations, there are three aspects of simplicity that require separate attention. First is the fact, observed earlier, that most modern complex sample surveys are multipurpose in character, meaning that there are many variables and statistics of interest, each of which requires an estimate of its corresponding variance. From the point of view of theoretical accuracy, this multitude of purposes may suggest a different preferred variance estimator for each of the survey statistics, or at least a different variance estimator for different classes of statistics. Such use of different variance estimators may be feasible in certain survey environments, where budget, professional staff, time, and computing resources are abundant. In many survey environments, however, these resources are scarce; this approach will not be feasible, and it will be necessary to use one, or at most a few, variance estimating methodologies. In this case compromises must be made, selecting a variance estimator that might not be optimal for any one statistic, but that involves a tolerable loss of accuracy for all, or at least the most important, survey statistics. The second aspect of simplicity involves the computer processing system used for the survey. As of this writing, the development of general purpose and portable software for the analysis of survey data, including variance estimation, is in its infancy. A few software packages are available, however, and they are discussed in Appendix E. When such packages are available, they are a boon to the processing of survey data, and there may be little concern with this aspect of the simplicity issue. When such packages are not available, however, special purpose computer programs must be written to process the data and estimate variances. In this case, one must give consideration to the abilities and skills of the computer programming staff. The specification of a variance estimating methodology must be commensurate with the staff's abilities to program that methodology. It may serve no purpose to specify an elaborate variance estimation scheme if the programming staff cannot devise the appropriate computer programs correctly. The third and final aspect of simplicity is concerned with the survey sponsor and users of the survey data. Often, the survey goals will be better served if simple estimation methods are used, ones that are readily understood by the survey sponsor and other users of the data. For statistically sophisticated sponsors and users, however, this should not be a concern.

Thus, the process of evaluating alternative variance estimators and selecting a specific estimator(s) for use in a particular application is a complicated one, involving both objective and subjective elements. In this process, the survey statistician must make intelligent trade-offs between the important, and often conflicting, considerations of accuracy, cost, timing, and simplicity.

1.2. The Scope and Organization of this Book

The main purpose of this book is to describe a number of the techniques for variance estimation that have been suggested in recent years, and to demonstrate how they may be used in the context of modern complex sample surveys. The various techniques to be described are widely scattered through the statistical literature; currently, there is no systematic treatment of this methodology that brings together the state of the art in one manuscript. One of the purposes of this book is to accomplish this objective.

Few fields of statistical study have such a variety of excellent texts as survey sampling. Examples include the very readable accounts by Cochran (1977), Deming (1950, 1960), Hansen, Hurwitz, and Madow (1953), Raj (1968), Sukhatme and Sukhatme (1970), and Yates (1949). Each of these texts discusses variance estimation for some of the basic estimators and sample designs. For convenience, we shall refer to these as the *standard* (or *textbook* or *customary*) variance estimators. Most of the text book discussions about the standard variance estimators emphasize unbiasedness and minimum mean square error. These discussions stop short of dealing with some of the important features of complex surveys, such as nonresponse, measurement errors, cost, and other operational issues.

In this book we consider certain *nonstandard* variance estimating techniques. As we shall see, these nonstandard estimators are not necessarily unbiased, but they are sufficiently flexible to accommodate most features of a complex survey. Except for a brief discussion in Section 1.4, we do not discuss the standard variance estimators, because they are adequately discussed elsewhere. In so doing we have tried to avoid duplication with the earlier sampling texts. The techniques we discuss overcome, to a large extent, some of the deficiencies in the standard estimators, such as the treatment of nonresponse, cost, and other operational issues.

Although the main area of application is the complex survey, part of the text is devoted to a description of the methods in the context of simple sampling designs and estimators. This approach is used to motivate the methods and to provide emphasis on the basic principles involved in applying the methods. It is important to emphasize the basic principles because, to some extent, each survey is different and it is nearly impossible in a moderately sized manuscript to describe appropriate variance estimating techniques for every conceivable survey situation.

Examples form an integral part of the effort to emphasize principles. Some are simple and used merely to acquaint the reader with the basics of a given technique. Others, however, are more elaborate, illustrating how the basic principles can be used to modify and adopt a variance estimating procedure to a complex problem.

Chapters 2, 3, and 4 describe three methods of variance estimation based on the concept of replication. The three methods—random groups, balanced half-samples, and jackknife—differ only in the way the replicates are formed. These chapters should be read in sequence, as each builds on concepts introduced in the preceding chapters.

The remaining chapters are largely self-contained and may be read in any suitable order. A minor exception, however, is that some of the examples used in later chapters draw on examples first introduced in Chapters 2, 3, or 4. To fully understand such examples, the reader would first need to study the background of the example in the earlier chapter.

The subject of Generalized Variance Functions (GVF) is introduced in Chapter 5. This method is applicable to surveys with an extraordinarily large publication schedule. The idea is simply to model an estimator's variance as a function of the estimator's expectation. To estimate the variance, one evaluates the function at the estimate; a separate variance computation is not required.

Taylor series (or linearization) methods are used in the basic sampling texts to obtain an estimator of variance for certain nonlinear estimators, e.g., the classical ratio estimator. In Chapter 6 a complete account of this methodology is given, showing how most nonlinear statistics can be linearized as a preliminary step in the process of variance estimation.

Chapter 7 discusses variance estimation for both equal and unequal probability systematic sampling. Although many of the estimators that are presented have been mentioned previously in the earlier sampling texts, little advice was given there about their usage. This chapter aims to provide some guidance about the tricky problem of variance estimation for this widely-used method of sampling.

A general summary of Chapters 2 through 6 is presented in Chapter 8. This chapter also makes some recommendations about the advantages, disadvantages, and appropriateness of the alternative variance estimation methodologies.

The book closes with five short appendices on special topics. Appendix A discusses the topic of Hadamard matrices, which are useful in implementing the ideas of balancing found in Chapter 3. Hadamard matrices are presented for all orders from 2 to 100. Appendix B discusses the asymptotic properties of the variance estimating methodologies presented in Chapters 2, 3, 4, and 6. Data transformations are discussed in Appendix C. This topic offers possibilities for improving the quality of survey-based interval estimates. Nonsampling errors are treated in Appendix D. Here, the notion of *total variance* is introduced, and the behavior of the variance estimators in the presence of measurement errors is discussed. The closing section, Appendix E, deals with computer software for variance estimation.

1.3. Notation and Basic Definitions

This section is devoted to some basic definitions and notation that shall be useful throughout the text. Many will find this material quite familiar. In any case, the reader is urged to look through this material before proceeding further because the basic framework (or foundations) of survey sampling is established herein. For a comprehensive treatment of this subject see Cassel, Särndal, and Wretman (1977).

1. We shall let $\mathcal{U} = \{1, \ldots, N\}$ denote a *finite population* of identifiable units, where $N < \infty$. N is the *size of the population*. In the case of multistage surveys, we shall use N to denote the number of primary units, and other symbols, such as M, to denote the number of second and successive stage units.

2. There are two definitions of the term *sample* that we shall find useful.
   a. A sample s is a finite sequence $\{i_1, i_2, \ldots, i_{n(s)}\}$ such that $i_j \in \mathcal{U}$ for $j = 1, 2, \ldots, n(s)$. In this case, the selected units are not necessarily distinct.
   b. A sample s is a nonempty subset of $\mathcal{U}$. In this case, the selected units are necessarily distinct.

3. In either definition of a sample s, we use $n(s)$ to denote the *sample size*. Many common sampling designs have a fixed sample size that does not vary from sample to sample, in which case we shall use the shorthand notation n.

4. For a given definition of the term sample, we shall let $\mathcal{S}$ denote the collection of all possible samples from $\mathcal{U}$.

5. A *probability measure* $\mathcal{P}$ is a nonnegative function defined over $\mathcal{S}$ such that the function values add to unity, i.e.

$$\mathcal{P}\{s\} \geq 0 \quad \text{and} \quad \sum_{s \in \mathcal{S}} \mathcal{P}\{s\} = 1.$$

Let S be the random variable taking values $s \in \mathcal{S}$.

6. We shall call the pair $(\mathcal{S}, \mathcal{P})$ a *sampling design*. It should be observed that it makes little conceptual difference whether we let $\mathcal{S}$ include all possible samples, or merely those samples with positive probability of being selected, i.e., $\mathcal{P}\{s\} > 0$.

7. For a given sampling design, the first order *inclusion probability* π_i is the probability of drawing the i-th unit into the sample

$$\pi_i = \sum_{s \supset i} \mathcal{P}\{s\},$$

for $i = 1, \ldots, N$, where $\sum_{s \supset i}$ stands for summation over all samples s that contain the i-th unit. The second order inclusion probability π_{ij} is the probability of drawing both the i-th and j-th units into the sample

$$\pi_{ij} = \sum_{s \supset i,j} \mathcal{P}\{s\},$$

for $i, j = 1, \ldots, N$.

8. Attached to each unit i in the population is the value Y_i of some *characteristic* of interest. Sometimes we may be interested in more than one characteristic, then denoting the values by $Y_{i1}, Y_{i2}, \ldots$, or by Y_i, $X_i, \ldots$.

9. A sampling design $(\mathcal{S}, \mathcal{P})$ is called *noninformative* if and only if the measure $\mathcal{P}\{\cdot\}$ does not depend on the values of the characteristics of interest. In this book we only consider noninformative designs.

10. As much as is feasible, we shall use upper case letters to indicate the values of units in the population and lower case letters to indicate the values of units in the sample. Thus, for example, we may write the sample mean based on a sample s as

$$\bar{y} = \sum_{i \in s} Y_i / n(s),$$

$$\bar{y} = \sum_{i=1}^{n(s)} y_i / n(s),$$

or

$$\bar{y} = \sum_{i=1}^{N} \alpha_i Y_i / n(s),$$

where

$$\alpha_i = \begin{cases} \text{number of times } i \text{ occurs in } s, & \text{if } i \in s, \\ 0, & \text{if } i \notin s. \end{cases}$$

11. In the case of a single characteristic of interest, we call the vector $\mathbf{Y} = (Y_1, \ldots, Y_N)$ the *population parameter*. In the case of multiple (r) characteristics, we let $\mathbf{Y}_i$ be a $(1 \times r)$ row vector composed of the values associated with unit i, and the matrix

$$\mathbf{Y} = \begin{pmatrix} \mathbf{Y}_1 \\ \vdots \\ \mathbf{Y}_N \end{pmatrix} \cdot$$

is the population parameter. We denote the parameter space by Ω. Usually, Ω is the N-dimensional Euclidean space in the single characteristic case (Nr-dimensional Euclidean space in the multiple characteristic case), or some subspace thereof.

12. Any real function on Ω is called a *parameter*. We shall often use the letter θ to denote an arbitrary parameter to be estimated. In the case of certain widely used parameters we may use special notation, such as

$$Y = \sum_{i=1}^{N} Y_i \qquad \qquad \text{population total}$$

$$R = Y/X \qquad \qquad \text{ratio of population totals}$$

$$D = Y/X - W/Z \qquad\qquad \text{difference between ratios}$$

$$\beta = \frac{\displaystyle\sum_{i=1}^{N} (Y_i - \bar{Y})(X_i' - \bar{X})}{\displaystyle\sum_{i=1}^{N} (X_i - \bar{X})^2} \qquad\qquad \text{regression coefficient}$$

$$\bar{Y} = Y/N \qquad\qquad \text{population mean per element.}$$

13. An *estimator* of θ will usually be denoted by $\hat{\theta}$. An estimator $\hat{\theta} = \theta(S, \mathbf{Y})$ is a real-valued statistic, thought to be good for estimating θ, such that for any given $s \in \mathcal{S}$, $\theta(s, \mathbf{Y})$ depends on $\mathbf{Y}$ only through those Y_i for which $i \in s$. In the case of the special parameters, we may use the notation

$$\hat{Y}$$
$$\hat{R}$$
$$\hat{D}$$
$$\hat{\beta}$$

and

$$\bar{y}.$$

We shall often adjoin subscripts to these symbols to indicate specific estimators.

14. The *expectation* and *variance* of $\hat{\theta}$ with respect to the design $(\mathcal{S}, \mathcal{P})$ shall be denoted by

$$\mathrm{E}\{\hat{\theta}\} = \sum_{s} \mathcal{P}\{s\}\theta(s, \mathbf{Y})$$

$$\mathrm{Var}\{\hat{\theta}\} = \sum_{s} \mathcal{P}\{s\}[\theta(s, \mathbf{Y}) - \mathrm{E}\{\hat{\theta}\}]^2.$$

15. In this book we shall be concerned almost exclusively with the estimation of the design-variance $\mathrm{Var}\{\hat{\theta}\}$. There are at least two other concepts of variability that arise in the context of survey sampling:

(i) In the prediction theory approach to survey sampling, it is assumed that the population parameter $\mathbf{Y}$ is itself a random variable with some distribution function $\xi(\cdot)$. Inferences about θ are based on the ξ-distribution rather than on the $\mathcal{P}$-distribution. In this approach, concern centers around the estimation of the ξ-variance

$$\mathcal{V}ar\{\hat{\theta} - \theta\} = \int [(\hat{\theta} - \theta) - \mathcal{E}\{\hat{\theta} - \theta\}]^2 \, d\xi,$$

where

$$\mathcal{E}\{\hat{\theta} - \theta\} := \int (\hat{\theta} - \theta) \, d\xi$$

is the ξ-expectation. The problem of estimating $\mathcal{V}a\imath\{\hat{\theta} - \theta\}$ is not treated in this book. For more information about ξ-variances, see Cassel, Särndal, and Wretman (1977) and Royall and Cumberland (1978, 1981a, 1981b).

(ii) In the study of measurement (or response) errors, it is assumed that the data

$$\{Y_i: i \in s\}$$

is unobservable, but rather

$$\{Y_i^0 = Y_i + e_i: i \in s\}$$

is observed, where e_i denotes an error of measurement. Such errors are particularly common in social and economic surveys, e.g., Y_i is the true income and Y_i^0 is the observed income of the i-th unit. In this context, the total variability of an estimator $\hat{\theta}$ arises from the sampling design, from the distribution of the errors e_i, and from any interaction between the design and error distributions. The problems of estimating total variability are treated briefly in Appendix D.

Henceforth, for convenience, the term "variance" shall refer strictly to "design-variance" unless otherwise indicated.

16. An *estimator of variance*, i.e., an estimator of Var$\{\hat{\theta}\}$, will usually be denoted by $v(\hat{\theta})$. We shall adjoint subscripts to indicate specific estimators.

17. Often, particularly in Chapters 2, 3, and 4, we shall be interested in estimation based on k subsamples (or replicates or pseudoreplicates) of the full samples. In such cases, we shall let $\hat{\theta}$ denote the estimator based on the full sample and $\hat{\theta}_\alpha$ the estimator, of the same functional form as $\hat{\theta}$, based on the α-th subsample, for $\alpha = 1, \ldots, k$. We shall use $\bar{\hat{\theta}}$ to denote the mean of the $\hat{\theta}_\alpha$, i.e.

$$\bar{\hat{\theta}} = \sum_{\alpha=1}^{k} \hat{\theta}_\alpha / k.$$

18. We shall often speak of the *Horvitz–Thompson* (1952) estimator of a population total Y. For an arbitrary sampling design with $\pi_i > 0$ for $i = 1, \ldots, N$, this estimator is

$$\hat{Y} = \sum_{i \in d(S)} Y_i / \pi_i,$$

where $\sum_{i \in d(S)}$ denotes a summation over the distinct units in S.

19. When speaking of unequal probability sampling designs, we shall use p_i to denote the *per draw selection probability* of the i-th unit ($i = 1, \ldots, N$). That is, in a sample of size one, p_i is the probability of drawing the i-th unit. If X is an auxiliary variable which is available

for all units in the population, then we may define

$$p_i = X_i / X,$$

where X is the population total of the X-variable.

20. Many common sampling designs will be discussed repeatedly throughout the text. To facilitate the presentation we shall employ the following abbreviations:

Sampling Design	Abbreviation
simple random sampling without replacement	srs wor
simple random sampling with replacement	srs wr
probability proportional to size sampling with replacement	pps wr
single-start, equal probability, systematic sample	sys

21. An unequal probability without replacement sampling design with $\pi_i = np_i$ and fixed sample size $n(s) = n$ shall be called a πps *sampling design*. Such designs arise frequently in practice; we shall discuss them further at various points in latter chapters.

1.4. Standard Sampling Designs and Estimators

Although it is not our intention to repeat in detail the standard theory and methods of variance estimation, it will be useful to review briefly some of this work. Such a review will serve to clarify the standard variance estimating formulae and to motivate the methods to be discussed in subsequent chapters.

We discuss nine basic sampling designs and associated estimators in the following paragraphs. All are discussed in the context of estimating a population total Y. These are basic sampling designs and estimators; they are commonly used in practice and form the basis for more complicated designs, also used in practice. The estimators are unbiased estimators of the total Y. The variance of each is given along with the standard unbiased estimator of variance. A reference is also given in case the reader desires a complete development of the theory.

1. *Design:*   srs wor of size n

     Estimator: $\hat{Y} = f^{-1} \sum\limits_{i=1}^{n} y_i$

                 $f = n/N$

Variance: $\mathrm{Var}\{\hat{Y}\} \doteq N^2(1-f)S^2/n$

$$S^2 = \sum_{i=1}^{N} (Y_i - \bar{Y})^2/(N-1)$$

$$\bar{Y} = \sum_{i=1}^{N} Y_i/N$$

Variance Estimator: $v(\hat{Y}) = N^2(1-f)s^2/n$

$$s^2 = \sum_{i=1}^{n} (y_i - \bar{y})^2/(n-1)$$

$$\bar{y} = \sum_{i=1}^{n} y_i/n$$

Reference: Cochran (1977), pp. 21–27.

2. *Design*:   srs wr of size n

Estimator: $\hat{Y} = N \sum_{i=1}^{n} y_i/n$

Variance: $\mathrm{Var}\{\hat{Y}\} = N^2\sigma^2/n$

$$\sigma^2 = \sum_{i=1}^{N} (Y_i - \bar{Y})^2/N$$

Variance Estimator: $v(\hat{Y}) = N^2 s^2/n$

$$s^2 = \sum_{i=1}^{n} (y_i - \bar{y})^2/(n-1)$$

Reference: Cochran (1977), pp. 29–30.

3. *Design*:   pps wr of size n

Estimator: $\hat{Y} = \sum_{i=1}^{n} y_i/np_i$

Variance: $\mathrm{Var}\{\hat{Y}\} = \sum_{i=1}^{N} p_i(Z_i - Y)^2/n$

$$Z_i = Y_i/p_i$$

Variance Estimator: $v(\hat{Y}) = \sum_{i=1}^{n} (z_i - \hat{Y})^2/n(n-1)$

Reference: Cochran (1977), pp. 252–255.

4. *Design*: πps of size n

Estimator: $\hat{Y} = \sum_{i=1}^{n} y_i/\pi_i$

(Horvitz–Thompson Estimator)

Variance: $\mathrm{Var}\{\hat{Y}\} = \sum\limits_{i=1}^{N} \sum\limits_{j>1}^{N} (\pi_i\pi_j - \pi_{ij})(Y_i/\pi_i - Y_j/\pi_j)^2$

Variance Estimator: $v(\hat{Y}) = \sum\limits_{i=1}^{n} \sum\limits_{j>1}^{n} [(\pi_i\pi_j - \pi_{ij})/\pi_{ij}](y_i/\pi_i - y_j/\pi_j)^2$

(Yates–Grundy Estimator)

Reference: Cochran (1977), pp. 259–261.

A two-stage sampling design is used in paragraphs 5, 6, and 7. In all cases N denotes the number of primary sampling units in the population, and M_i denotes the number of elementary units within the i-th primary unit. The symbols n and m_i denote the first and second stage sample sizes, respectively, and Y_{ij} denotes the value of the j-th elementary unit within the i-th primary unit. The population total is now

$$Y_{..} = \sum_{i=1}^{N} \sum_{j=1}^{M_i} Y_{ij}.$$

5. *Design:*   srs wor at both the first and second stage of sampling

Estimator: $\hat{Y}_{..} = (N/n) \sum\limits_{i=1}^{n} M_i\bar{y}_{i\cdot}$

$$\bar{y}_{i\cdot} = \sum_{j=1}^{m_i} y_{ij}/m_i$$

Variance:

$$\mathrm{Var}\{\hat{Y}_{..}\} = N^2(1-f_1)(1/n) \sum_{i=1}^{N} (Y_{i\cdot} - Y_{..}/N)^2/(N-1)$$

$$+ (N/n) \sum_{i=1}^{N} M_i^2(1-f_{2i})S_i^2/m_i$$

$$Y_{i\cdot} = \sum_{j=1}^{M_i} Y_{ij}$$

$$\bar{Y}_{i\cdot} = Y_{i\cdot}/M_i$$

$$S_i^2 = \sum_{j=1}^{M_i} (Y_{ij} - \bar{Y}_{i\cdot})^2/(M_i-1)$$

$$f_1 = n/N$$

$$f_{2i} = m_i/M_i$$

Variance Estimator:

$$v(\hat{Y}..) = N^2(1 - f_1)(1/n) \sum_{i=1}^{n} (M_i\bar{y}_{i\cdot} - \hat{Y}../N)^2/(n-1)$$

$$+ (N/n) \sum_{i=1}^{n} M_i^2(1 - f_{2i})s_i^2/m_i$$

$$s_i^2 = \sum_{j=1}^{n} (y_{ij} - \bar{y}_{i\cdot})^2/(m_i - 1)$$

Reference: Cochran (1977), pp. 300–303.

6. *Design*:   pps wr at the first stage of sampling; srs wor at the second stage

Estimator: $\hat{Y}.. = \sum_{i=1}^{n} M_i\bar{y}_{i\cdot}/np_i$

Variance:

$$\text{Var}\{\hat{Y}..\} = \sum_{i=1}^{n} p_i(Y_{i\cdot}/p_i - Y..)^2/n$$

$$+ \sum_{i=1}^{N} (1/np_i)M_i^2(1 - f_{2i})S_i^2/m_i$$

Variance Estimator:

$$v(\hat{Y}..) = \sum_{i=1}^{n} (M_i\bar{y}_{i\cdot}/p_i - \hat{Y}..)^2/n(n-1)$$

Reference: Cochran (1977), pp. 306–308.

7. *Design*: πps at the first stage of sampling; srs wor at the second stage

Estimator: $\hat{Y}.. = \sum_{i=1}^{n} M_i\bar{y}_{i\cdot}/\pi_i$

$\pi_i = np_i$ is the probability that the
i-th primary unit is selected

Variance:

$$\text{Var}\{\hat{Y}..\} = \sum_{i=1}^{N} \sum_{j>i}^{N} (\pi_i\pi_j - \pi_{ij})(Y_{i\cdot}/\pi_i - Y_{j\cdot}/\pi_j)^2$$

$$+ \sum_{i=1}^{N} (1/\pi_i)M_i^2(1 - f_{2i})S_i^2/m_i$$

π_{ij} is the joint probability that the *i*-th and *j*-th primary units are selected.

Variance Estimator:

$$v(\hat{Y}..) = \sum_{i=1}^{n} \sum_{j>i}^{n} [(\pi_i \pi_j - \pi_{ij})/\pi_{ij}](M_i \bar{y}_{i.}/\pi_i - M_j \bar{y}_{j.}/\pi_j)^2$$

$$+ \sum_{i=1}^{n} (1/\pi_i) M_i^2 (1 - f_{2i}) s_i^2 / m_i$$

Reference: Cochran (1977), pp. 308–310.

Any of the above sampling designs may be used within strata in a stratified sampling design.

8. *Design*: L strata; sample size n_h in the h-th stratum ($h = 1, \ldots, L$); the sampling design within the strata is one of those described in paragraphs 1, 2, ..., 7.

Estimator: $\hat{Y} = \sum_{h=1}^{L} \hat{Y}_h$

$\hat{Y}_h$ = estimator of the total in the h-th stratum; corresponds to the given within stratum sampling design.

Variance: $\text{Var}\{\hat{Y}\} = \sum_{h=1}^{L} \text{Var}\{\hat{Y}_h\}$

$\text{Var}\{\hat{Y}_h\}$ corresponds to the given estimator and sampling design; see paragraphs 1, 2, ... or 7.

Variance Estimator: $v(\hat{Y}) = \sum_{h=1}^{L} v(\hat{Y}_h)$

$v(\hat{Y}_h)$ corresponds to the given estimator and sampling design; see paragraphs 1, 2, ... or 7.

Reference: Cochran (1977), pp. 91–96 and Raj (1968), pp. 61–64.

Finally, we illustrate the concept of double sampling designs.

9. *Design*:   The first sample is a srs wor of size n'; this sample is classified into L strata; the second sample is a stratified random subsample of size n; the subsample size within the h-th stratum is $n_h = \nu_h n'_h$, where $0 < \nu_h \leq 1$, ν_h is specified in advance of sampling, and n'_h is the number of units from the first sample that were classified in the h-th stratum.

Estimator: $\hat{Y} = N \sum_{h=1}^{L} w_h \bar{y}_h$

$w_h = n'_h / n'$

$\bar{y}_h$ is the sample mean of the simple random subsample from the h-th stratum.

Variance: $\text{Var}\{\hat{Y}\} = N^2(1 - f')S^2/n' + N^2 \sum_{h=1}^{L} (W_h S_h^2/n')(1/v_h - 1)$

$f = n'/N$

$S^2 = \sum_{h=1}^{L} \sum_{i=1}^{N_h} (Y_{hi} - \bar{Y})^2/(N - 1)$

$\bar{Y} = \sum_{h=1}^{L} \sum_{i=1}^{N_h} Y_{hi}/N$

$\bar{Y}_h = \sum_{i=1}^{N_h} Y_{hi}/N_h$

N_h is the size of the population in the h-th stratum.

$S_h^2 = \sum_{i=1}^{N_h} (Y_{hi} - \bar{Y}_h)^2/(N_h - 1)$

Variance Estimator:

$$v(\hat{Y}) = N^2[n'(N - 1)/(n' - 1)N]\left[\sum_{h=1}^{L} w_h s_h^2(1/n'v_h - 1/N) \right.$$

$$+ [(N - n')/n'(N - 1)] \sum_{h=1}^{L} s_h^2(w_h/N - 1/n'v_h)$$

$$+ \left. [(N - n')/n'(N - 1)] \sum_{h=1}^{L} w_h(\bar{y}_h - \hat{Y}/N)^2 \right]$$

$s_h^2 = \sum_{i=1}^{n_h} (y_{hi} - \bar{y}_h)^2/(n_h - 1)$

Reference: Cochran (1977), pp. 327–335.

In what follows we refer frequently to the standard or textbook or customary estimators of variance. It will be understood, unless specified otherwise, that these are the estimators of variance discussed here in paragraphs 1–9.

1.5. Linear Estimators

Linear estimators play a central role in survey sampling, and we shall often discuss special results about estimators of their variance. Indeed, we shall often provide motivation for variance estimators by discussing them in the context of estimating the variance of a linear estimator, even though their real utility may be in the context of estimating the variance of a nonlinear estimator. Furthermore, in Chapter 6 we shall show that the problem of estimating the variance of a nonlinear estimator may be tackled by estimating the variance of an appropriate linear approximation.

There is little question about the meaning of the term *linear estimator* when dealing in the context of random samples from an infinite population. In finite population sampling, however, there are several meanings which may be ascribed to this term.

Horvitz and Thompson (1952) devised three classes of linear estimators for without replacement sampling designs:

1.
$$\hat{\theta} = \sum_{i \in S} \beta_i Y_i,$$

where $\sum_{i \in S}$ is a summation over the units in the sample S and β_i is defined in advance of the survey, for $i = 1, \ldots, N$, and is associated with the i-th unit in the population.

2.
$$\hat{\theta} = \sum_{i=1}^{n(S)} \beta_i y_i,$$

where β_i is defined in advance of the survey, for $i = 1, \ldots, n(S)$, and is associated with the unit selected at the i-th draw.

3.
$$\hat{\theta} = \beta_S \sum_{i \in S} Y_i,$$

where β_S is defined in advance of the survey for all possible samples S.

Certain linear estimators are members of all three classes, such as the sample mean for srs wor. Other estimators belong to only one or two of these classes. If we use srs wor within each of $L \geq 2$ strata, then the usual estimator of the population total

$$\hat{Y} = \sum_{h=1}^{L} (N_h/n_h) \sum_{i \in S(h)} Y_i,$$

where $S(h)$ denotes the sample from the h-th stratum and N_h and n_h denote the sizes of the population and sample respectively in that stratum, belongs only to classes 1 and 2, unless proportional allocation is used. An example of a class 3 estimator is the classical ratio estimator

$$\hat{Y}_R = \sum_{i \in S} Y_i \sum_{i=1}^{N} X_i / \sum_{i \in S} X_i, \tag{1.00}$$

where X is an auxiliary variable and

$$\beta_S = \sum_{i=1}^{N} X_i / \sum_{i \in S} X_i$$

is assumed known in advance of the survey.

Some linear estimators do not fit into any of Horvitz and Thompson's classes, most notably those associated with replacement sampling designs. An easy example is the estimator of the population total

$$\hat{Y}_{\text{pps}} = (1/n) \sum_{i=1}^{n} y_i / p_i$$

for pps wr sampling, where $\{p_i\}_{i=1}^N$ is the sequence of selection prob-abilities and $n(S) = n$ is the sample size. To include this estimator and other possibilities, Godambe (1955) suggested that the most general linear estimator may be written as

$$\hat{\theta} = \sum_{i \in S} \beta_{Si} Y_i, \qquad (1.01)$$

where the β_{si} are defined in advance of the survey for all samples $s \in \mathscr{S}$ and for all units $i \in s$. Cassel, Särndal, and Wretman (1977) define linear estimators to be of the form

$$\hat{\theta} = \beta_{S0} + \sum_{i \in S} \beta_{Si} Y_i. \qquad (1.02)$$

They call estimators of the form (1.01) linear homogeneous estimators.

If multiple characteristics are involved in the estimator, then even (1.02) does not exhaust the supply of linear estimators. For example, for srs wor we wish to regard the difference estimator

$$\bar{y}_d = \sum_{i \in S} Y_i/n + \beta \left(\sum_{i=1}^N X_i/N - \sum_{i \in S} X_i/n \right)$$

(β a known constant) as a linear estimator, yet it does not fit the form of (1.02). In view of these considerations, we shall use the following definition in this book:

Definition 1.1. Let $\hat{\theta}(1), \ldots, \hat{\theta}(p)$ denote p statistics in the form of (1.02), possibly based on different characteristics, and let $\{\gamma_j\}_{j=0}^p$ denote a sequence of fixed real numbers. An estimator $\hat{\theta}$ is said to be a *linear estimator* if it is expressible as

$$\hat{\theta} = \gamma_0 + \gamma_1 \hat{\theta}(1) + \ldots + \gamma_p \hat{\theta}(p).$$

This definition is sufficiently general for our purpose. Sometimes, however, it is more general than we require, in which case we shall consider restricted classes of linear estimators. For example, we may eliminate the ratio estimator (1.00) from certain discussions because it lacks certain properties possessed by other estimators we wish to consider.

CHAPTER 2

The Method of Random Groups

2.1. Introduction

The random group method of variance estimation consists in selecting two or more samples from the population, usually using the same sampling design for each sample; constructing a separate estimate of the population parameter of interest from each sample and an estimate from the combination of all samples; and computing the sample variance among the several estimates. Historically, this was one of the first techniques developed to simplify variance estimation for complex sample surveys. It was introduced in jute acreage surveys in Bengal by Mahalanobis (1939, 1946), who called the various samples *interpenetrating samples*. Deming (1956), the United Nations Subcommission on Statistical Sampling (1949), and others proposed the alternative term *replicated samples*. Hansen, Hurwitz, and Madow (1953) referred to the *ultimate cluster* technique in multistage surveys, and to the *random group* method in general survey applications. All of these terms have been used in the literature by various authors and all refer to the same basic method. Although the same basic method is involved, the various terminologies arose originally from somewhat different applications, e.g., interpenetrating samples in the case of the measurement of correlated response errors and ultimate cluster in the case of multistage sampling.

We will employ the term *random group* when referring to this general method of variance estimation.

There are two fundamental variations of the random group method. The first case is where the samples or random groups are mutually independent, while the second case arises when there is a dependency of some kind between the random groups. The case of independent samples is treated in Section 2.2. Although this variation of the method is not frequently

employed in practice, there is exact theory regarding the properties of the
estimators whenever the samples are independent. In practical applications,
the samples are often dependent in some sense, and exact statistical theory
for the estimators is generally not available. This case is discussed beginning
in Section 2.3.

2.2. The Case of Independent Random Groups

Mutual independence of the various samples (or more properly, of
estimators derived from the samples) arises when one sample is replaced
into the population before selecting the next sample. To describe this in
some detail, we assume there exists a well-defined finite population and
that it is desired to estimate some parameter θ of the population. We place
no restrictions on the form of the parameter θ. It may be a linear statistic
such as a population mean or total, or nonlinear such as a ratio of totals
or a regression or correlation coefficient.

The overall sampling scheme that is required may be described as follows:

(i) A sample, s_1, of units is drawn from the finite population according
    to a well-defined sampling design $(\mathcal{S}, \mathcal{P})$. No restrictions are placed
    on the design: it may involve multiple frames, multiple stages, fixed
    or random sample size, double sampling, stratification, equal or
    unequal selection probabilities, or with or without replacement selec-
    tion, but the design is not restricted to these features.

(ii) Following the selection of the first sample, s_1 is replaced into the
     population, and a second sample, s_2, is drawn according to the same
     sampling design $(\mathcal{S}, \mathcal{P})$.

(iii) This process is repeated until $k \geq 2$ samples, $s_1, \ldots, s_k$, are obtained,
      each being selected according to the common sampling design and
      each being replaced before the selection of the next sample. We shall
      call these k samples *random groups*.

In most applications of the random group method, there is a common
estimation procedure and a common *measurement process* that is applied to
each of the k random groups. Here, the terminology *estimation procedure*
is used in a broad sense to include all of the steps in the processing of the
survey data that occur after that data has been put in machine readable
form. Included are such steps as the editing procedures, adjustments for
nonresponse and other missing data, large observation or outlier procedures,
and the computation of the estimates themselves. In applying the random
group method there are no restrictions on any of these steps beyond those
of good survey practice. The terminology *measurement process* is used in
an equally broad sense to include all of the steps in the conduct of the
survey up to and including the conversion of the survey responses to machine

readable form. This includes all of the field work (whether it be by mail, telephone, or personal visit), call backs for nonresponse, clerical screening and coding of the responses, and transcription of the data to machine readable form. There are no restrictions on any of these steps either, with one possible exception (see Appendix D).[1]

Application of the common measurement process and estimation procedure results in k estimators of θ, which we shall denote by $\hat{\theta}_\alpha$, $\alpha = 1, \ldots, k$. Then, the main theorem that describes the random group estimator of variance may be stated as follows:

Theorem 2.2.1. *Let $\hat{\theta}_1, \ldots, \hat{\theta}_k$ be uncorrelated random variables with common expectation* $E\{\hat{\theta}_1\} = \mu$. *Let $\hat{\bar{\theta}}$ be defined by*

$$\hat{\bar{\theta}} = \sum_{\alpha=1}^{k} \hat{\theta}_\alpha / k.$$

Then $E\{\hat{\bar{\theta}}\} = \mu$ *and an unbiased estimator of* $\mathrm{Var}\{\hat{\bar{\theta}}\}$ *is given by*

$$v(\hat{\bar{\theta}}) = \sum_{\alpha=1}^{k} (\hat{\theta}_\alpha - \hat{\bar{\theta}})^2 / k(k-1). \tag{2.2.1}$$

PROOF. It is obvious that $E\{\hat{\bar{\theta}}\} = \mu$. The variance estimator $v(\hat{\bar{\theta}})$ may be written as

$$v(\hat{\bar{\theta}}) = \left[\sum_{\alpha=1}^{k} \hat{\theta}_\alpha^2 - k\hat{\bar{\theta}}^2 \right] / k(k-1).$$

Since the $\hat{\theta}_\alpha$ are uncorrelated we have

$$E\{v(\hat{\bar{\theta}})\} = \left[\sum_{\alpha=1}^{k} (\mathrm{Var}\{\hat{\theta}_\alpha\} + \mu^2) - k(\mathrm{Var}\{\hat{\bar{\theta}}\} + \mu^2) \right] \Big/ k(k-1)$$

$$= (k^2 - k) \, \mathrm{Var}\{\hat{\bar{\theta}}\} / k(k-1)$$

$$= \mathrm{Var}\{\hat{\bar{\theta}}\}. \qquad \square$$

The statistic $\hat{\bar{\theta}}$ may be used as the overall estimator of θ while $v(\hat{\bar{\theta}})$ is the *random group estimator* of its variance $\mathrm{Var}\{\hat{\bar{\theta}}\}$.

---

[1] In the study of measurement (or response) errors, it is assumed that the characteristic of interest Y cannot be observed. Rather, Y^0 is observed, where Y^0 is the characteristic Y plus an additive error e. If care is not taken, correlations between the various random groups can occur because of correlations between the errors associated with selected units in different random groups. An important example is where the errors are introduced by interviewers, and an interviewer's assignment covers selected units from two or more random groups. To avoid such correlation, interviewer assignments should be arranged entirely within a single random group. Error might also be introduced by other clerical operations, such as in coding survey responses on occupation, in which case clerical work assignments should be nested within random groups.

In many survey applications, the parameter of interest θ is the same as the expectation μ,

$$E\{\hat{\bar{\theta}}\} = \mu = \theta, \qquad (2.2.1a)$$

or at least approximately so. In survey sampling it has been traditional to employ design unbiased estimators, and this practice tends to guarantee (2.2.1a), at least in cases where $\hat{\theta}_\alpha(\alpha = 1, \ldots, k)$ is a linear estimator and where θ is a linear function of the population parameter $\mathbf{Y}$. When θ and $\hat{\theta}_\alpha$ are nonlinear, then μ and θ may be unequal because of a small technical bias arising from the nonlinear form.[2]

It is interesting to observe that Theorem 2.2.1 does not require that the variances of the random variables $\hat{\theta}_\alpha$ be equal. Thus, the samples (or random groups) s_α could be generated by different sampling designs and the estimators $\hat{\theta}_\alpha$ could have different functional forms, yet the theorem would remain valid so long as the $\hat{\theta}_\alpha$ are uncorrelated with common expectation μ. In spite of this additional generality of Theorem 2.2.1, we will continue to assume the samples s_α are each generated from a common sampling design and the estimators $\hat{\theta}_\alpha$ from a common measurement process and estimation procedure.

The assumption of uncorrelated random variables also has important implications. While Theorem 2.2.1 only requires that the random variables $\hat{\theta}_\alpha$ be uncorrelated, the procedure of replacing sample $s_{\alpha-1}$ prior to selecting sample s_α tends to induce independence among the $\hat{\theta}_\alpha$. Thus, the method of sampling described by (i)–(iii) seems to more than satisfy the requirements of the theorem. However, the presence of measurement errors, as noted earlier, can introduce a correlation between the $\hat{\theta}_\alpha$ unless different sets of interviewers and different processing facilities are employed in the various samples. Certain nonresponse and poststratification adjustments may also introduce a correlation between the $\hat{\theta}_\alpha$ if they are not applied individually within each sample. This topic is discussed further in Section 2.7.

Inferences about the parameter θ are usually based on normal theory or on Student's t theory. The central mathematical result is given in the following well-known theorem, which we state without proof.

Theorem 2.2.2. *Let* $\hat{\theta}_1, \ldots, \hat{\theta}_k$ *be independent and identically distributed normal* (θ, σ^2) *random variables. Then*
 (i) *the statistic*

$$z = (\hat{\bar{\theta}} - \theta)/\sqrt{\sigma^2/k}.$$

is distributed as a normal $(0, 1)$ *random variable; and*
 (ii) *the statistic*

$$t = (\hat{\bar{\theta}} - \theta)/\sqrt{v(\hat{\bar{\theta}})}$$

is distributed as Student's t with $k - 1$ degrees of freedom. □

---

[2] In most simple examples, this bias is at most of order n^{-1}, where n is the sample size. Such biases are usually unimportant in large samples.

If the variance of $\hat{\hat{\theta}}$ is essentially known without error or if k is very large, then a $(1 - \alpha)$ 100 percent confidence interval for θ is

$$(\hat{\hat{\theta}} - z_{\alpha/2}\sqrt{v(\hat{\hat{\theta}})}, \hat{\hat{\theta}} + z_{\alpha/2}\sqrt{v(\hat{\hat{\theta}})}),$$

where $z_{\alpha/2}$ is the upper $\alpha/2$ percentage point of the $N(0, 1)$ distribution. When the variance of $\hat{\hat{\theta}}$ is not known or when k is not large, the confidence interval takes the form

$$(\hat{\hat{\theta}} - t_{k-1,\alpha/2}\sqrt{v(\hat{\hat{\theta}})}, \hat{\hat{\theta}} + t_{k-1,\alpha/2}\sqrt{v(\hat{\hat{\theta}})}),$$

where $t_{k-1,\alpha/2}$ is the upper $\alpha/2$ percentage point of the t_{k-1} distribution. In like manner, statistical tests of hypotheses about θ may be based on Theorem 2.2.2.

The assumptions required in Theorem 2.2.2 are stronger than those in Theorem 2.2.1, and may not be strictly satisfied in sampling from finite populations. First, the random variables $\hat{\theta}_\alpha$ must now be independent and identically distributed (θ, σ^2) random variables, whereas before they were only held to be uncorrelated with common mean μ. These assumptions, as noted before do not cause serious problems, because the overall sampling scheme (i)-(iii) and the common estimation procedure and measurement process tend to ensure that the more restrictive assumptions are satisfied. There may be concern about a bias $\mu - \theta \neq 0$ for nonlinear estimators, but such biases are usually unimportant in the large samples encountered in modern complex sample surveys. Second, Theorem 2.2.2 assumes normality of the $\hat{\theta}_\alpha$, and this is never satisfied exactly in finite population sampling. Asymptotic theory for survey sampling, however, suggests that the $\hat{\theta}_\alpha$ may be approximately normally distributed in large samples. A discussion of some of the relevant asymptotic theory is presented in Appendix B.

Notwithstanding these failures of the stated assumptions, Theorem 2.2.2 has historically formed the basis for inference in complex sample surveys, largely because of the various asymptotic results.

Many of the important applications of the random group technique concern nonlinear estimators. In such applications efficiency considerations may suggest an estimator $\hat{\theta}$ computed from the combination of all random groups, rather than $\hat{\hat{\theta}}$. For certain linear estimators $\hat{\theta}$ and $\hat{\hat{\theta}}$ are identical, whereas for nonlinear estimators they are generally not identical. This point is discussed further in Section 2.8. The difference between $\hat{\theta}$ and $\hat{\hat{\theta}}$ is illustrated in the following example.

EXAMPLE 2.2.1. Suppose that it is desired to estimate the ratio

$$\theta = Y/X,$$

where Y and X denote the population totals of two of the survey characteristics. Let $\hat{Y}_\alpha$ and $\hat{X}_\alpha$ $(\alpha = 1, \ldots, k)$ denote estimators of Y and X derived from the α-th random group. In practice, these are often linear estimators and unbiased for Y and X, respectively. Then,

$$\hat{\theta}_\alpha = \hat{Y}_\alpha / \hat{X}_\alpha,$$

$$\hat{\bar{\theta}} = (1/k) \sum_{\alpha=1}^{k} \hat{Y}_\alpha / \hat{X}_\alpha,$$

and

$$\hat{\theta} = \frac{\sum_{\alpha=1}^{k} \hat{Y}_\alpha / k}{\sum_{\alpha=1}^{k} \hat{X}_\alpha / k}.$$

In general, $\hat{\bar{\theta}} \neq \hat{\theta}$.   □

There are two random group estimators of the variance of $\hat{\theta}$ which are used in practice. First, one may use

$$v_1(\hat{\theta}) = \sum_{\alpha=1}^{k} (\hat{\theta}_\alpha - \hat{\bar{\theta}})^2 / k(k-1), \tag{2.2.2}$$

which is the same as $v(\hat{\bar{\theta}})$. Thus, $v(\hat{\bar{\theta}}) = v_1(\hat{\theta})$ is used not only to estimate $\mathrm{Var}\{\hat{\bar{\theta}}\}$, which it logically estimates, but also to estimate $\mathrm{Var}\{\hat{\theta}\}$. However, straightforward application of the Cauchy–Schwarz inequality gives

$$0 \leq [\sqrt{\mathrm{Var}\{\hat{\bar{\theta}}\}} - \sqrt{\mathrm{Var}\{\hat{\theta}\}}]^2 \leq \mathrm{Var}\{\hat{\bar{\theta}} - \hat{\theta}\} \tag{2.2.3}$$

and $\mathrm{Var}\{\hat{\bar{\theta}} - \hat{\theta}\}$ is generally small relative to both $\mathrm{Var}\{\hat{\bar{\theta}}\}$ and $\mathrm{Var}\{\hat{\theta}\}$. In fact, using Taylor series expansions (see Chapter 6) it is possible to show that $\mathrm{Var}\{\hat{\bar{\theta}} = \hat{\theta}\}$ is of smaller order than $\mathrm{Var}\{\hat{\bar{\theta}}\}$ or $\mathrm{Var}(\hat{\theta})$. Thus, the two variances are usually of similar magnitude and $v_1(\hat{\theta})$ should be a reasonable estimator of $\mathrm{Var}\{\hat{\theta}\}$.

The second random group estimator is

$$v_2(\hat{\theta}) = \sum_{\alpha=1}^{k} (\hat{\theta}_\alpha - \hat{\theta})^2 / k(k-1). \tag{2.2.4}$$

When the estimator of θ is linear, then clearly $\hat{\bar{\theta}} = \hat{\theta}$ and $v_1(\hat{\theta}) = v_2(\hat{\theta})$. For nonlinear estimators we have

$$\sum_{\alpha=1}^{k} (\hat{\theta}_\alpha - \hat{\theta})^2 = \sum_{\alpha=1}^{k} (\hat{\theta} - \hat{\bar{\theta}})^2 + k(\hat{\bar{\theta}} - \hat{\theta})^2. \tag{2.2.5}$$

Thus,

$$v_1(\hat{\theta}) \leq v_2(\hat{\theta})$$

and $v_2(\hat{\theta})$ will be preferred when a conservative estimate of variance is desired. However, as observed before, the expectation of the squared difference $(\hat{\bar{\theta}} - \hat{\theta})^2$ will be unimportant in many complex survey applications and there should be little difference between v_1 and v_2. If an important difference does occur between v_1 and v_2 (or between $\tilde{\hat{\theta}}$ and $\hat{\theta}$), then this could signal either a computational error or a bias associated with small sample sizes.

Little else can be said by way of recommending $v_1(\hat{\theta})$ or $v_2(\hat{\theta})$. Using second-order Taylor series expansions, Dippo (1981) has shown that the bias of v_1 as an estimator of $\text{Var}\{\hat{\theta}\}$ is less than or equal to the bias of v_2. To the same order of approximation, Dippo shows that the variances of v_1 and v_2 are identical. Neither of these results, however, has received any extensive empirical testing. And in general, we feel that it is an open question as to which of v_1 or v_2 is the more accurate estimator of the variance of $\hat{\theta}$.

Before considering the case of nonindependent random groups, we present a simple, artificial example of the method of sample selection discussed in this section.

EXAMPLE 2.2.2. Suppose that a sample of households is to be drawn using a multistage sampling design. Two random groups are desired. An areal frame exists, and the target population is divided into two strata (defined, say, on the basis of geography). Stratum I contains N_1 clusters (or primary sampling units (PSU)); stratum II consists of one PSU that is to be selected with certainty. Sample s_1 is selected according to the following plan:

 (i) Two PSUs are selected from stratum I using some πps sampling design. From each selected PSU, an equal probability, single-start, systematic sample of m_1 households is selected and enumerated.
(ii) The certainty PSU is divided into well-defined units, say city blocks, with the block size varying between 10 and 15 households. An unequal probability systematic sample of m_2 blocks is selecting with probabilities proportional to the block sizes. All households in selected blocks are enumerated.

After drawing sample s_1 according to this plan, s_1 is replaced and the second random group, s_2, is selected by independently repeating steps (i) and (ii).

Separate estimates, $\hat{\theta}_1$ and $\hat{\theta}_2$, of a population parameter of interest are computed from the two random groups. An overall estimator of the parameter and the random group estimator of its variance are

$$\bar{\hat{\theta}} = (\hat{\theta}_1 + \hat{\theta}_2)/2$$

and

$$v(\bar{\hat{\theta}}) = \frac{1}{2(1)} \sum_{\alpha=1}^{2} (\hat{\theta}_\alpha - \bar{\hat{\theta}})^2$$

$$= (\hat{\theta}_1 - \hat{\theta}_2)^2/4$$

respectively.   □

The example nicely illustrates the simplifications that result from proper use of random groups. Had we not employed the random group methodology, variance estimation would have been considerably more difficult, particularly for designs that do not admit an unbiased estimator of variance, e.g., systematic sampling designs.

2.3. Example: A Survey of AAA Motels[3]

We now illustrate the use of independent random groups with a real survey. The example is concerned with a survey of motel operators affiliated with the American Automobile Association (AAA). The purpose of the survey was to determine whether the operators were in favor of instituting a system whereby motorists could make reservations in advance of their arrival.

The survey frame was a file of cards maintained in the AAA's central office. It consisted of 172 file drawers, 64 cards per drawer. Each card represented one of the following kinds of establishments:

> A contract motel
> > 1 to 10 rooms
> > 11 to 24 rooms
> > 25 rooms and over
> A hotel
> A restaurant
> A special attraction
> A blank card.

The sampling unit was the card (or equivalently, the establishment operator).

The sampling design for the survey consisted of the following key features:

1. Each of the 172 drawers was augmented by 6 blank cards, so that each drawer now contained 70 cards. This augmentation was based on 1) the prior belief that there were approximately 5000 contract motels in the population and 2) the desire to select about 700 of them into the overall sample. Thus, an overall sampling fraction of about one in $5000/700 \doteq 7$ was required.
2. A sample of 172 cards was chosen by selecting one card at random from each drawer. Sampling was independent in different drawers.
3. Nine additional samples were selected according to the procedure in step 2. Each of the samples was selected after replacing the previous sample. Thus, estimators derived from the 10 samples (or random groups) may be considered to be independent.

[3] This example is from Deming (1960, Chapter 7). The permission of the author is gratefully acknowledged.

4. Eight hundred and fifty-four motels were drawn into the overall sample, and each was mailed a questionnaire. The 866 remaining units were not members of the domain for which estimates were desired (i.e., contract motels). Although the random groups were selected with replacement, no duplicates were observed.
5. At the end of 10 days, a second notice was sent to delinquent cases, and at the end of another week, a third notice. Every case that had not reported by the end of 24 days was declared a nonrespondent.
6. Nonrespondents were then ordered by random group number, and from each consecutive group of three, one was selected at random. The selected nonrespondents were enumerated via personal interview. In this sampling, nonrespondents from the end of one random group were tied to those at the beginning of the next random group, thus abrogating, to a minor degree, the condition of independence of the random group estimators. This fact, however, is ignored in the ensuing development of the example.

Table 2.3.1 gives the results to the question, "How frequently do people ask you to make reservations for them?" after 24 days. The results of the 1 in 3 subsample of nonrespondents are contained in Table 2.3.2.

Estimates for the domain of contract motels may be computed by noting that the probability of a given sampling unit being included in any one of the random groups is 1/70, and the conditional probability of being included in the subsample of nonrespondents is 1/3. Thus, for example, the estimated

Table 2.3.1. Number of Replies to Each Category of the Question "How Frequently Do People Ask You to Make Reservations for Them?" After 24 Days

Random Group	Frequently	Rarely	Never	Ambiguous Answer	No Reply Yet	Total
1	16	40	17	2	19	94
2	20	30	17	3	15	85
3	18	35	16	1	15	85
4	17	31	14	2	16	80
5	14	32	15	3	18	82
6	15	32	12	4	16	79
7	19	30	17	3	17	86
8	13	37	11	3	18	82
9	19	39	19	2	14	93
10	17	39	15	2	15	88
Total	168	345	153	25	163	854

Source: Table 3, Deming (1960, Chapter 7).

Table 2.3.2. Results of Personal Interviews with Subsample of
Nonrespondents

Random Group	Frequently	Rarely	Never	Temporarily Closed (vacation, sick, etc.)	Total
1	1	2	2	1	6
2	1	2	1	1	5
3	2	2	0	1	5
4	2	1	2	0	5
5	1	3	1	2	7
6	2	2	0	1	5
7	1	3	1	1	6
8	1	2	1	2	6
9	2	2	1	0	5
10	1	2	0	2	5
Total	14	21	9	11	55

Source: Table 4, Deming (1960, Chapter 7).

total number of contract motels from the first random group is

$$\hat{X}_1 = 70 \sum_{i=1}^{172} X_{1i}$$

$$= 70(94)$$

$$= 6580,$$

where

$$X_{1i} = \begin{cases} 1, & \text{if the } i\text{-th selected unit in the first} \\ & \text{random group is a contract motel} \\ 0, & \text{if not} \end{cases}$$

and the estimated total over all random groups is

$$\hat{\bar{X}} = \sum_{\alpha=1}^{10} \hat{X}_\alpha / 10 = 5978.$$

Since the estimator is linear, $\hat{X}$ and $\hat{\bar{X}}$ are identical. The corresponding estimate of variance is

$$v(\hat{\bar{X}}) = \frac{1}{10(9)} \sum_{\alpha=1}^{10} (\hat{X}_\alpha - \hat{\bar{X}})^2 = 12,652.9.$$

Estimated totals for each of the categories of the question "How frequently do people ask you to make reservations for them?" are given in Table 2.3.3. For example, the estimate from random group 1 of the total number of

Table 2.3.3. Estimated Totals

Random Group	Frequently	Rarely	Never	Ambiguous Answer	Temporarily Closed
1	1330	3220	1610	140	210
2	1610	2520	1400	210	210
3	1680	2870	1120	70	210
4	1610	2380	1400	140	0
5	1190	2870	1260	210	420
6	1470	2660	840	280	210
7	1540	2730	1400	210	210
8	1120	3010	980	210	420
9	1750	3150	1540	140	0
10	1400	3150	1050	140	420
Parent Sample	1470	2856	1260	175	231

motels that would respond "frequently" is

$$\hat{Y}_1 = 70\left(\sum_{i \in r_1} Y_{1i} + 3 \sum_{i \in nr_1} Y_{1i} \right)$$

$$= 70(16 + 3 \cdot 1)$$

$$= 1330,$$

where $\sum_{i \in r_1}$ and $\sum_{i \in nr_1}$ denote summations over the respondents and the subsample of nonrespondents, respectively, in the first random group, and

$$Y_{1i} = \begin{cases} 1, & \text{if the } i\text{-th selected unit in the first random group is a} \\ & \text{contract motel and reported "frequently"} \\ 0, & \text{if not.} \end{cases}$$

All of the estimates in Table 2.3.3 may be computed in this manner.

Various nonlinear statistics may also be prepared from these data. The estimate from the first random group of the ratio of the "rarely" plus "never" to the "frequently" plus "rarely" plus "never" is

$$\hat{R}_1 = \frac{3220 + 1610}{1330 + 3220 + 1610}$$

$$= 0.784.$$

The estimate of this ratio over all random groups is

$$\hat{\bar{R}} = \sum_{\alpha=1}^{10} \hat{R}_\alpha / 10$$

$$= 0.737$$

with corresponding variance estimate

$$v(\hat{\bar{R}}) = \frac{1}{10(9)} \sum_{\alpha=1}^{10} (\hat{R}_\alpha - \hat{\bar{R}})^2$$

$$= 0.0001139.$$

Since the ratio is a nonlinear statistic, we may use the alternative estimate

$$\hat{R} = \frac{2856 + 1260}{1470 + 2856 + 1260} = 0.737.$$

The two random group estimates of Var$\{\hat{R}\}$ are

$$v_1(\hat{R}) = v(\hat{\bar{R}})$$

$$= 0.0001139$$

and

$$v_2(\hat{R}) = \frac{1}{10(9)} \sum_\alpha^{10} (\hat{R}_\alpha - \hat{R})^2 = 0.0001139.$$

Clearly, there is little difference between $\hat{\bar{R}}$ and $\hat{R}$, and $v_1(\hat{R})$ and $v_2(\hat{R})$ for these data. Other nonlinear statistics and corresponding variance estimates may be prepared in similar manner.

2.4. The Case of Nonindependent Random Groups

In practical applications, survey samples are often selected as a whole using some form of sampling without replacement, rather than by selecting a series of independent random groups. Even in these circumstances the random group estimator of variance is used widely, at least within government statistical bureaus where large-scale surveys are conducted with small sampling fractions. Random groups are now formed following sample selection by randomly dividing the parent sample into k groups. Estimates are computed from each random group, and the variance is estimated using an expression of the form of (2.2.1). The random group estimators $\hat{\theta}_\alpha$ are no longer uncorrelated because sampling is performed without replacement, and the result of Theorem 2.2.1 is no longer strictly valid. The random group estimator now incurs a bias. In the remainder of this section we describe the formation of random groups in some detail, and then investigate the magnitude and sign of the bias for some simple (but popular) examples.

2.4.1. The Formation of Random Groups

To ensure that the random group estimator of variance has acceptable statistical properties, the random groups must not be formed in a purely arbitrary fashion. Rather, *the principal requirement is that they be formed so that each random group has essentially the same sampling design as the parent sample.* This requirement can be satisfied for most survey designs by adhering to the following rules:

(i) If a single-stage sample of size n is selected by either srs wor or pps wor, then the random groups should be formed by dividing the parent sample at random. This means that the first random group (RG) is obtained by drawing a simple random sample without replacement (srs wor) of size $m = [n/k]$ from the parent sample; the second RG by drawing a srs wor of size m from the remaining $n - m$ units in the parent sample; and so forth. If n/k is not an integer, i.e., $n = km + q$ with $0 < q < k$, then the q excess units may be left out of the k random groups. An alternative method of handling excess units is to add one of the units to each of the first q RGs.

(ii) If a single-start systematic sample of size n is selected with either equal or unequal probabilities, then the random groups should be formed by dividing the parent sample systematically. This may be accomplished by generating a random integer between 1 and k, say α^*. The first unit in the parent sample is then assigned to random group α^*, the second to random group $\alpha^* + 1$, and so forth in a modulo k fashion. Variance estimation for systematic sampling is discussed more fully in Chapter 7.

(iii) In multistage sampling, the random groups should be formed by dividing the *ultimate clusters*, i.e., the aggregate of all elementary units selected from the same primary sampling unit (PSU), into k groups. That is, all second, third, and successive stage units selected from the PSU must be treated as a single unit when forming RGs. The actual division of the ultimate clusters into random groups is made according to either rule (i) or (ii), depending upon the nature of the first-stage sampling design. If this design is srs wor or pps wor, then rule (i) should be used, whereas for systematic sampling designs rule (ii) should be used. Particular care is required when applying the ultimate cluster principal to so-called self-representing PSUs, where terminology may cause considerable confusion.[4] From the point of view of variance estimation, a self-representing PSU should be considered a separate stratum, and the units used at the first stage of subsampling are the basis for the formation of random groups.

(iv) In stratified sampling, two options are available. First, if we wish to estimate the variance within a given stratum, then we invoke either

---

[4] A self-representing PSU is a PSU selected with probability one.

rule (i), (ii), or (iii) depending upon the nature of the within stratum sampling design. For example, rule (iii) is employed if a multistage design is used within the stratum. Second, if we wish to estimate the total variance across all strata then each random group must itself be a stratified sample comprised of units from each stratum. In this case, the first RG is obtained by drawing a srs wor of size $m_h = n_h/k$ from the parent sample n_h in the h-th stratum, for $h = 1, \ldots, L$. The second RG is obtained in the same fashion by selecting from the remaining $n_h - m_h$ units in the h-th stratum. The remaining RGs are formed in like manner. If excess observations remain in any of the strata, i.e., $n_h = km_h + q_h$, they may be left out of the k random groups or added, one each, to the first q_h RGs. If the parent sample is selected systematically within strata, then the random groups must also be formed in a systematic fashion. In other words, each random group must be comprised of a systematic subsample from the parent sample in each stratum.

(v) If an estimator is to be constructed for some double sampling scheme, such as double sampling for stratification or double sampling for the ratio estimator (see Cochran (1977, Chapter 12)), then the n' sampling units selected into the initial sample should be divided into the k random groups. The division should be made randomly for srs wor and pps wor designs and systematically for systematic sampling designs. When n' is not an integer multiple of k, either of the procedures given in rule (i) for dealing with excess units may be used. The second phase sample, say of size n, is divided into random groups according to the division of the initial sample. In other words, a given selected unit i is assigned the same random group number as it was assigned in the initial sample. This procedure is used when both initial and second phase samples are selected in advance of the formation of random groups. Alternatively, in some applications it may be possible to form the random groups after selection of the initial sample but before selection of the second phase sample. In this case, the sample n' is divided into the k random groups and the second phase sample is obtained by independently drawing $m = n/k$ units from each random group.

These rules, or combinations thereof, should cover many of the popular sampling designs used in modern, large-scale sample surveys. The rules will, of course, have to be used in combination with one another in many situations. An illustration is where a multistage sample is selected within each of $L \geq 2$ strata. For this case rules (iii) and (iv) must be used in combination. The ultimate clusters are the basis for the formation of random groups, and each random group is composed of ultimate clusters from each stratum. Another example is where a double sampling scheme for the ratio estimator is used within each of $L \geq 2$ strata. For this case rules (iv) and

(v) are used together. Some exotic sampling designs may not be covered by any combination of the rules. In such cases, the reader should attempt to form the random groups by adhering to the principal requirement that each group has essentially the same design as the parent sample.

2.4.2. A General Estimation Procedure

In general, the estimation methodology for a population parameter θ is the same as in the case of independent random groups (cf. Section 2.2). We let $\hat{\theta}$ denote the estimator of θ obtained from the parent sample, $\hat{\theta}_\alpha$ the estimator obtained from the α-th RG, and $\hat{\bar{\theta}} = \sum_{\alpha=1}^{k} \hat{\theta}_\alpha/k$. The random group estimator of $\mathrm{Var}\{\hat{\bar{\theta}}\}$ is then given by

$$v(\hat{\bar{\theta}}) = \frac{1}{k(k-1)} \sum_{\alpha=1}^{k} (\hat{\theta}_\alpha - \hat{\bar{\theta}})^2 \qquad (2.4.1)$$

which is identical to (2.2.1). We estimate the variance of $\hat{\theta}$ by either

$$v_1(\hat{\theta}) = v(\hat{\bar{\theta}}) \qquad (2.4.2)$$

or

$$v_2(\hat{\theta}) = \frac{1}{k(k-1)} \sum_{\alpha=1}^{k} (\hat{\theta}_\alpha - \hat{\theta})^2, \qquad (2.4.3)$$

which are identical to (2.2.2) and (2.2.4). When a conservative estimator of $\mathrm{Var}\{\hat{\theta}\}$ is sought, $v_2(\hat{\theta})$ is preferred to $v_1(\hat{\theta})$. The estimators $\hat{\theta}_\alpha$ should be prepared by application of a common measurement process and estimation procedure to each random group. This process was described in detail in Section 2.2.

Because the random group estimators are not independent, $v(\hat{\bar{\theta}})$ is not an unbiased estimator of the variance of $\hat{\bar{\theta}}$. The following theorem describes some of the properties of $v(\hat{\bar{\theta}})$:

Theorem 2.4.1. *Let $\hat{\theta}_\alpha$ be defined as above and let $\mu_\alpha = \mathrm{E}\{\hat{\theta}_\alpha\}$, where μ_α is not necessarily equal to θ. Then,*

$$\mathrm{E}\{\hat{\bar{\theta}}\} = \sum_{\alpha=1}^{k} \mu_\alpha/k \overset{(say)}{=} \bar{\mu},$$

and the expectation of the random group estimator of $\mathrm{Var}\{\hat{\bar{\theta}}\}$ is given by

$$\mathrm{E}\{v(\hat{\bar{\theta}})\} = \mathrm{Var}\{\hat{\bar{\theta}}\} + \frac{1}{k(k-1)} \sum_{\alpha=1}^{k} (\mu_\alpha - \bar{\mu})^2$$

$$- 2 \sum_{\alpha=1}^{k} \sum_{\beta>\alpha}^{k} \mathrm{Cov}\{\hat{\theta}_\alpha, \hat{\theta}_\beta\}/\{k(k-1)\}.$$

Further, if each RG *is the same size, then*

$$\mu_\alpha = \bar{\mu}(\alpha = 1, \ldots, k),$$

$$E\{\hat{\bar{\theta}}\} = \bar{\mu},$$

and

$$E\{v(\hat{\bar{\theta}})\} = \text{Var}\{\hat{\bar{\theta}}\} - \text{Cov}\{\hat{\theta}_1, \hat{\theta}_2\}.$$

PROOF. It is obvious that

$$E\{\hat{\bar{\theta}}\} = \bar{\mu}.$$

The random group estimator of variance may be reexpressed as

$$v(\hat{\bar{\theta}}) = \hat{\bar{\theta}}^2 - 2 \sum_\alpha^k \sum_{\beta > \alpha}^k \hat{\theta}_\alpha \hat{\theta}_\beta / k(k-1).$$

The conclusion follows because

$$E\{\hat{\bar{\theta}}^2\} = \text{Var}\{\hat{\bar{\theta}}\} + \bar{\mu}^2$$

and

$$E\{\hat{\theta}_\alpha \hat{\theta}_\beta\} = \text{Cov}\{\hat{\theta}_\alpha, \hat{\theta}_\beta\} + \mu_\alpha \mu_\beta. \qquad \square$$

Theorem 2.4.1 displays the bias in $v(\hat{\bar{\theta}})$ as an estimator of $\text{Var}\{\hat{\bar{\theta}}\}$. For large populations and small sampling fractions, however, $2 \sum_\alpha^k \sum_{\beta > \alpha}^k \text{Cov}\{\hat{\theta}_\alpha, \hat{\theta}_\beta\}/\{k(k-1)\}$ will tend to be relatively small and negative. The quantity

$$\frac{1}{k(k-1)} \sum_{\alpha=1}^k (\mu_\alpha - \bar{\mu})^2$$

will tend to be relatively small when $\mu_\alpha \doteq \bar{\mu}(\alpha = 1, \ldots, k)$. Thus, the bias of $v(\hat{\bar{\theta}})$ will be unimportant in many large-scale surveys, and tending to be slightly positive.

When the estimator $\hat{\theta}$ is linear, various special results and estimators are available. This topic is treated in Subsections 2.4.3 and 2.4.4.

When the estimator $\hat{\theta}$ is nonlinear, little else is known regarding the exact bias properties of (2.4.1), (2.4.2), or (2.4.3). Some empirical investigations of the variance estimators have been encouraging in the sense that their bias is often found to be unimportant. See Frankel (1971b) for one of the largest such studies to date. The evidence as of this writing suggests that the bias of the random group estimator of variance is often small and decreases as the size of the groups increase (or equivalently as the number of groups decreases). This result occurs because the variability among the $\hat{\theta}_\alpha$ is close to the variability in $\hat{\theta}$ when the sample size involved in $\hat{\theta}_\alpha$ is close to that involved in $\hat{\theta}$. See Dippo and Wolter (1984) for a recent discussion of this evidence.

It is possible to approximate the bias properties of the variance estimators by working with a linear approximation to $\hat{\theta}$ (see Chapter 6) and then applying known results for linear estimators. Such approximations generally suggest that the bias is unimportant in the context of large samples with small sampling fractions. See Dippo (1981) for discussion of second-order approximations.

2.4.3. Linear Estimators with Random Sampling of Elementary Units

In this subsection we show how the general estimation procedure applies to a rather simple estimator and sampling design. Specifically, we consider the problem of variance estimation for linear estimators where the parent sample is selected in $L \geq 1$ strata. Within the h-th stratum we assume that a simple random sample without replacement (srs wor) of n_h elementary units is selected. Without essential loss of generality, the ensuing development will be given for the case of estimating the population mean $\theta = \bar{Y}$.

The standard unbiased estimator of θ is given by

$$\hat{\theta} = \bar{y}_{st} = \sum_{h=1}^{L} W_h \bar{y}_h$$

where $W_h = N_h/N$, $\bar{y}_h = \sum_{j=1}^{n_h} y_{hj}/n_h$, N_h denotes the number of units in the h-th stratum and $N = \sum_{h=1}^{L} N_h$. If n_h is an integer multiple of k (i.e., $n_h = km_h$) for $h = 1, \ldots, L$, then we form the k random groups as described by rule (iv) of Subsection 2.4.1 and the estimator of θ from the α-th RG is

$$\hat{\theta}_\alpha = \bar{y}_{st,\alpha} = \sum_{h=1}^{L} W_h \bar{y}_{h,\alpha}, \tag{2.4.4}$$

where $\bar{y}_{h,\alpha}$ is the sample mean of the m_h units in stratum h that were selected into the α-th RG. Because the estimator $\hat{\theta}$ is linear and since $n_h = km_h$ for $h = 1, \ldots, L$, it is clear that $\hat{\theta} = \hat{\bar{\theta}}$.

The random group estimator of the variance $\text{Var}\{\hat{\theta}\}$ is

$$v(\hat{\theta}) = \frac{1}{k(k-1)} \sum_{\alpha=1}^{k} (\bar{y}_{st,\alpha} - \bar{y}_{st})^2.$$

where it is clear that (2.4.1), (2.4.2), and (2.4.3) are identical in this case.

Theorem 2.4.2. *When $n_h = km_h$ for $h = 1, \ldots, L$, the expectation of the random group estimator of* $\text{Var}\{\bar{y}_{st}\}$ *is given by*

$$E\{v(\hat{\theta})\} = \sum_{h=1}^{L} W_h^2 S_h^2/n_h,$$

where

$$S_h^2 = \frac{1}{N_h - 1} \sum_{j=1}^{N_h} (Y_{hj} - \bar{Y}_h)^2.$$

PROOF. By definition

$$E\{\bar{y}_{\mathrm{st},\alpha}^2\} = \mathrm{Var}\{\bar{y}_{\mathrm{st},\alpha}\} + E^2\{\bar{y}_{\mathrm{st},\alpha}\}$$

$$= \sum_{h=1}^{L} W_h^2 \left(\frac{1}{m_h} - \frac{1}{N_h} \right) S_h^2 + \bar{Y}^2$$

and

$$E\{\bar{y}_{\mathrm{st}}^2\} = \mathrm{Var}\{\bar{y}_{\mathrm{st}}\} + E^2\{\bar{y}_{\mathrm{st}}\}$$

$$= \sum_{h=1}^{L} W_h^2 \left(\frac{1}{n_h} - \frac{1}{N_h} \right) S_h^2 + \bar{Y}^2.$$

The result follows by writing

$$v(\hat{\theta}) = \frac{1}{k(k-1)} \left(\sum_{\alpha=1}^{k} \bar{y}_{\mathrm{st},\alpha}^2 - k\bar{y}_{\mathrm{st}}^2 \right). \qquad \square$$

The reader will recall that

$$\mathrm{Var}\{\hat{\theta}\} = \sum_{h=1}^{L} W_h^2 (1 - f_h) S_h^2 / n_h,$$

where $f_h = n_h / N_h$ is the sampling fraction. Thus, Theorem 2.4.2 shows that $v(\hat{\theta})$ is essentially unbiased whenever the sampling fractions f_h are negligible. If some of the f_h are not negligible, then $v(\hat{\theta})$ is conservative in the sense that it will tend to overestimate $\mathrm{Var}\{\hat{\theta}\}$.

If some of the sampling fractions f_h are important, then they may be included in the variance computations by working with

$$W_h^* = W_h \sqrt{1 - f_h}$$

in place of W_h. Under this procedure, we define the random group estimators by

$$\hat{\theta}_\alpha^* = \bar{y}_{\mathrm{st}} + \sum_{h=1}^{L} W_h^* (\bar{y}_{h,\alpha} - \bar{y}_h), \qquad (2.4.5)$$

$$\hat{\bar{\theta}}^* = \frac{1}{k} \sum_{\alpha=1}^{k} \hat{\theta}_\alpha^*,$$

and

$$v(\hat{\bar{\theta}}^*) = \frac{1}{k(k-1)} \sum_{\alpha=1}^{k} (\hat{\theta}_\alpha^* - \hat{\bar{\theta}}^*)^2.$$

It is clear that

$$\hat{\bar{\theta}}^* = \hat{\theta} = \bar{y}_{st}.$$

Corollary 2.4.1. *Let $v(\hat{\bar{\theta}}^*)$ be defined as $v(\hat{\theta})$ with $\hat{\theta}^*_\alpha$ in place of $\hat{\theta}_\alpha$. Then, given the conditions of Theorem 2.4.2,*

$$E\{v(\hat{\bar{\theta}}^*)\} = \sum_{h=1}^{L} W_h^{*2} S_h^2 / n_h$$

$$= \mathrm{Var}\{\hat{\theta}\}. \qquad \square$$

This corollary shows that an unbiased estimator of variance, including the finite population corrections, can be achieved by exchanging the weights W_h^* for W_h.

Next, we consider the general case where the stratum sample sizes are not integer multiples of the number of random groups. Assume $n_h = km_h + q_h$ for $h = 1, \ldots, L$, with $0 \le q_h < k$. A straightforward procedure for estimating the variance of $\hat{\theta} = \bar{y}_{st}$ is to leave the q_h excess observations out of the k random groups. The random group estimators are defined as before, but now

$$\hat{\bar{\theta}} = \sum_{\alpha=1}^{k} \bar{y}_{st,\alpha} / k \ne \hat{\theta}$$

because the excess observations are in $\hat{\theta}$ but not in $\hat{\bar{\theta}}$. The expectation of the random group estimator $v(\hat{\bar{\theta}})$ is described by Theorem 2.4.2, where the n_h in the denominator is replaced by $n_h - q_h = km_h$.

Corollary 2.4.2. *Let $v(\hat{\bar{\theta}})$ be defined as in Theorem 2.4.2 with q_h excess observations omitted from the k random groups. Then,*

$$E\{v(\hat{\bar{\theta}})\} = \sum_{h=1}^{L} W_h^2 S_h^2 / km_h. \qquad \square$$

The reader will immediately observe that $v(\hat{\bar{\theta}})$ tends to overestimate $\mathrm{Var}\{\hat{\theta}\}$, not only due to possibly nonnegligible f_h, but also because $km_h \le n_h$. As before, if some of the f_h are important, they may be accounted for by working with $\hat{\theta}^*_\alpha$ instead of $\hat{\theta}_\alpha$. If both the f_h and the q_h are important, they may be accounted for by replacing W_h by

$$W_h'' = W_h \sqrt{(1 - f_h) \frac{km_h}{n_h}}$$

and by defining

$$\hat{\theta}''_\alpha = \bar{y}_{st} + \sum_{h=1}^{L} W_h'' \left(\bar{y}_{h,\alpha} - \frac{1}{k} \sum_{\beta=1}^{k} \bar{y}_{h,\beta} \right).$$

Then $\hat{\bar{\theta}}'' = \hat{\theta}$ and $v(\hat{\bar{\theta}}'')$ is an unbiased estimator of $\mathrm{Var}\{\hat{\theta}\}$.

An alternative procedure, whose main appeal is that it does not omit observations, is to form q_h random groups of size $m_h + 1$ and $k - q_h$ of size m_h. Letting

$$a_{h,\alpha} = \begin{cases} k(m_h + 1)/n_h, & \text{if the } \alpha\text{-th RG contains } m_h + 1 \\ & \text{units from the } h\text{-th stratum} \\ km_h/n_h, & \text{if the } \alpha\text{-th RG contains } m_h \text{ units} \\ & \text{from the } h\text{-th stratum,} \end{cases}$$

we define the α-th random group estimator by

$$\tilde{\theta}_\alpha = \sum_{h=1}^{L} W_h a_{h,\alpha} \bar{y}_{h,\alpha}.$$

It is important to note that

$$E\{\tilde{\theta}_\alpha\} = \sum_{h=1}^{L} W_h a_{h,\alpha} \bar{Y}_h$$

$$\neq \bar{Y}. \tag{2.4.6}$$

However, because

$$\tilde{\bar{\theta}} = \sum_{\alpha=1}^{k} \tilde{\theta}_\alpha / k = \bar{y}_{st} = \hat{\theta}$$

and $E\{\hat{\theta}\} = \theta$, $\tilde{\bar{\theta}}$ is an unbiased estimator of θ even though the individual $\tilde{\theta}_\alpha$ are not unbiased. The random group estimator of $\text{Var}\{\hat{\theta}\}$ is now given by

$$v(\tilde{\bar{\theta}}) = \frac{1}{k(k-1)} \sum_{\alpha=1}^{k} (\tilde{\theta}_\alpha - \tilde{\bar{\theta}})^2$$

$$= \frac{1}{k(k-1)} \sum_{\alpha=1}^{k} (\tilde{\theta}_\alpha - \hat{\theta})^2. \tag{2.4.7}$$

Theorem 2.4.3. *When $n_h = km_h + q_h$ for $h = 1, \ldots, L$, the expectation of the random group estimator $v(\tilde{\bar{\theta}})$ is given by*

$$E\{v(\tilde{\bar{\theta}})\} = \text{Var}\{\bar{y}_{st}\} + \frac{1}{k(k-1)} \sum_{\alpha=1}^{k} (E\{\tilde{\theta}_\alpha\} - \theta)^2.$$

$$-2 \sum_{\alpha=1}^{k} \sum_{\beta>\alpha}^{k} \left[\sum_{h=1}^{L} W_h^2 a_{h,\alpha} a_{h,\beta} (-S_h^2/N_h) \right] \bigg/ k(k-1). \tag{2.4.8}$$

PROOF. Follows directly from Theorem 2.4.1 by noting that

$$\text{Cov}\{\tilde{\theta}_\alpha, \tilde{\theta}_\beta\} = \sum_{h=1}^{L} W_h^2 a_{h,\alpha} a_{h,\beta} (-S_h^2/N_h)$$

whenever $\alpha \neq \beta$. $\square$

The reader will observe that $v(\tilde{\bar{\theta}})$ is a biased estimator of $\text{Var}\{\bar{y}_{st}\}$ with the bias being given by the second and third terms on the right side of

(2.4.8). When the f_h's are negligible, the contribution of the third term will be unimportant. The contribution of the second term will be unimportant whenever

$$E\{\tilde{\theta}_\alpha\} = \sum_{h=1}^{L} W_h a_{h,\alpha} \bar{Y}_h \doteq \bar{Y}$$

for $\alpha = 1, \ldots, k$. Thus, in many surveys the bias of the random group estimator $v(\tilde{\bar{\theta}})$ will be unimportant.

It is an open question as to whether the estimator $v(\tilde{\bar{\theta}})$ has better or worse statistical properties than the estimator obtained by leaving the excess observations out of the random groups.

2.4.2. Linear Estimators with Clustered Samples

In the last subsection, we discussed the simple situation where a srs wor of elementary units is selected within each of L strata. We now turn to the case where a sample of n clusters (or PSUs) is selected and possibly several stages of subsampling occur independently within the selected PSUs. To simplify the presentation, we initially discuss the case of $L = 1$ stratum. The reader will be able to connect the results of Subsection 2.4.3 with the results of this subsection to handle cluster sampling within $L \geq 2$ strata. We continue to discuss linear estimators, only now focus our attention on estimators of the population total $\theta = Y$.

Let N denote the number of PSUs in the population, Y_i the total in the i-th primary, and $\hat{Y}_i$ the estimator of Y_i due to subsampling at the second and successive stages. The method of subsampling is left unspecified, but, for example, it may involve systematic sampling or other sampling designs which ordinarily do not admit an unbiased estimator of variance. We assume that n PSUs are selected according to some πps scheme, so that the i-th unit is included in sample with probability

$$\pi_i = np_i,$$

where $0 < np_i < 1$, $\sum_{i=1}^{N} p_i = 1$, and p_i is proportional to some auxiliary measure X_i. The reader will observe that srs wor at the first stage is simply a special case of this sampling framework, with $\pi_i = n/N$.

We consider estimators of the population total of the form

$$\hat{\theta} = \sum_{i=1}^{n} \hat{Y}_i / \pi_i = \sum_{i=1}^{n} \hat{Y}_i / np_i.^5$$

---

[5] The material presented here extends easily to estimators of the more general form $\hat{\theta} = \sum_{i=1}^{N} a_{is} \hat{Y}_i$, where the a_{is} are defined in advance for each sample s and satisfy $E\{a_{is}\} = 1$. See Durbin (1953), Raj (1966), and Rao (1975) for discussion of unbiased variance estimation for such estimators.

After forming $k = n/m$ (m an integer) random groups, the α-th random group estimator is given by

$$\hat{\theta}_\alpha = \sum_{i=1}^{m} \hat{Y}_i/mp_i$$

where the sum is taken over all PSUs selected into the α-th RG. Because $\hat{\theta}$ is linear in the $\hat{Y}_i$, it is clear that $\hat{\theta} = \hat{\bar{\theta}} = \sum_{\alpha=1}^{k} \hat{\theta}_\alpha/k$. We define the RG estimator of $\text{Var}\{\hat{\theta}\}$ by

$$v(\hat{\theta}) = \frac{1}{k(k-1)} \sum_{\alpha=1}^{k} (\hat{\theta}_\alpha - \hat{\theta})^2. \tag{2.4.9}$$

Some of the properties of $v(\hat{\theta})$ are given in the following theorems:

Theorem 2.4.4. *Let $v_k(\hat{\theta})$ be the estimator defined in (2.4.9) based on k random groups, and let $n = km$ (m an integer). Then,*

(i) $E_2\{v_k(\hat{\theta})\} = v_n(\hat{\theta})$
(ii) $\text{Var}\{v_k(\hat{\theta})\} \geq \text{Var}\{v_n(\hat{\theta})\}$,

where E_2 denotes the conditional expectation over all possible choices of k random groups for a given parent sample.

PROOF. Part (i) follows immediately from Theorem 2.4.2 by letting $L = 1$. Part (ii) follows from part (i) since

$$\text{Var}\{v_k(\hat{\theta})\} = \text{Var}\{E_2\{v_k(\hat{\theta})\}\} + E\{\text{Var}_2\{v_k(\hat{\theta})\}\}$$
$$\geq \text{Var}\{E_2\{v_k(\hat{\theta})\}\}. \qquad \square$$

Theorem 2.4.4 shows that $v_k(\hat{\theta})$ has the same expectation regardless of the value of k, so long as n is an integer multiple of k. However, the choice $k = n$ minimizes the variance of the RG estimator. The reader will recall that $v_n(\hat{\theta})$ is the standard unbiased estimator of the variance given pps wr sampling at the first stage.

Theorem 2.4.5. *Let $v_k(\hat{\theta})$ be the estimator defined in (2.4.9) based on k random groups and let*

$$Y_i = E\{\hat{Y}_i|i\}$$
$$\sigma_{2i}^2 = \text{Var}\{\hat{Y}_i|i\},$$

and $n = km$ (m an integer). Then, the expectation of $v_k(\hat{\theta})$ is given by

$$E\{v_k(\hat{\theta})\} = E\left\{ \frac{1}{n(n-1)} \sum_{i=1}^{n} \left(Y_i/p_i - \sum_{j=1}^{n} Y_j/np_j \right)^2 \right\} + \sum_{i=1}^{N} \sigma_{2i}^2/np_i.$$
$$\tag{2.4.10}$$

PROOF. The result follows from Theorem 2.4.4 and the fact that

$$v_n(\hat{\theta}) = \frac{1}{n(n-1)}\left(\sum_{i=1}^{n} \frac{\hat{Y}_i^2}{p_i^2} - n\hat{Y}^2\right)$$

and that

$$E\{\hat{Y}_i^2|i\} = Y_i^2 + \sigma_{2i}^2 . \qquad\qquad \square$$

Remarkably, (2.4.10) shows that the RG estimator completely includes the within component of variance since, the reader will recall,

$$\text{Var}\{\hat{\theta}\} = \text{Var}\left\{\sum_{i=1}^{n} Y_i/np_i\right\} + \sum_{i=1}^{N} \sigma_{2i}^2/np_i . \qquad (2.4.11)$$

Thus, the bias in $v_k(\hat{\theta})$ arises only in the between component, i.e., the difference between the first terms on the right side of (2.4.10) and (2.4.11). In surveys where the between component is a small portion of the total variance, we would anticipate that the bias in $v_k(\hat{\theta})$ would be unimportant. The bias is discussed further in Subsection 2.4.5 where it is related to the efficiency of πps sampling vis-à-vis pps wr sampling.

Now let us see how these results apply in the case of srs wor at the first stage. We continue to make no assumptions about the sampling designs at the second and subsequent stages, except we do require that such subsampling be independent from one PSU to another.

Corollary 2.4.3 *Suppose the n PSUs are selected at random and without replacement. Then, the expectation of $v_k(\hat{\theta})$ is given by*

$$E\{v_k(\hat{\theta})\} = N^2 S_b^2/n + (N^2/n)\sigma_w^2, \qquad (2.4.12)$$

where

$$S_b^2 = (N-1)^{-1} \sum_{i=1}^{N} \left(Y_i - N^{-1}\sum_{j=1}^{N} Y_j\right)^2$$

and

$$\sigma_w^2 = N^{-1} \sum_{i=1}^{N} \sigma_{2i}^2 . \qquad\qquad \square$$

The true variance of $\hat{\theta}$ given srs wor at the first stage is

$$\text{Var}\{\hat{\theta}\} = N^2(1 - n/N)S_b^2/n + (N^2/n)\sigma_w^2,$$

and thus $v_k(\hat{\theta})$ tends to overestimate the variance. Not surprisingly, the problem is with the finite population correction $(1 - n/N)$. One may attempt to adjust for the problem by working with the modified estimator $(1 - n/N)v_k(\hat{\theta})$, but this modification is downward biased by the amount

$$\text{Bias}\{(1 - n/N)v_k(\hat{\theta})\} = -N\sigma_w^2 .$$

Of course, when n/N is negligible $v_k(\hat{\theta})$ is essentially unbiased.

In the case of sampling within $L \geq 2$ strata, similar results are available. Let the estimator be of the form

$$\hat{\theta} = \sum_{h=1}^{L} \hat{Y}_h = \sum_{h=1}^{L} \sum_{i=1}^{n_h} \frac{\hat{Y}_{hi}}{\pi_{hi}} ,$$

where $\pi_{hi} = n_h p_{hi}$ is the inclusion probability associated with the (h, i)-th PSU. Two random group methodologies were discussed for this problem in Subsection 2.4.1. The first method works *within* strata and the estimator of $\text{Var}\{\hat{Y}\}$ is of the form

$$v(\hat{\theta}) = \sum_{h=1}^{L} v(\hat{Y}_h) ,$$

$$v(\hat{Y}_h) = \frac{1}{k(k-1)} \sum_{\alpha=1}^{k} (\hat{Y}_{h(\alpha)} - \hat{Y}_h)^2, \tag{2.4.13}$$

$$\hat{Y}_{h(\alpha)} = \sum_{i=1}^{m_h} \frac{\hat{Y}_{hi}}{m_h p_{hi}} ,$$

where the latter sum is over the m_h units that were drawn into the α-th random group within the h-th stratum. The second method works *across* strata and the estimator is of the form

$$v(\hat{\theta}) = \frac{1}{k(k-1)} \sum_{\alpha=1}^{k} (\hat{\theta}_\alpha - \hat{\theta})^2$$

$$\hat{\theta}_\alpha = \sum_{h=1}^{L} \hat{Y}_{h(\alpha)} \tag{2.4.14}$$

$$\hat{Y}_{h(\alpha)} = \sum_{i=1}^{m_h} \frac{\hat{Y}_{hi}}{m_h p_{hi}} ,$$

where this latter sum is over the m_h units from the h-th stratum that were assigned to the α-th random group. It is easy to show that both (2.4.13) and (2.4.14) have the same expectation, namely

$$E\{v(\hat{\theta})\} = \sum_{h=1}^{L} E\left\{ \frac{1}{n_h(n_h-1)} \sum_{i=1}^{n_h} \left(\frac{Y_{hi}}{p_{hi}} - \frac{1}{n_h} \sum_{j=1}^{n_h} \frac{Y_{hj}}{p_{hj}} \right)^2 \right\} + \sum_{h=1}^{L} \sum_{i=1}^{N_h} \frac{\sigma_{2hi}^2}{n_h p_{hi}} .$$

Since the true variance of $\hat{Y}$ is

$$\text{Var}\{\hat{\theta}\} = \sum_{h=1}^{L} \text{Var}\left\{ \sum_{i=1}^{n_h} \frac{Y_{hi}}{n_h p_{hi}} \right\} + \sum_{h=1}^{L} \sum_{i=1}^{N_h} \frac{\sigma_{2hi}^2}{n_h p_{hi}} ,$$

we see once again that the random group estimator incurs a bias in the between PSU component of variance.

It may be possible to reduce the bias by making adjustments to the estimator analogous to those made in Subsection 2.4.3. Let

$$\text{Var}\left\{ \left(\sum_{i=1}^{n_h} \frac{Y_{hi}}{n_h p_{hi}} \right)_{\text{wr}} \right\}$$

denote the between PSU variance given pps wr sampling within strata, and let

$$\mathrm{Var}\left\{ \sum_{i=1}^{n_h} \frac{Y_{hi}}{n_h p_{hi}} \right\}$$

denote the between variance given πps sampling. Suppose that it is possible to obtain a measure of the factor

$$R_h = \frac{n_h}{n_h - 1} \left(\frac{\left\{ \mathrm{Var}\left(\sum_{j=1}^{n_h} \frac{Y_{hi}}{n_h p_{hi}} \right)_{\mathrm{wr}} \right\}}{\mathrm{Var}\left\{ \sum_{i=1}^{n_h} \frac{\hat{Y}_{hi}}{n_h p_{hi}} \right\}} - \frac{\mathrm{Var}\left\{ \sum_{i=1}^{n_h} \frac{Y_{hi}}{n_h p_{hi}} \right\}}{\mathrm{Var}\left\{ \sum_{i=1}^{n_h} \frac{\hat{Y}_{hi}}{n_h p_{hi}} \right\}} \right),$$

either from previous census data, computations on an auxiliary variable, or by professional judgment. Then define the adjusted estimators

$$\hat{\theta}_\alpha^* = \hat{\theta} + \sum_{h=1}^{L} A_h^{-1/2} (\hat{Y}_{h(\alpha)} - \hat{Y}_h).$$

$$A_h = 1 + R_h.$$

It is clear that

$$\hat{\bar{\theta}}^* = \sum_{\alpha=1}^{k} \hat{\theta}_\alpha^* / k = \hat{\theta}$$

since the estimators are linear in the $\hat{Y}_{hi}$. If the A_h are known without error, then the usual random group estimator of variance

$$v(\hat{\bar{\theta}}^*) = \frac{1}{k(k-1)} \sum_{\alpha=1}^{k} (\hat{\theta}_\alpha^* - \hat{\bar{\theta}}^*)^2$$

is unbiased for $\mathrm{Var}\{\hat{\theta}\}$. The estimator $v(\hat{\bar{\theta}}^*)$ will be biased to the extent that the measures of R_h are erroneous. So long as the measures of R_h are reasonable, however, the bias of $v(\hat{\bar{\theta}}^*)$ should be reduced relative to that of $v(\hat{\theta})$.

The results of this subsection were developed given the assumption that n (or n_h) is an integer multiple of k. If $n = km + q$, with $0 < q < k$, then either of the techniques discussed in Subsection 2.4.3 may be used to obtain the variance estimator.

2.4.5. A General Result About Variance Estimation for Without Replacement Sampling

As was demonstrated in Theorem 2.4.4, the random group estimator tends to estimate the variance as if the sample were selected with replacement, even though it may in fact have been selected without replacement. The price we pay for this practice is a bias in the estimator of variance, although the bias is probably not very important in modern, large-scale surveys. An advantage of this method, in addition to the obvious computational advantages, is that the potentially troublesome calculation of the joint inclusion probabilities π_{ij} is avoided. In the case of srs wor, we can adjust for the bias by applying the usual finite population correction. For unequal probability sampling without replacement, there is no general correction to the variance estimator that accounts for the without replacement feature. In the case of multistage sampling, we know (cf., Theorem 2.4.5) that the bias occurs only in the between PSU component of variance.

This section is devoted to a general result that relates the bias of the random group variance estimator to the efficiency of without replacement sampling. We shall make repeated use of this result in future chapters. To simplify the discussion, we assume initially that a single-stage sample of size n is drawn from a finite population of size N, where Y_i is the value of the i-th unit in the population and p_i is the corresponding nonzero selection probability, possibly based upon some auxiliary measure of size X_i. We shall consider two methods of drawing the sample: (1) pps wr sampling and (2) an arbitrary πps scheme. The reader will recall that a πps scheme is a without replacement sampling design with inclusion probabilities $\pi_i = np_i$. Let

$$\hat{Y}_{\pi ps} = \sum_{i=1}^{n} y_i / \pi_i$$

denote the Horvitz–Thompson estimator of the population total given the πps scheme, where $\pi_i = np_i$ for $i = 1, \ldots, N$, and let

$$\hat{Y}_{wr} = (1/n) \sum_{i=1}^{n} y_i / p_i$$

denote the customary estimator of the population total Y given the pps wr scheme. Let $\text{Var}\{\hat{Y}_{\pi ps}\}$ and $\text{Var}\{\hat{Y}_{wr}\}$ denote the variances of $\hat{Y}_{\pi ps}$ and $\hat{Y}_{wr}$, respectively. Further, let

$$v(\hat{Y}_{wr}) = \{1/n(n-1)\} \sum_{i=1}^{n} (y_i/p_i - \hat{Y}_{wr})^2$$

be the usual unbiased estimator of $\text{Var}\{\hat{Y}_{wr}\}$. Then we have the following:

Theorem 2.4.6. *Suppose that we use the estimator $v(\hat{Y}_{wr})$ to estimate $\text{Var}\{\hat{Y}_{\pi ps}\}$ given the πps sampling design. Then, the bias of $v(\hat{Y}_{wr})$ is given by*

$$\text{Bias}\{v(\hat{Y}_{wr})\} = \frac{n}{n-1}(\text{Var}\{\hat{Y}_{wr}\} - \text{Var}\{\hat{Y}_{\pi ps}\}). \qquad (2.4.12)$$

PROOF. The variances of $\hat{Y}_{\pi ps}$ and $\hat{Y}_{wr}$ are

$$\text{Var}\{\hat{Y}_{\pi ps}\} = \sum_{i=1}^{N} \pi_i(1 - \pi_i)\left(\frac{Y_i}{\pi_i}\right)^2 + 2 \sum_{i=1}^{N} \sum_{j>i}^{N} (\pi_{ij} - \pi_i\pi_j)\left(\frac{Y_i}{\pi_i}\right)\left(\frac{Y_j}{\pi_j}\right) \qquad (2.4.13)$$

and

$$\text{Var}\{\hat{Y}_{wr}\} = \frac{1}{n}\sum_i^N p_i\left(\frac{Y_i}{p_i} - Y\right)^2,$$

respectively. Thus

$$\text{Var}\{\hat{Y}_{wr}\} - \text{Var}\{\hat{Y}_{\pi ps}\} = \frac{n-1}{n}Y^2 - 2\sum_{i=1}^N\sum_{j>i}^N \pi_{ij}\left(\frac{Y_i}{\pi_i}\right)\left(\frac{Y_j}{\pi_j}\right).$$

The variance estimator may be expressed as

$$v(\hat{Y}_{wr}) = \sum_{i=1}^n \left(\frac{y_i}{np_i}\right)^2 - \frac{2}{n-1}\sum_{i=1}^n\sum_{j>i}^n \left(\frac{y_i}{np_i}\right)\left(\frac{y_j}{np_j}\right)$$

with expectation (given the πps sampling design)

$$\text{E}\{v(\hat{Y}_{wr})\} = \sum_{i=1}^N \pi_i\left(\frac{Y_i}{np_i}\right)^2 - \frac{2}{n-1}\sum_{i=1}^N\sum_{j>1}^N \pi_{ij}\left(\frac{y_i}{np_i}\right)\left(\frac{y_j}{np_j}\right). \quad (2.4.14)$$

Combining (2.4.13) and (2.4.14) gives

$$\text{Bias}\{v(\hat{Y}_{wr})\} = Y^2 - \frac{2n}{n-1}\sum_{i=1}^N\sum_{j>1}^N \pi_{ji}\left(\frac{Y_i}{\pi_i}\right)\left(\frac{Y_j}{\pi_j}\right),$$

from which the result follows immediately. □

Theorem 2.4.6, originally due to Durbin (1953), implies that when we use the pps wr estimator $v(\hat{Y}_{wr})$ we tend to overestimate the variance of $\hat{Y}_{\pi ps}$ whenever that variance is smaller than the variance of $\hat{Y}_{wr}$ given pps wr sampling. Conversely, when pps wr sampling is more efficient than the πps scheme, the estimator $v(\hat{Y}_{wr})$ tends to underestimate $\text{Var}\{\hat{Y}_{\pi ps}\}$. Thus, we say that $v(\hat{Y}_{wr})$ is a conservative estimator of $\text{Var}\{\hat{Y}_{\pi ps}\}$ for the useful applications of πps sampling.

This result extends easily to the case of multistage sampling, and as before, the bias occurs only in the between PSU component of variance. To see this, consider estimators of the population total Y of the form

$$\hat{Y}_{\pi ps} = \sum_{i=1}^n \frac{\hat{y}_i}{\pi_i}$$

$$\hat{Y}_{wr} = \frac{1}{n}\sum_{i=1}^n \frac{\hat{y}_i}{p_i}$$

for πps sampling and pps wr sampling at the first stage, where $\hat{y}_i$ is an estimator of the total in the i-th selected PSU due to sampling at the second and subsequent stages. Consider the pps wr estimator of variance

$$v(\hat{Y}_{wr}) = \frac{1}{n(n-1)}\sum_{i=1}^n \left(\frac{\hat{y}_i}{p_i} - \hat{Y}_{wr}\right)^2. \quad (2.4.15)$$

Assuming that sampling at the second and subsequent stages is independent from one PSU selection to the next, $v(\hat{Y}_{wr})$ is an unbiased estimator of $\text{Var}\{\hat{Y}_{wr}\}$ given pps wr sampling. When used for estimating the variance of $\hat{Y}_{\pi ps}$ given πps sampling, $v(\hat{Y}_{wr})$ incurs the bias

$$\text{Bias}\{v(\hat{Y}_{wr})\} = Y^2 - \frac{2n}{n-1} \sum_{i=1}^{N} \sum_{j>i}^{N} \pi_{ij}\left(\frac{Y_i}{\pi_i}\right)\left(\frac{Y_j}{\pi_j}\right)$$

$$= \frac{n}{n-1}(\text{Var}\{\hat{Y}_{wr}\} - \text{Var}\{\hat{Y}_{\pi ps}\}),$$

(2.4.16)

where Y_i is the total of the i-th PSU. See Durbin (1953). This result confirms that the bias occurs only in the between PSU component of variance. The bias will be unimportant in many practical applications, particularly those where the between variance is a small fraction of the total variance.

Usage of (2.4.15) for estimating the variance of $\hat{Y}_{\pi ps}$ given πps sampling not only avoids the troublesome calculation of the joint inclusion probabilities π_{ij}, but also avoids the computation of estimates of the components of variance due to sampling at the second and subsequent stages. This is a particularly nice feature because the sampling designs used at these stages often do not admit unbiased estimators of the variance (e.g., systematic sampling).

The above results have important implications for the random group estimator of variance with $n = km$. By Theorem 2.4.4 we know that the random group estimator of $\text{Var}\{\hat{Y}_{\pi ps}\}$, given πps sampling and based upon $k \le n/2$ random groups, has the same expectation as, but equal or larger variance than, the estimator $v(\hat{Y}_{wr})$. In fact, $v(\hat{Y}_{wr})$ is the random group estimator for the case $(k, m) = (n, 1)$. Thus the expressions for bias given in Theorem 2.4.6 and in (2.4.16) apply to the random group estimator regardless of the number of groups k. Furthermore, all of these results may be extended to the case of sampling within $L \ge 2$ strata.

Finally, it is of some interest to investigate the properties of the estimator

$$v(\hat{Y}_{\pi ps}) = \sum_{i=1}^{n} \sum_{j>i}^{n} \frac{\pi_i \pi_j - \pi_{ij}}{\pi_{ij}}\left(\frac{\hat{y}_i}{\pi_i} - \frac{\hat{y}_j}{\pi_j}\right)^2$$

(2.4.17)

given a πps sampling design. This is the Yates–Grundy estimator of variance applied to the estimated PSU totals $\hat{y}_i$, and is the first term in the standard unbiased estimator of $\text{Var}\{\hat{Y}_{\pi ps}\}$. See Cochran (1977, pp. 300–302). The estimator (2.4.17) is not as simple as the estimator $v(\hat{Y}_{wr})$ (or the random group estimator with $k \le n/2$) because it requires computation of the joint inclusion probabilities π_{ij}. However, it shares the desirable feature that the calculation of estimates of the within variance components is avoided. Its expectation is easily established.

Theorem 2.4.7. *Suppose that we use $v(\hat{Y}_{\pi ps})$ to estimate $\text{Var}\{\hat{Y}_{\pi ps}\}$ given a*

πps *sampling design. Then*

$$\text{Bias}\{v(\hat{Y}_{\pi\text{ps}})\} = -\sum_{i=1}^{N} \sigma_{2i}^2,$$

where

$$\sigma_{2i}^2 = \text{Var}\{\hat{Y}_i | i\}$$

is the conditional variance of $\hat{Y}_i$ *due to sampling at the second and subsequent stages given that* i-th PSU *is in the sample.*

PROOF. Follows directly from Cochran's (1977) Theorem 11.2. □

It is interesting to contrast this theorem with earlier results. Contrary to the random group estimator (or $v(\hat{Y}_{\text{wr}})$), the estimator $v(\hat{Y}_{\pi\text{ps}})$ is always *downward* biased and the bias is in the *within* PSU component of variance. Since the bias of this estimator is in the opposite direction (in the useful applications of πps sampling) from that of $v(\hat{Y}_{\text{wr}})$, the interval

$$(v(\hat{Y}_{\pi\text{ps}}), v(\hat{Y}_{\text{wr}}))$$

may provide useful bounds on the variance $\text{Var}\{\hat{Y}_{\pi\text{ps}}\}$. Of course, the random group estimator may be substituted for $v(\hat{Y}_{\text{wr}})$ in this expression.

We speculate that in many cases the absolute biases will be in the order

$$|\text{Bias}\{v(\hat{Y}_{\text{wr}})\}| \le |\text{Bias}\{v(\hat{Y}_{\pi\text{ps}})\}|.$$

This is because the within variance dominates the total variance $\text{Var}\{\hat{Y}_{\pi\text{ps}}\}$ in many modern, large-scale surveys and the bias of $v(\hat{Y}_{\pi\text{ps}})$ is in that component. Also, the within component of $\text{Var}\{\hat{Y}_{\pi\text{ps}}\}$ is $\sum_{i=1}^{N} \sigma_{2i}^2/\pi_i$ so that the relative bias of $v(\hat{Y}_{\pi\text{ps}})$ is bounded by

$$\frac{|\text{Bias}\{v(\hat{Y}_{\pi\text{ps}})\}|}{\text{Var}\{\hat{Y}_{\pi\text{ps}}\}} \le \frac{\displaystyle\sum_{i=1}^{N} \sigma_{2i}^2}{\displaystyle\sum_{i=1}^{N} \sigma_{2i}^2/\pi_i}$$

$$\le \max_i \pi_i.$$

This bound may not be very small in large-scale surveys when sampling PSUs within strata. In any case, the random group estimator (or $v(\hat{Y}_{\text{wr}})$) will be preferred when a conservative estimator of variance is desired.

2.5. The Collapsed Stratum Estimator

Considerations of efficiency sometimes lead to sampling designs which call for a single primary unit (PSU) per stratum. In such cases an unbiased estimator of the variance is not available, not even for linear statistics. Nor is a consistent estimator available. It is possible, however, to give an estimator which tends towards an overestimate of the variance. This is the *collapsed stratum estimator*, and it is closely related to the random group estimator discussed elsewhere in this chapter. Generally speaking, the collapsed stratum estimator is applicable only to problems of estimating the variance of linear statistics. In the case of nonlinear estimators, the variance may be estimated by a combination of collapsed stratum and Taylor series methodology (see Chapter 6).

We suppose that it is desired to estimate a population total, say Y, using an estimator of the form

$$\hat{Y} = \sum_{h=1}^{L} \hat{Y}_h, \qquad (2.5.1)$$

where L denotes the number of strata and $\hat{Y}_h$ an estimator of the total in the h-th stratum, Y_h, resulting from sampling within the h-th stratum. We assume that one primary unit is selected independently from each of the L strata, that any subsampling is independent from one primary to the next, but otherwise leave unspecified the nature of the subsampling schemes within primaries. Note that this form (2.5.1) includes as special cases the Horvitz–Thompson estimator and such nonlinear estimators as the separate ratio and regression estimators. It does not include the combined ratio and regression estimators. To estimate the variance of $\hat{Y}$, we combine the L strata into G groups of at least two strata each. In practice, each group often contains exactly two strata.

Let us begin by considering the simple case where L is even and each group contains precisely two of the original strata. Then the estimator of total may be expressed as

$$\hat{Y} = \sum_{g=1}^{G} \hat{Y}_g = \sum_{g=1}^{G} (\hat{Y}_{g1} + \hat{Y}_{g2}),$$

where $L = 2G$ and $\hat{Y}_{gh}(h = 1, 2)$ denotes the estimator of total for the h-th stratum in the g-th group (or collapsed stratum). If we ignore the original stratification within groups and treat $(g, 1)$ and $(g, 2)$ as independent selections from the g-th group, for $g = 1, \ldots, G$, then the natural estimator of the variance of $\hat{Y}_g$ is

$$v_{cs}(\hat{Y}_g) = (\hat{Y}_{g1} - \hat{Y}_{g2})^2.$$

The corresponding estimator of the variance of $\hat{Y}$ is

$$v_{cs}(\hat{Y}) = \sum_{g=1}^{G} v_{cs}(\hat{Y}_g) \qquad (2.5.2)$$

$$= \sum_{g=1}^{G} (\hat{Y}_{g1} - \hat{Y}_{g2})^2.$$

In fact, the expectation of this estimator given the original sampling design is

$$E\{v_{cs}(\hat{Y})\} = \sum_{g=1}^{G} (\sigma_{g1}^2 + \sigma_{g2}^2) + \sum_{g=1}^{G} (\mu_{g1} - \mu_{g2})^2,$$

where $\sigma_{gh}^2 = \text{Var}\{\hat{Y}_{gh}\}$ and $\mu_{gh} = E\{\hat{Y}_{gh}\}$. Since the variance of $\hat{Y}$ is

$$\text{Var}\{\hat{Y}\} = \sum_{g=1}^{G} (\sigma_{g1}^2 + \sigma_{g2}^2),$$

the collapsed stratum estimator is biased by the amount

$$\text{Bias}\{v_{cs}(\hat{Y})\} = \sum_{g=1}^{G} (\mu_{g1} - \mu_{g2})^2. \tag{2.5.3}$$

This, of course, tends to be an upward bias since the right side of (2.5.3) is nonnegative.

Equation (2.5.3) suggests a strategy for grouping the original strata so as to minimize the bias of the collapsed stratum estimator. The strategy is to form groups so that the means μ_{gh} are as alike as possible within groups, i.e., the differences $|\mu_{g1} - \mu_{g2}|$ are as small as possible. If the estimator $\hat{Y}_{gh}$ is unbiased for the stratum total Y_{gh}, or approximately so, then essentially $\mu_{gh} = Y_{gh}$ and the formation of groups is based upon similar stratum totals, i.e., small values of $|Y_{g1} - Y_{g2}|$.

Now suppose that a known auxiliary variable A_{gh} is available for each stratum and that this variable is well correlated with the expected values μ_{gh} (i.e., essentially well-correlated with the stratum totals Y_{gh}). For this case Hansen, Hurwitz, and Madow (1953) give the following estimator which is intended to reduce the bias term (2.5.3):

$$v_{cs}(\hat{Y}) = \sum_{g=1}^{G} v_{cs}(\hat{Y}_g) \tag{2.5.4}$$

$$= \sum_{g=1}^{G} 4(P_{g2}\hat{Y}_{g1} - P_{g1}\hat{Y}_{g2})^2,$$

where $P_{gh} = A_{gh}/A_g$, $A_g = A_{g1} + A_{g2}$. When $P_{gh} = 1/2$ for all (g, h) then this estimator is equivalent to the simple estimator (2.5.2). The bias of this estimator contains the term

$$\sum_{g=1}^{G} 4(P_{g2}\mu_{g1} - P_{g1}\mu_{g2})^2, \tag{2.5.5}$$

which is analogous to (2.5.3). If the measure of size is such that $\mu_{gh} = \beta A_{gh}$ (or approximately so) for all (g, h), then the bias component (2.5.5) vanishes (or approximately so). On this basis Hansen, Hurwitz, and Madow's estimator might be preferred uniformly to the simple estimator (2.5.2). Unfortunately this may not always be the case because two additional terms appear in the bias of (2.5.4) that did not appear in the bias of the simple estimator (2.5.2). These terms, which we shall display formally in Theorem 2.5.1, essentially have to do with the variability of the A_{gh} within groups. When this variability is small relative to unity, these components of bias should be small, and otherwise not. Thus the choice of (2.5.2) or (2.5.4) involves some judgment about which of several components of bias dominates a particular application.

In many real applications of the collapsed stratum estimator, A_{gh} is taken simply to be the number of elementary units within the (g, h)-th stratum. For example, Section 2.12 presents a survey concerned with estimating total consumer expenditures on certain products, and A_{gh} is the population of the (g, h)-th stratum.

The estimators generalize easily to the case of more than two strata per group. Once again let G denote the number of groups, and let L_g denote the number of original strata in the g-th group. The estimator of the total is

$$\hat{Y} = \sum_{g=1}^{G} \hat{Y}_g = \sum_{g=1}^{G} \sum_{h=1}^{L_g} \hat{Y}_{gh} \tag{2.5.6}$$

and the Hansen, Hurwitz, and Madow's estimator of variance is

$$v_{cs}(\hat{Y}) = \sum_{g=1}^{G} [L_g/(L_g - 1)] \sum_{h}^{L_g} (\hat{Y}_{gh} - P_{gh}\hat{Y}_g)^2, \tag{2.5.7}$$

where $P_{gh} = A_{gh}/A_g$ and

$$A_g = \sum_{h=1}^{L_g} A_{gh}.$$

If we take $A_{gh}/A_g = 1/L_g$ for $g = 1, \ldots, G$, then the estimator reduces to

$$v_{cs}(\hat{Y}) = \sum_{g=1}^{G} [L_g/(L_g - 1)] \sum_{h=1}^{L_g} (\hat{Y}_{gh} - \hat{Y}_g/L_g)^2, \tag{2.5.8}$$

which is the generalization of the simple collapsed stratum estimator (2.5.2). If the $\hat{Y}_{gh}$ were a random sample from the g-th collapsed stratum, then the reader would recognize (2.5.8) as the random group estimator. In fact though, the $\hat{Y}_{gh}$ do not constitute a random sample from the g-th group, and $v_{cs}(\hat{Y})$ is a biased and inconsistent estimator of $\mathrm{Var}\{\hat{Y}\}$.

Theorem 2.5.1. *Let $\hat{Y}$ and $v_{cs}(\hat{Y})$ be defined by (2.5.6) and (2.5.7), respectively. Let the sampling be conducted independently in each of the L strata, so that the $\hat{Y}_{gh}$ are independent random variables. Then,*

$$E\{v_{cs}(\hat{Y})\} = \sum_{g=1}^{G} \frac{L_g - 1 + V^2_{A(g)} - 2V_{A(g), \sigma(g)}}{L_g - 1} \sigma_g^2$$

$$+ \sum_{g=1}^{G} \frac{L_g}{L_g - 1} \sum_{h=1}^{L_g} (\mu_{gh} - P_{gh}\mu_g)^2,$$

$$\mathrm{Var}\{\hat{Y}\} = \sum_{g=1}^{G} \sigma_g^2,$$

and

$$\text{Bias}\{v_{cs}(\hat{Y})\} = \sum_{g=1}^{G} \frac{V_{A(g)}^2 - 2V_{A(g),\sigma(g)}}{L_g - 1} \sigma_g^2$$

$$+ \sum_{g=1}^{G} \frac{L_g}{L_g - 1} \sum_{h=1}^{L_g} (\mu_{gh} - P_{gh}\mu_g)^2,$$

where

$$\mu_{gh} = \text{E}\{\hat{Y}_{gh}\}, \mu_g = \sum_{h=1}^{L_g} \mu_{gh}$$

$$\sigma_{gh}^2 = \text{Var}\{\hat{Y}_{gh}\},$$

$$\sigma_g^2 = \sum_{h=1}^{L_g} \sigma_{gh}^2 ,$$

$$V_{A(g)}^2 = \sum_{h=1}^{L_g} A_{gh}^2 / L_g \bar{A}_g^2 - 1,$$

$$V_{A(g),\sigma(g)} = \sum_{h=1}^{L_g} A_{gh}\sigma_{gh}^2 / \bar{A}_g \sigma_g^2 - 1,$$

and

$$\bar{A}_g = \sum_{h=1}^{L_g} A_{gh} / L_g .$$

PROOF. See Hansen, Hurwitz, and Madow (1953), Volume II, Chapter 9.
□

Corollary. *If $P_{gh} = 1/L_g$ for all h and g, then*

$$\text{Bias}\{v_{cs}(\hat{Y})\} = \sum_{g=1}^{G} \frac{L_g}{L_g - 1} \sum_{h=1}^{L_g} (\mu_{gh} - \mu_g/L_g)^2.$$
□

Corollary. *If $\mu_{gh} = \beta_g A_{gh}$ for all h and g, then*

$$\text{Bias}\{v_{cs}(\hat{Y})\} = \sum_{g=1}^{G} \frac{V_{A(g)}^2 - 2V_{A(g),\sigma(g)}}{L_g - 1} \sigma_g^2 .$$
□

Corollary. *If both $P_{gh} = 1/L_g$ and $\mu_{gh} = \mu_g/L_g$ for all g and h, then $v_{cs}(\hat{Y})$ is an unbiased estimator of* $\text{Var}\{\hat{Y}\}$.
□

Theorem 2.5.1 gives the expectation and bias of the collapsed stratum estimator. It is clear that $v_{cs}(\hat{Y})$ tends to give an overestimate of the variance whenever the A_{gh} are similar within each group. If the A_{gh} are dissimilar within groups so that $V_{A(g)}^2$ and $V_{A(g),\sigma(g)}$ are large relative to $L_g - 1$, the bias could be in either direction. To reduce the bias, one may group strata so that the expected values μ_{gh} (essentially the stratum totals Y_{gh}), are similar within each group, choose $A_{gh} \doteq \beta_g^{-1}\mu_{gh}$ for some constant β_g, or both, as is evident from the corollaries. As was noted earlier for the special case $L_g = 2$, the choice between equal $P_{gh} = 1/L_g$ and other alternatives involves some judgment about which components of bias dominate and how closely the available measures of size are to being proportional to the μ_{gh} (or Y_{gh}).

A word of caution regarding the grouping of strata is in order. While it is true that strata should be grouped so that the μ_{gh} (or the totals Y_{gh}) are alike, the grouping must be performed prior to looking at the observed data. If one groups on the basis of similar $\hat{Y}_{gh}$, a severe downward bias may result. Another problem to be avoided is the grouping of a self-representing (SR) primary with a nonself-representing (NSR) primary.[6] Since the SR PSU is selected with probability one, it contributes only to the within PSU component of variance, not to the between component. The collapsed stratum estimator, however, would treat such a PSU as contributing to both components of variance, thus increasing the overstatement of the total variance.

One of the conditions of Theorem 2.5.1 is that sampling be conducted independently in each of the strata. Strictly speaking, this eliminates sampling schemes such as controlled selection, where a dependency exists between the selections in different strata. See, e.g., Goodman and Kish (1950) and Ernst (1981). Nevertheless, because little else is available, the collapsed stratum estimator is often used to estimate the variance for controlled selection designs. The theoretical properties of this practice are not known, although Brooks (1977) has investigated them empirically. Using 1970 Census data on labor force participation, school enrollment, and income, the bias of the collapsed stratum estimator was computed for the Census Bureau's Current Population Survey (CPS). The reader will note that in this survey the PSUs were selected using a controlled selection procedure (see Hanson (1978)). In almost every instance the collapsed stratum estimator resulted in an overstatement of the variance. For estimates concerned with Blacks, the ratios of the expected value of the variance estimator to the true variance were almost always between 1.0 and 2.0, while for Whites the ratios were between 3.0 and 4.0.

Finally, in a recent unpublished paper, Bev Causey (1982) has shown that the collapsed stratum estimator may be seriously upward biased for characteristics related to labor force status for a sampling design where counties are the primary sampling units. Both the Brooks and Causey results support the view that the more effective the stratification (or controlled selection) is in reducing the true variance of estimate, the greater the bias in the collapsed stratum estimator of variance.

---

[6] A SR primary is one selected with probability one. A NSR primary is one selected with probability less than one from a stratum containing two or more primaries.

In discussing the collapsed stratum estimator, we have presented the relatively simple situation where one unit is selected from each stratum. We note, however, that mixed strategies for variance estimation are available, and even desirable, depending upon the nature of the sampling design. To illustrate a mixed strategy involving the collapsed stratum estimator, suppose that there are $L = L' + L''$ strata, where one unit is selected independently from each of the first L' strata and two units are selected from each of the remaining L'' strata. Suppose that the sampling design used in these latter strata is such that it permits an unbiased estimator of the within stratum variance. Then, for an estimator of the form

$$\hat{Y} = \sum_{h=1}^{L} \hat{Y}_h$$

$$= \sum_{h=1}^{L'} \hat{Y}_h + \sum_{h=L'+1}^{L} \hat{Y}_h$$

$$= \sum_{g=1}^{G'} \sum_{h=1}^{L'_g} \hat{Y}_{gh} + \sum_{h=L'+1}^{L} \hat{Y}_h$$

$$L' = \sum_{g=1}^{G'} L'_g,$$

we may estimate the variance by

$$v(\hat{Y}) = \sum_{g=1}^{G'} \frac{L'_g}{L'_g - 1} \sum_{h=1}^{L'_g} (\hat{Y}_{gh} - P_{gh}\hat{Y}_g)^2 + \sum_{h=L'+1}^{L} v(\hat{Y}_h) \qquad (2.5.9)$$

where $v(\hat{Y}_h)$, for $h = L'+1, \ldots, L$, denotes an unbiased estimator of the variance of $\hat{Y}_h$ based upon sampling within the h-th stratum. In this illustration, the collapsed stratum estimator is only used for those strata where one primary unit is sampled and another, presumably unbiased, variance estimator is used for those strata where more than one primary unit is sampled. Another illustration of a mixed strategy occurs in the case of self-representing (SR) primary units (i.e., the stratum contains only one primary and it is selected with certainty). Suppose now that the L'' strata each contain one SR primary, and that the L' are as before. The variance estimator is again of the form (2.5.9), where the $v(\hat{Y}_h)$ now represent estimators of the variance due to sampling within the self-representing primaries. Of course, the $v(\hat{Y}_h)$ will themselves be collapsed stratum estimators if the SR primary is stratified internally to the point where one secondary unit is selected per stratum.

As we have seen, the collapsed stratum estimator is usually, though not necessarily, upward biased, depending upon the measure of size A_{gh} and its relation to the stratum totals Y_{gh}. In an effort to reduce the size of the bias several authors have suggested alternative variance estimators for one-per-stratum sampling designs:

(a) The method of Hartley, Rao, and Kiefer (1969) relies upon a linear model connecting the Y_h with one or more known measures of size. No collapsing of strata is required. Since the Y_h are unknown, the model is fit using the $\hat{Y}_h$. Estimates $\hat{\sigma}_h^2$ of the within stratum variances σ_h^2 are then prepared from the regression residuals, and the overall variance is estimated by

$$v(\hat{Y}) = \sum_{h=1}^{L} \hat{\sigma}_h^2 .$$

The bias of this statistic as an estimator of $\text{Var}\{\hat{Y}\}$ is a function of the error variance of the true Y_h about the assumed regression line.

(b) Fuller's (1970) method depends on the notion that the stratum boundaries are chosen by a random process prior to sample selection. This preliminary stage of randomization yields nonnegative joint inclusion probabilities for sampling units in the same stratum. Without this randomization, such joint inclusion probabilities are zero. The Yates–Grundy (1953) estimator may then be used for estimating $\text{Var}\{\hat{Y}\}$, where the inclusion probabilities are specified by Fuller's scheme. The estimator incurs a bias in situations where the stratum boundaries are not randomized in advance.

(c) The original collapsed stratum estimator (2.5.2) was derived via the supposition that the primaries were selected with replacement from within the collapsed strata. Alternatively, numerous variance estimators may be derived by hypothesizing some without replacement sampling scheme within collapsed strata. A simple possibility is

$$v(\hat{Y}) = \sum_{g=1}^{G} (1 - 2/N_g)(\hat{Y}_{g1} - \hat{Y}_{g2})^2,$$

where N_g is the number of primary units in the g-th collapsed stratum. For this estimator srs wor is assumed and $(1 - 2/N_g)$ is the finite population correction. Shapiro and Bateman (1978) have suggested another possibility, where Durbin's (1967) sampling scheme is hypothesized within collapsed strata. The authors suggest using the Yates and Grundy (1953) variance estimator with the values of the inclusion probabilities specified by Durbin's scheme. The motivation behind all such alternatives is that variance estimators derived via without replacement assumptions should be less biased for one-per-stratum designs than estimators derived via *with replacement* assumptions.

The above methods appear promising for one-per-stratum designs. In fact, each of the originating authors gives an example where the new method is less biased than the collapsed stratum estimator. More comparative studies of the methods are needed though before a definitive recommendation can be made about preferences for the various estimators.

2.6. Stability of the Random Group Estimator of Variance

In most descriptive surveys, emphasis is placed on estimating parameters θ such as a population mean, a population total, a ratio of two totals, and so on. An estimator of variance is needed at the analysis stage in order to interpret the survey results and to make statistical inferences about θ. The variance of the survey estimator $\hat{\theta}$ is also of importance at the design stage, where the survey statistician is attempting to optimize the survey design and to choose a large enough sample to produce the desired levels of precision for $\hat{\theta}$. A subordinate problem in most surveys, though still a problem of great importance, is the stability or precision of the variance estimator. A related question in the context of the present chapter is "How many random groups are needed?"

One general criterion for assessing the stability of the random group estimator $v(\hat{\bar{\theta}})$ is its coefficient of variation,

$$\mathrm{CV}\{v(\hat{\bar{\theta}})\} = [\mathrm{Var}\{v(\hat{\bar{\theta}})\}]^{1/2}/\mathrm{Var}\{\hat{\bar{\theta}}\}.$$

We shall explore the CV criterion in this section. Another general criterion is the proportion of intervals

$$(\hat{\bar{\theta}} - c\{v(\hat{\bar{\theta}})\}^{1/2},\ \hat{\bar{\theta}} + c\{v(\hat{\bar{\theta}})\}^{1/2})$$

that contain the true population parameter θ, where c is a constant, often based upon normal or Student's t theory. This criterion will be addressed in Appendix C. Finally, the quality of a variance estimator may be assessed by its use in other statistical analyses, though no results about such criteria are presented in this book.

With respect to the CV criterion, we begin with the following theorem.

Theorem 2.6.1. *Let $\hat{\theta}_1, \ldots, \hat{\theta}_k$ be independent and identically distributed random variables, and let $v(\hat{\bar{\theta}})$ be defined by (2.2.1). Then*

$$\mathrm{CV}\{v(\hat{\bar{\theta}})\} = \left\{ \frac{\beta_4(\hat{\theta}_1) - (k-3)/(k-1)}{k} \right\}^{1/2}, \tag{2.6.1}$$

where

$$\beta_4(\hat{\theta}_1) = \frac{\mathrm{E}\{(\hat{\theta}_1 - \mu)^4\}}{[\mathrm{E}\{(\hat{\theta}_1 - \mu)^2\}]^2}$$

$$\mu = \mathrm{E}\{\hat{\theta}_1\}.$$

PROOF. Since the $\hat{\theta}_\alpha$ are independent we have

$$\mathrm{E}\{v^2(\hat{\bar{\theta}})\} = \frac{1}{k^4} \sum_{\alpha=1}^{k} \kappa_4(\hat{\theta}_\alpha)$$

$$+ \frac{2}{k^4}\left(1 + \frac{2}{(k-1)^2}\right) \sum_{\alpha=1}^{k} \sum_{\beta>\alpha}^{k} \kappa_2(\hat{\theta}_\alpha)\kappa_2(\hat{\theta}_\beta),$$

where

$$\kappa_4(\hat{\theta}_\alpha) = \mathrm{E}\{(\hat{\theta}_\alpha - \mu)^4\},$$

$$\kappa_2(\hat{\theta}_\alpha) = \mathrm{E}\{(\hat{\theta}_\alpha - \mu)^2\}.$$

And by the identically distributed condition,

$$\mathrm{Var}\{v(\hat{\bar{\theta}})\} = \frac{1}{k^3}\kappa_4(\hat{\theta}_1) + \frac{k-1}{k^3}\frac{k^2 - 2k + 3}{(k-1)^2}\kappa_2^2(\hat{\theta}_1) - \mathrm{E}^2\{v(\hat{\bar{\theta}})\}.$$

The result follows by the definition of the coefficient of variation.   □

From this theorem we see that the CV of the variance estimator depends upon both the kurtosis $\beta_4(\hat{\theta}_1)$ of the estimator and the number of groups k. If k is small, the CV will be large and the variance estimator of low precision. If the distribution of $\hat{\theta}_1$ has an excess of values near the mean and in the tails, the kurtosis $\beta_4(\hat{\theta}_1)$ will be large and the variance estimator of low precision. When k is large, we see that the CV^2 is approximately inversely proportional to the number of groups

$$\mathrm{CV}^2\{v(\hat{\bar{\theta}})\} \doteq \frac{\beta_4(\hat{\theta}_1) - 1}{k}.$$

Theorem 2.6.1 can be sharpened for specific estimators and sampling designs. Two common cases are treated in the following corollaries, which we state without proof.

Corollary. *A simple random sample with replacement is divided into k groups of $m = n/k$ units each. Let $\hat{\theta}_\alpha$ denote the sample mean based on the α-th group; let $\hat{\bar{\theta}} = \sum_{\alpha=1}^{k} \hat{\theta}_\alpha/k = \sum_{i=1}^{n} y_i/n$; and let $v(\hat{\bar{\theta}})$ be defined by (2.2.1). Then,*

$$\mathrm{CV}\{v(\hat{\bar{\theta}})\} = \left\{ \frac{\beta_4(\hat{\theta}_1) - (k-3)/(k-1)}{k} \right\}^{1/2}, \qquad (2.6.2)$$

where

$$\beta_4(\hat{\theta}_1) = \beta_4/m + 3(m-1)/m$$

$$\beta_4 = \frac{\displaystyle\sum_{i=1}^{N}(Y_i - \bar{Y})^4/N}{\left\{\displaystyle\sum_{i=1}^{N}(Y_i - \bar{Y})^2/N\right\}^2}.$$ □

Corollary. *A pps wr sample is divided into k groups of $m = n/k$ units each. Let*

$$\hat{\theta}_\alpha = \frac{1}{m}\sum_{i=1}^{m} y_i/p_i$$

denote the usual estimator of the population total based on the α-th group; let

$$\hat{\bar{\theta}} = \sum_{\alpha=1}^{k} \hat{\theta}_{\alpha}/k = \frac{1}{n} \sum_{i=1}^{n} y_i/p_i;$$

and let $v(\hat{\bar{\theta}})$ be defined by (2.2.1). Then,

$$\mathrm{CV}\{v(\hat{\bar{\theta}})\} = \left\{ \frac{\beta_4(\hat{\theta}_1) - (k-3)/(k-1)}{k} \right\}^{1/2}, \qquad (2.6.3)$$

where

$$\beta_4(\hat{\theta}_1) = \beta_4/m + 3(m-1)/m$$

$$\beta_4 = \frac{\sum\limits_{i=1}^{N} (Z_i - \bar{Z})^4/N}{\left\{ \sum\limits_{i=1}^{N} (Z_i - \bar{Z})^2/N \right\}^2}$$

$$Z_i = Y_i/p_i. \qquad \square$$

Both corollaries work with an estimator $\hat{\bar{\theta}}$ that is in the form of a sample mean, first for the y-variable and second for the z-variable. Correspondingly, the first corollary expresses the CV as a function of the kurtosis of the y-variable, while the second corollary expresses the CV as a function of the kurtosis of the z-variable. In this latter case, it is the distribution of z that is important, and when there is an excess of observations in the tail of this distribution, then β_4 is large, making $\beta_4(\hat{\theta}_1)$ large and the precision of $v(\hat{\bar{\theta}})$ low. Both corollaries are important in practical applications because it may be easier to interpret the kurtosis of y or z than that of $\hat{\theta}_1$.

As has been observed, $\mathrm{CV}\{v(\hat{\bar{\theta}})\}$ is an increasing function of $\beta_4(\hat{\theta}_1)$ and a decreasing function of k. The size of the random groups m exerts a minor influence on $\beta_4(\hat{\theta}_1)$, and thus on the $\mathrm{CV}\{v(\hat{\bar{\theta}})\}$. Because the kurtosis $\beta_4(\hat{\theta}_1)$ is essentially of the form $a/m + b$, where a and b are constants, it will decrease significantly as m increases initially from 1. As m becomes larger and larger, however, a law of diminishing returns takes effect, and the decrease in the kurtosis $\beta_4(\hat{\theta}_1)$ becomes less important. The marginal decrease in $\beta_4(\hat{\theta}_1)$ for larger and larger m is not adequate to compensate for the necessarily decreased k. Thus, the number of groups k has more of an impact on decreasing the $\mathrm{CV}\{v(\hat{\bar{\theta}})\}$ and increasing the precision of the variance estimate than does the size of the groups m.

While Theorem 2.6.1. and its corollaries were for the case of independent random groups, we may regard these results as approximate in the more common situation of without replacement sampling, particularly in large populations with small sampling fractions. This is demonstrated by Hansen,

Hurwitz, and Madow's (1953) result that

$$
\begin{aligned}
\mathrm{CV}\{v(\hat{\bar{\theta}})\} = \frac{(N-1)}{N(n-1)} \Bigg\{ & \Bigg\{ \frac{(n-1)^2}{n} - \frac{n-1}{n(N-1)}[(n-2)(n-3)-(n-1)] \\
& - \frac{4(n-1)(n-2)(n-3)}{n(N-1)(N-2)} - \frac{6(n-1)(n-2)(n-3)}{n(N-1)(N-2)(N-3)} \Bigg\} \beta_4 \\
& + \Bigg\{ \frac{(n-1)N}{n(N-1)}[(n-1)^2+2] \\
& + \frac{2(n-1)(n-2)(n-3)N}{n(N-1)(N-2)} + \frac{3(n-1)(n-2)(n-3)N}{n(N-1)(N-2)(N-3)} \\
& - \frac{N^2(n-1)^2}{(N-1)^2} \Bigg\} \Bigg\}^{1/2}
\end{aligned}
$$

$$(2.6.4)$$

for srs wor with $\hat{\bar{\theta}} = \bar{y}$ and $k = n$, $m = 1$. Clearly, when N and n are large and the sampling fraction n/N is small, (2.6.4) and (2.6.2) are approximately equal to one another. Based on this result we suggest that Theorem 2.6.1 and its corollaries may be used to a satisfactory approximation in studying the stability of the random group variance estimator for modern complex sample surveys, even for without replacement designs and nonindependent random groups.

Theorem 2.6.1 and its corollaries may be used to address two important questions: "How many random groups should be used?" and "What values of m and k are needed to meet a specified level of precision (i.e., a specified level of $\mathrm{CV}\{v(\hat{\bar{\theta}})\}$, say CV*)"?

The question about the number of random groups k involves many considerations, including both precision and cost. From a precision perspective, as has been noted, we would like to choose k as large as possible. From a cost perspective, however, increasing k implies increasing computational costs. Thus, the optimum value of k will be one that compromises and balances the cost and precision requirements. These requirements will, of course, vary from one survey to another. In one case, the goals of the survey may only seek to obtain a rough idea of the characteristics of a population, and cost considerations may outweigh precision considerations, suggesting that the optimum value of k is low. On the other hand, if major policy decisions are to be based on the survey results, precision considerations may prevail, suggesting a large value of k.

To show that a formal analysis of the cost-precision trade-off is possible, consider the simple case of srs wor where the sample mean $\hat{\bar{\theta}} = \bar{y}$ is used to estimate the population mean $\bar{Y}$. If the random group estimator of variance is computed according to the relation

$$
v(\hat{\bar{\theta}}) = \frac{1}{k(k-1)} \sum_{\alpha=1}^{k} (\hat{\theta}_\alpha - \hat{\bar{\theta}})(\hat{\theta}_\alpha - \hat{\bar{\theta}}),
$$

then $(m + 3)k + 1$ addition or subtraction instructions and $2k + 3$ multiplication or division instructions are used. Now suppose that C dollars are available in the survey budget for variance calculations and c_1 and c_2 are the per unit costs of an addition or subtraction instruction and a multiplication or division instruction, respectively. Then, m and k should be chosen to minimize $CV\{v(\hat{\bar{\theta}})\}$ subject to the cost constraint

$$\{(m + 3)k + 1\}c_1 + (2k + 3)c_2 \leq C. \tag{2.6.5}$$

As has been observed, however, $CV\{v(\hat{\bar{\theta}})\}$ is, to a good approximation, a decreasing function of k, and thus the approximate optimum is the largest value of k (and the corresponding m) that satisfies the constraint (2.6.5). In multipurpose surveys where two or more statistics are to be published, the objective function may be a linear combination of the individual CVs or the CV associated with the most important single statistic. Although the above analysis was for the simple case of srs wor and the sample mean, it may be extended in principal to complex survey designs and estimators.

To answer the second question, we suggest setting (2.6.1), (2.6.2), or (2.6.3) equal to the desired level of precision CV*. Then one can study the values of (m, k) needed to meet the precision constraint. Usually, many alternative values of (m, k) will be satisfactory, and the one which is preferred will be the one that minimizes total costs. In terms of the formal analysis, the cost model (2.6.5) is now the objective function to be minimized subject to the constraint on the coefficient of variation

$$CV\{v(\hat{\bar{\theta}})\} \leq CV^*.$$

In practical applications, some knowledge of the kurtosis $\beta_4(\hat{\theta}_1)$ (or β_4) will be necessary in addressing either of these two questions. If data from a prior survey is available, either from the same or similar populations, then such data should be used to estimate $\beta_4(\hat{\theta}_1)$. In the absence of such data, some form of subjective judgment or expert opinion will have to be relied upon.

In this connection, Tables 2.6.1 through 2.6.11 present information about 11 families of distributions, some discrete, some continuous, and one of a mixed type. Each table contains six properties of the associated distribution:

  (i) the density function,
 (ii) the constraints on the parameters of the distribution,
(iii) plots of the distribution for alternative values of the parameters,
 (iv) the mean,
  (v) the variance, and
 (vi) the kurtosis.

These tables may be useful in developing some idea of the magnitude of the kurtosis $\beta_4(\hat{\theta}_1)$ (or β_4). Simply choose the distribution that best represents the finite population under study and read the theoretical value of the kurtosis for that distribution. The choice of distribution may well be a

Table 2.6.1. Bernoulli Distribution

$$f(x) = Q \qquad , x = 0$$
$$= P \qquad , x = 1$$
$$= 0 \qquad , \text{otherwise.}$$

$$P + Q = 1$$
$$0 \le P \le 1$$

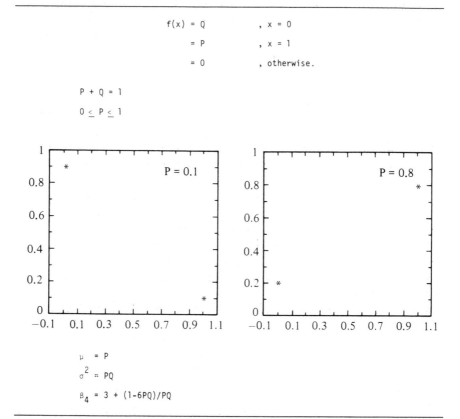

$$\mu = P$$
$$\sigma^2 = PQ$$
$$\beta_4 = 3 + (1-6PQ)/PQ$$

Table 2.6.2. Discrete Uniform Distribution

$$
\begin{aligned}
f(x) &= (K+1)^{-1} & , \ x &= 0 \\
&= (K+1)^{-1} & , \ x &= K^{-1} \\
&= (K+1)^{-1} & , \ x &= 2K^{-1} \\
&\quad \vdots & &\quad \vdots \\
&= (K+1)^{-1} & , \ x &= 1 \\
&= 0 & , \ &\text{otherwise}
\end{aligned}
$$

$$K \quad = 1, 2, \ldots$$

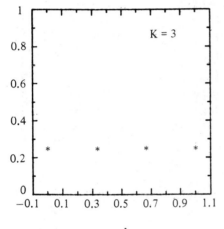

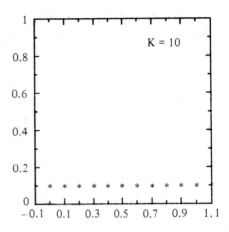

$$
\mu = \frac{1}{2}
$$

$$
\sigma^2 = \frac{1}{12} + \frac{1}{6K}
$$

$$
\beta_4 = 3\{3(K+1)^2 - 7\}/5\{(K+1)^2 - 1\}
$$

Table 2.6.3. Poisson Distribution

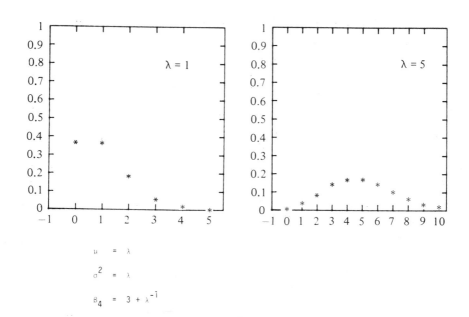

$$f(x) = e^{-\lambda} \lambda^x / x! \qquad , x = 0, 1, 2, \ldots$$
$$= 0 \qquad , \text{otherwise}$$

$$\lambda > 0$$

$$\mu = \lambda$$
$$\sigma^2 = \lambda$$
$$\beta_4 = 3 + \lambda^{-1}$$

Table 2.6.4. Logarithmic Series Distribution

$$f(x) \quad = \quad \alpha\theta^x/x \qquad\qquad , \; x = 1, 2, \ldots$$
$$= \; 0 \qquad\qquad\quad , \; \text{otherwise}$$

$$0 < \theta < 1$$
$$\alpha = -1/\log(1-\theta)$$

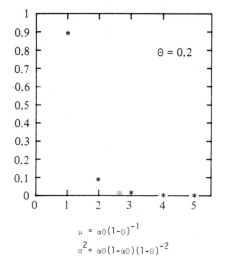 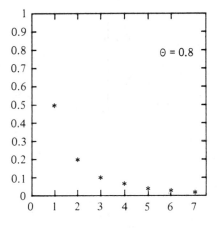

$$\mu \; = \; \alpha\theta(1-\theta)^{-1}$$
$$\sigma^2 = \; \alpha\theta(1-\alpha\theta)(1-\theta)^{-2}$$

$$\beta_4 = \frac{1+4\theta+\theta^2-4\alpha\theta(1+\quad)+6\alpha^2\theta^2-3\alpha^3\theta^3}{\alpha\theta(1-\alpha\theta)^2(1-\theta)^{-4}}$$

Table 2.6.5. Uniform Distribution

f(x) = 1   , $0 \le x \le 1$

     = 0   , otherwise

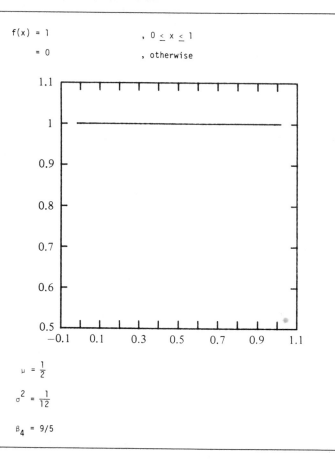

$\mu = \frac{1}{2}$

$\sigma^2 = \frac{1}{12}$

$\beta_4 = 9/5$

Table 2.6.6. Beta Distribution

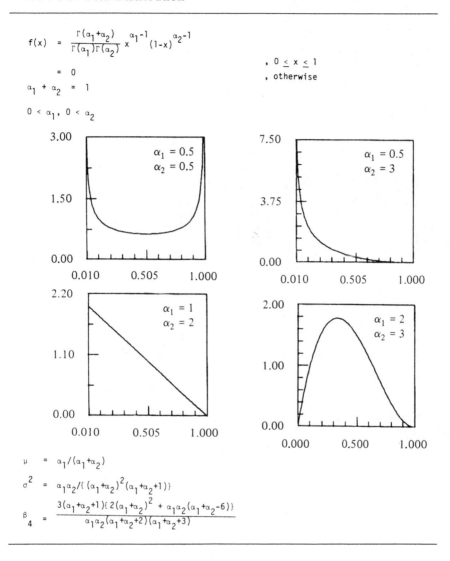

$$f(x) = \frac{\Gamma(\alpha_1+\alpha_2)}{\Gamma(\alpha_1)\Gamma(\alpha_2)} x^{\alpha_1-1} (1-x)^{\alpha_2-1} \qquad , \; 0 \le x \le 1$$

$$= 0 \qquad , \; \text{otherwise}$$

$$\alpha_1 + \alpha_2 = 1$$

$$0 < \alpha_1, \; 0 < \alpha_2$$

$$\mu = \alpha_1/(\alpha_1+\alpha_2)$$

$$\sigma^2 = \alpha_1\alpha_2/\{(\alpha_1+\alpha_2)^2(\alpha_1+\alpha_2+1)\}$$

$$\beta_4 = \frac{3(\alpha_1+\alpha_2+1)\{2(\alpha_1+\alpha_2)^2 + \alpha_1\alpha_2(\alpha_1+\alpha_2-6)\}}{\alpha_1\alpha_2(\alpha_1+\alpha_2+2)(\alpha_1+\alpha_2+3)}$$

Table 2.6.7. Triangular Distribution

$f(x)$ = $(2/P)x$, $0 \le x \le P$

        = $-(2/Q)x + (2/Q)$, $P \le x \le 1$

        = 0 , otherwise

$P + Q = 1$

$0 \le P \le 1$

$0 \le Q \le 1$

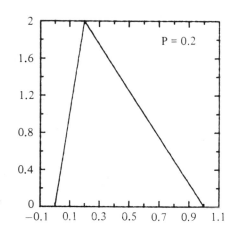
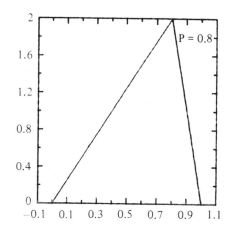

μ = $(1+P)/3$

σ^2 = $(1-PQ)/18$

β_4 = $12/5$

Table 2.6.8. Standard Normal Distribution

$$f(x) = (2\pi)^{-\frac{1}{2}} e^{-x^2/2} \; ; \; -\infty < x < \infty$$

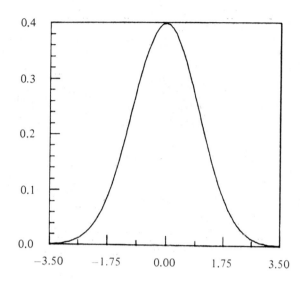

$\mu = 0$

$\sigma^2 = 1$

$\beta_4 = 3$

Table 2.6.9. Gamma Distribution

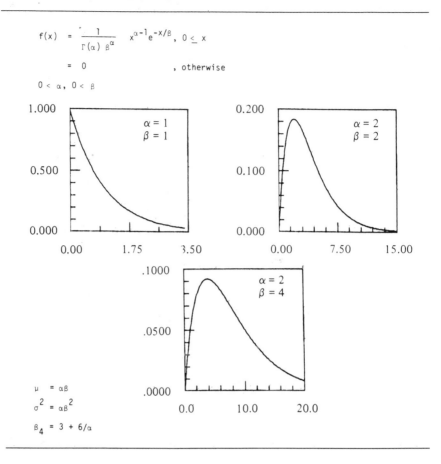

$$f(x) = \frac{1}{\Gamma(\alpha)\,\beta^{\alpha}}\; x^{\alpha-1}e^{-x/\beta},\; 0 \leq x$$

$$= 0 \qquad\qquad\qquad , \text{otherwise}$$

$$0 < \alpha,\; 0 < \beta$$

$\alpha = 1$
$\beta = 1$

$\alpha = 2$
$\beta = 2$

$\alpha = 2$
$\beta = 4$

$$\mu = \alpha\beta$$
$$\sigma^2 = \alpha\beta^2$$
$$\beta_4 = 3 + 6/\alpha$$

Table 2.6.10. Standard Weibull Distribution

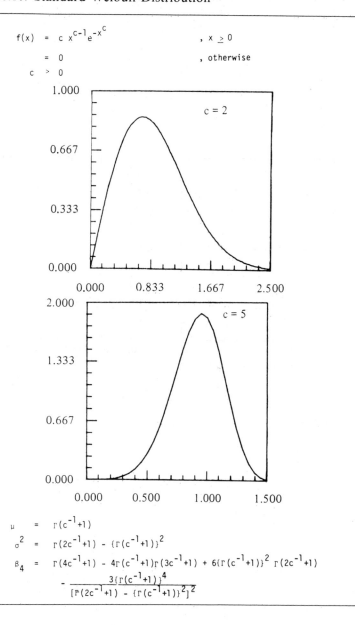

$$f(x) = c\,x^{c-1}e^{-x^c} \qquad , x \geq 0$$

$$ = 0 \qquad\qquad\qquad , \text{otherwise}$$

$$c > 0$$

$$\mu = \Gamma(c^{-1}+1)$$

$$\sigma^2 = \Gamma(2c^{-1}+1) - \{\Gamma(c^{-1}+1)\}^2$$

$$\beta_4 = \Gamma(4c^{-1}+1) - 4\Gamma(c^{-1}+1)\Gamma(3c^{-1}+1) + 6\{\Gamma(c^{-1}+1)\}^2\,\Gamma(2c^{-1}+1)$$

$$- \frac{3\{\Gamma(c^{-1}+1)\}^4}{[\Gamma(2c^{-1}+1) - \{\Gamma(c^{-1}+1)\}^2]^2}$$

Table 2.6.11. Mixed Uniform Distribution

$$f(x) = P \qquad\qquad , x = 0$$
$$\quad = Q \qquad\qquad , 0 < x \le 1$$
$$\quad = 0 \qquad\qquad , \text{otherwise}$$

$$P + Q = 1$$
$$0 \le P, \ 0 \le Q$$

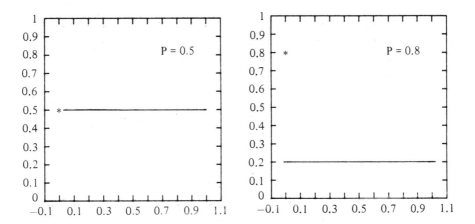

$$\mu = Q/2$$

$$\sigma^2 = Q(1+3P)/12$$

$$\beta_4 = \frac{9\{Q^2(25+15P) + 40P - 24\}}{5Q(1+3P)^2}$$

Table 2.6.12. $\mathrm{CV}\{v(\hat{\theta})\}$ for $n = 1000$

			β_4	
m	k	9/5	3	10
500	2	1.41379	1.41421	1.41669
250	4	0.81576	0.81650	0.82077
200	5	0.70626	0.70711	0.71204
125	8	0.53340	0.53452	0.54103
100	10	0.47013	0.47140	0.47877
50	20	0.32259	0.32444	0.33506
40	25	0.28659	0.28868	0.30056
20	50	0.19904	0.20203	0.21867
10	100	0.13785	0.14213	0.16493
8	125	0.12218	0.12700	0.15208
5	200	0.09408	0.10025	0.13058
4	250	0.08266	0.08962	0.12261
2	500	0.05299	0.06331	0.10492
1	1000	0.02832	0.04474	0.09488

highly subjective one, in which case the kurtosis could only be regarded as an approximation to the true kurtosis for the population under study. Such approximations, however, may well be adequate for purposes of planning the survey estimators and variance estimators. If the chosen distribution is intended to represent the distribution of the estimator $\hat{\theta}_1$, then the tabular kurtosis is $\beta_4(\hat{\theta}_1)$, whereas if the distribution represents the unit values Y_i or Z_i, then the tabular kurtosis is β_4.

The kurtosis $\beta_4(\hat{\theta}_1)$ (or β_4) is invariant under linear transformations of $\hat{\theta}_1$ (or Y_i or Z_i). Therefore if a given finite population can be represented by a linear transformation of one of the 11 distributions, then the kurtosis of the original distribution applies. If the population cannot be represented by any of the 11 distributions or by linear transformations thereof, then see Johnson and Kotz (1969, 1970a, 1970b) for discussion of a wide range of distributions.

To illustrate the utility of the tables, suppose a given survey is concerned with three variables, where the unit values are distributed approximately as a uniform, a normal, and a $\Gamma(6/7, 1)$ random variable, respectively. The corresponding values of β_4 are then 9/5, 3, and 10. Table 2.6.12 gives the corresponding values of $\mathrm{CV}\{v(\hat{\theta})\}$ for srs wor of size $n = 1,000$, where sample means are used to estimate population means. If $\mathrm{CV}\{v(\hat{\theta})\}$ is to be no larger than 15% for each of the three survey variables, then at least $k = 200$ random groups are needed.

2.7. Estimation Based on Order Statistics

In view of the computational simplicity of the random group estimator, one may question the need for still "quicker" estimators of the variance of $\hat{\bar{\theta}}$ or $\hat{\theta}$. There are, however, circumstances where the additional simplicity of an estimator based, say, on the range may be useful. An example is in field work where a quick computation of variance is desired but only a primitive desk calculator is available. Also, estimators based on the range or on quasiranges may be robust in some sense, and may be used to identify errors in the calculation of the random group estimator. This section, then, is devoted to a brief discussion of such estimators. We do not attempt a complete discussion of this topic, nor do we assign priority to specific authors. For additional information the reader is referred to David (1970). The estimators described here have received little previous attention in the survey sampling literature.

The specific problem that we shall address is that of estimating

$$\sigma \stackrel{[\text{defn}]}{=} [k \cdot \text{Var}\{\hat{\bar{\theta}}\}]^{1/2}.$$

Given an estimator $\hat{\sigma}$ of σ, we may estimate the standard error of $\hat{\bar{\theta}}$ or $\hat{\theta}$ by $\hat{\sigma}/k^{1/2}$.

As before we let $\hat{\theta}_1, \ldots, \hat{\theta}_k$ denote the k random group estimators of θ, and we let

$$\hat{\theta}_{(1)}, \ldots, \hat{\theta}_{(k)}$$

denote the observations ordered from smallest to largest. We define the *range*

$$W = \hat{\theta}_{(k)} - \hat{\theta}_{(1)},$$

and the *i*-th *quasirange*

$$W_{(i)} = \hat{\theta}_{(k+1-i)} - \hat{\theta}_{(i)}$$

for $1 \leq i \leq [k/2]$, where the notation $[x]$ signifies the largest integer $< x$. Note that the range and the first quasirange are identical.

The utility of the range for checking calculations is easily seen. Letting $v(\hat{\bar{\theta}})$ denote the random group estimator, the ratio $W^2/v(\hat{\bar{\theta}})$ is algebraically bounded by

$$W^2/v(\hat{\bar{\theta}}) \leq 2k(k-1)$$

$$W^2/v(\hat{\bar{\theta}}) \geq \begin{cases} 4(k-1), & k \text{ even,} \\ 4k^2/(k+1), & k \text{ odd.} \end{cases}$$

The upper bound results from a sample configuration with $k-2$ observations at $\hat{\bar{\theta}}$ and $\hat{\theta}_{(1)}$ and $\hat{\theta}_{(k)}$ at equal distances below and above $\hat{\bar{\theta}}$. The lower bound corresponds to half the observations at one extreme $\hat{\theta}_{(k)}$ and half (plus 1 if k is odd) at the other $\hat{\theta}_{(1)}$. Consequently, if the computed value of $v(\hat{\hat{\theta}})$ is larger than its upper bound

$$W^2/4(k-1), \qquad k \text{ even},$$

$$W^2(k+1)/4k^2, \qquad k \text{ odd},$$

or smaller than its lower bound

$$W^2/2k(k-1),$$

then an error has been made either in the computation of the random group estimator or in the computation of the range.

Ranges and quasiranges form the basis for some extremely simple estimators of σ. The first of these is

$$\hat{\sigma}_1 = W/d_k, \qquad (2.7.1)$$

where $d_k = \mathscr{E}\{W/\sigma\}$ and the expectation operator, $\mathscr{E}$, is with respect to an assumed parent distribution for the $\hat{\theta}_\alpha$. For the time being, we shall assume that the $\hat{\theta}_\alpha (\alpha = 1, \ldots, k)$ comprise a random sample from the $N(\theta, \sigma^2)$ distribution. For this normal parent, values of d_k are given in the second column of Table 2.7.1 for $k = 2, 3, \ldots, 100$. The efficiency of $\hat{\sigma}_1$ is very good for $k \le 12$.

In normal samples, however, the efficiency of $\hat{\sigma}_1$ declines with increasing k, and at a certain point estimators based on the quasiranges will do better. It has been shown that $\hat{\sigma}_1$ is more efficient than any quasirange for $k \le 17$, but thereafter a multiple of $W_{(2)}$ is more efficient, to be in turn replaced by a multiple of $W_{(3)}$ for $k \ge 32$, and so on. Table 2.7.1. also presents the appropriate divisors for $W_{(i)}$ for $i = 2, 3, \ldots, 9$ and $k = 2, 3, \ldots, 100$.

Very efficient estimators can be constructed from *thickened ranges*, e.g., $W + W_{(2)} + W_{(4)}$, and other linear combinations of quasiranges. A typical estimator is

$$\hat{\sigma}_2 = (W + W_{(2)} + W_{(4)})/e_k, \qquad (2.7.2)$$

where $e_k = \mathscr{E}\{(W + W_{(2)} + W_{(4)})/\sigma\}$ may be obtained by summing the appropriate elements of Table 2.7.1. For $k = 16$, the estimator $\hat{\sigma}_2$ has efficiency 97.5%.

Table 2.7.2 presents several unbiased estimators of σ and their associated efficiencies for $k = 2, 3, \ldots, 100$. Column 2 gives the most efficient estimator, say $\hat{\sigma}_3$, based on a single quasirange ($i \le 9$); Column 3 gives its corresponding efficiency; Column 4 gives the most efficient estimator, say $\hat{\sigma}_4$, based on a linear combination of two quasiranges (i and $i' \le 9$); and Column 5 gives its corresponding efficiency. The efficiencies,

$$\text{eff}\{\hat{\sigma}_3\} = \text{Var}\{\tilde{\sigma}\}/\text{Var}\{\hat{\sigma}_3\}$$

$$\text{eff}\{\hat{\sigma}_4\} = \text{Var}\{\tilde{\sigma}\}/\text{Var}\{\hat{\sigma}_4\},$$

74   2. The Method of Random Groups

Table 2.7.1. Denominator d_k of Unbiased Estimator of σ Based on the i-th Quasirange for Samples of k from $N(\theta, \sigma^2)$

k	$i = 1$	$i = 2$	$i = 3$	$i = 4$	$i = 5$	$i = 6$	$i = 7$	$i = 8$	$i = 9$
2	1.128								
3	1.692								
4	2.058	0.594							
5	2.325	0.990							
6	2.534	1.283	0.403						
7	2.704	1.514	0.705						
8	2.847	1.704	0.945	0.305					
9	2.970	1.864	1.143	0.549					
10	3.077	2.002	1.312	0.751	0.245				
11	3.172	2.123	1.457	0.923	0.449				
12	3.258	2.231	1.585	1.073	0.624	0.205			
13	3.335	2.328	1.699	1.205	0.776	0.381			
14	3.406	2.415	1.802	1.323	0.911	0.534	0.176		
15	3.471	2.495	1.895	1.429	1.031	0.670	0.330		
16	3.531	2.569	1.980	1.526	1.140	0.792	0.467	0.154	
17	3.587	2.637	2.058	1.614	1.238	0.902	0.590	0.291	
18	3.640	2.700	2.131	1.696	1.329	1.003	0.701	0.415	0.137
19	3.688	2.759	2.198	1.771	1.413	1.095	0.803	0.527	0.261
20	3.734	2.815	2.261	1.841	1.490	1.180	0.896	0.629	0.373
21	3.778	2.867	2.320	1.907	1.562	1.259	0.982	0.724	0.476
22	3.819	2.916	2.376	1.969	1.630	1.333	1.063	0.811	0.571
23	3.858	2.962	2.428	2.027	1.693	1.402	1.137	0.892	0.659
24	3.895	3.006	2.478	2.081	1.753	1.467	1.207	0.967	0.740
25	3.930	3.048	2.525	2.133	1.810	1.528	1.273	1.038	0.817
26	3.964	3.088	2.570	2.182	1.863	1.585	1.335	1.105	0.888
27	3.996	3.126	2.612	2.229	1.914	1.640	1.394	1.168	0.956
28	4.027	3.162	2.653	2.273	1.962	1.692	1.450	1.227	1.019
29	4.057	3.197	2.692	2.316	2.008	1.741	1.503	1.284	1.079
30	4.085	3.231	2.729	2.357	2.052	1.788	1.553	1.337	1.136
31	4.112	3.263	2.765	2.396	2.094	1.833	1.601	1.388	1.190
32	4.139	3.294	2.799	2.433	2.134	1.876	1.647	1.437	1.242
33	4.164	3.324	2.832	2.469	2.173	1.918	1.691	1.484	1.291
34	4.189	3.352	2.864	2.503	2.210	1.957	1.733	1.528	1.339
35	4.213	3.380	2.895	2.537	2.245	1.995	1.773	1.571	1.384
36	4.236	3.407	2.924	2.569	2.280	2.032	1.812	1.612	1.427
37	4.258	3.433	2.953	2.600	2.313	2.067	1.849	1.652	1.469
38	4.280	3.458	2.981	2.630	2.345	2.101	1.886	1.690	1.509
39	4.301	3.482	3.008	2.659	2.376	2.134	1.920	1.726	1.547
40	4.321	3.506	3.034	2.687	2.406	2.166	1.954	1.762	1.585
41	4.341	3.529	3.059	2.714	2.435	2.197	1.986	1.796	1.621
42	4.360	3.551	3.083	2.740	2.463	2.227	2.018	1.829	1.655
43	4.379	3.573	3.107	2.766	2.491	2.256	2.048	1.861	1.689
44	4.397	3.594	3.130	2.791	2.517	2.284	2.078	1.892	1.721
45	4.415	3.614	3.153	2.815	2.543	2.311	2.107	1.922	1.753

Table 2.7.1. (*Cont.*)

k	i = 1	i = 2	i = 3	i = 4	i = 5	i = 6	i = 7	i = 8	i = 9
46	4.432	3.634	3.175	2.839	2.568	2.337	2.135	1.951	1.784
47	4.449	3.654	3.196	2.862	2.592	2.363	2.162	1.980	1.813
48	4.466	3.673	3.217	2.884	2.616	2.388	2.188	2.007	1.842
49	4.482	3.691	3.237	2.906	2.639	2.413	2.214	2.034	1.870
50	4.498	3.709	3.257	2.927	2.662	2.436	2.238	2.060	1.897
51	4.513	3.727	3.276	2.948	2.684	2.460	2.263	2.086	1.924
52	4.528	3.744	3.295	2.968	2.705	2.482	2.286	2.110	1.950
53	4.543	3.761	3.313	2.988	2.726	2.504	2.310	2.135	1.975
54	4.557	3.778	3.331	3.007	2.746	2.526	2.325	2.158	1.999
55	4.571	3.794	3.349	3.026	2.766	2.547	2.354	2.181	2.023
56	4.585	3.810	3.366	3.044	2.786	2.567	2.376	2.204	2.047
57	4.599	3.825	3.383	3.062	2.805	2.587	2.397	2.226	2.069
58	4.612	3.840	3.400	3.080	2.824	2.607	2.417	2.247	2.092
59	4.625	3.855	3.416	3.097	2.842	2.626	2.437	2.268	2.113
60	4.638	3.870	3.432	3.114	2.860	2.645	2.457	2.288	2.135
61	4.651	3.884	3.447	3.131	2.878	2.663	2.476	2.308	2.156
62	4.663	3.898	3.463	3.147	2.895	2.681	2.495	2.328	2.176
63	4.675	3.912	3.478	3.163	2.912	2.699	2.513	2.347	2.196
64	4.687	3.926	3.492	3.179	2.928	2.716	2.532	2.366	2.215
65	4.699	3.939	3.507	3.194	2.944	2.733	2.549	2.385	2.235
66	4.710	3.952	3.521	3.209	2.960	2.750	2.567	2.403	2.253
67	4.721	3.965	3.535	3.224	2.976	2.767	2.584	2.420	2.272
68	4.733	3.977	3.549	3.239	2.991	2.783	2.601	2.438	2.290
69	4.743	3.990	3.562	3.253	3.006	2.798	2.617	2.455	2.308
70	4.754	4.002	3.575	3.267	3.021	2.814	2.633	2.472	2.325
71	4.765	4.014	3.588	3.281	3.036	2.829	2.649	2.488	2.342
72	4.775	4.026	3.601	3.294	3.050	2.844	2.665	2.504	2.359
73	4.785	4.037	3.613	3.308	3.064	2.859	2.680	2.520	2.375
74	4.796	4.049	3.626	3.321	3.078	2.873	2.695	2.536	2.391
75	4.805	4.060	3.638	3.334	3.091	2.887	2.710	2.551	2.407
76	4.815	4.071	3.650	3.346	3.105	2.901	2.724	2.566	2.423
77	4.825	4.082	3.662	3.359	3.118	2.915	2.739	2.581	2.438
78	4.834	4.093	3.673	3.371	3.131	2.929	2.753	2.596	2.453
79	4.844	4.103	3.685	3.383	3.144	2.942	2.767	2.610	2.468
80	4.853	4.114	3.696	3.395	3.156	2.955	2.780	2.624	2.483
81	4.862	4.124	3.707	3.407	3.169	2.968	2.794	2.638	2.497
82	4.871	4.134	3.718	3.419	3.181	2.981	2.807	2.652	2.511
83	4.880	4.144	3.729	3.430	3.193	2.993	2.820	2.665	2.525
84	4.889	4.154	3.740	3.442	3.205	3.006	2.833	2.679	2.539
85	4.897	4.164	3.750	3.453	3.216	3.018	2.845	2.692	2.553
86	4.906	4.173	3.760	3.464	3.228	3.030	2.858	2.705	2.566
87	4.914	4.183	3.771	3.474	3.239	3.042	2.870	2.717	2.579
88	4.923	4.192	3.781	3.485	3.250	3.053	2.882	2.730	2.592
89	4.931	4.201	3.791	3.496	3.261	3.065	2.894	2.742	2.605

Table 2.7.1. (*Cont.*)

k	i = 1	i = 2	i = 3	i = 4	i = 5	i = 6	i = 7	i = 8	i = 9
90	4.939	4.211	3.801	3.506	3.272	3.076	2.906	2.754	2.617
91	4.947	4.220	3.810	3.516	3.283	3.087	2.918	2.766	2.630
92	4.955	4.228	3.820	3.526	3.294	3.098	2.929	2.778	2.642
93	4.963	4.237	3.829	3.536	3.304	3.109	2.940	2.790	2.654
94	4.970	4.246	3.839	3.546	3.314	3.120	2.951	2.802	2.666
95	4.978	4.254	3.848	3.556	3.325	3.131	2.963	2.813	2.678
96	4.985	4.263	3.857	3.566	3.335	3.141	2.973	2.824	2.689
97	4.993	4.271	3.866	3.575	3.345	3.152	2.984	2.835	2.701
98	5.000	4.280	3.875	3.585	3.355	3.162	2.995	2.846	2.712
99	5.007	4.288	3.884	3.594	3.364	3.172	3.005	2.857	2.723
100	5.015	4.296	3.892	3.603	3.374	3.182	3.016	2.868	2.734

Source: Table 1 of Harter (1959). Harter's tables are given to six decimal places. We have truncated (not rounded) his figures to three decimal places.

Table 2.7.2. Most Efficient Unbiased Estimators of σ Based on Quasiranges for Samples of k from $N(\theta, \sigma^2)$

k	Based on One Quasirange		Based on a Linear Combination of Two Quasiranges Among Those with $i < i' \leq 9$	
	Estimate	Eff(%)	Estimate	Eff(%)
2	$0.886W_1$	100.00		
3	$0.590W_1$	99.19		
4	$0.485W_1$	97.52	$0.453(W_1 + 0.242W_2)$	98.92
5	$0.429W_1$	95.48	$0.372(W_1 + 0.363W_2)$	98.84
6	$0.394W_1$	93.30	$0.318(W_1 + 0.475W_2)$	98.66
7	$0.369W_1$	91.12	$0.279(W_1 + 0.579W_2)$	98.32
8	$0.351W_1$	89.00	$0.250(W_1 + 0.675W_2)$	97.84
9	$0.336W_1$	86.95	$0.227(W_1 + 0.765W_2)$	97.23
10	$0.324W_1$	84.99	$0.209(W_1 + 0.848W_2)$	96.54
11	$0.315W_1$	83.13	$0.194(W_1 + 0.927W_2)$	95.78
12	$0.306W_1$	81.36	$0.211(W_1 + 0.923W_3)$	95.17
13	$0.299W_1$	79.68	$0.198(W_1 + 1.001W_3)$	95.00
14	$0.293W_1$	78.09	$0.187(W_1 + 1.076W_3)$	94.77
15	$0.288W_1$	76.57	$0.177(W_1 + 1.147W_3)$	94.50
16	$0.283W_1$	75.13	$0.168(W_1 + 1.216W_3)$	94.18
17	$0.278W_1$	73.76	$0.160(W_1 + 1.281W_3)$	93.82
18	$0.370W_2$	72.98	$0.153(W_1 + 1.344W_3)$	93.43
19	$0.362W_2$	72.98	$0.147(W_1 + 1.405W_3)$	93.02
20	$0.355W_2$	72.91	$0.141(W_1 + 1.464W_3)$	92.59

Table 2.7.2. (*Cont.*)

k	Based on One Quasirange		Based on a Linear Combination of Two Quasiranges Among Those with $i < i' \leq 9$	
	Estimate	Eff(%)	Estimate	Eff(%)
21	$0.348W_2$	72.77	$0.136(W_1 + 1.520W_3)$	92.14
22	$0.342W_2$	72.59	$0.146(W_1 + 1.529W_4)$	91.78
23	$0.337W_2$	72.37	$0.141(W_1 + 1.588W_4)$	91.61
24	$0.332W_2$	72.11	$0.136(W_1 + 1.644W_4)$	91.42
25	$0.328W_2$	71.82	$0.132(W_1 + 1.699W_4)$	91.21
26	$0.323W_2$	71.52	$0.128(W_1 + 1.752W_4)$	90.98
27	$0.319W_2$	71.20	$0.124(W_1 + 1.805W_4)$	90.73
28	$0.316W_2$	70.86	$0.121(W_1 + 1.855W_4)$	90.48
29	$0.312W_2$	70.51	$0.118(W_1 + 1.905W_4)$	90.21
30	$0.309W_2$	70.15	$0.115(W_1 + 1.953W_4)$	89.93
31	$0.306W_2$	69.78	$0.112(W_1 + 2.000W_4)$	89.63
32	$0.357W_3$	69.57	$0.109(W_1 + 2.046W_4)$	89.35
33	$0.353W_3$	69.58	$0.115(W_1 + 2.067W_5)$	89.11
34	$0.349W_3$	69.57	$0.112(W_1 + 2.115W_5)$	88.97
35	$0.345W_3$	69.53	$0.110(W_1 + 2.161W_5)$	88.82
36	$0.341W_3$	69.48	$0.107(W_1 + 2.207W_5)$	88.66
37	$0.338W_3$	69.41	$0.105(W_1 + 2.252W_5)$	88.48
38	$0.335W_3$	69.32	$0.103(W_1 + 2.296W_5)$	88.31
39	$0.332W_3$	69.21	$0.101(W_1 + 2.339W_5)$	88.12
40	$0.329W_3$	69.10	$0.099(W_1 + 2.381W_5)$	87.92
41	$0.326W_3$	68.97	$0.097(W_1 + 2.423W_5)$	87.73
42	$0.324W_3$	68.83	$0.095(W_1 + 2.464W_5)$	87.52
43	$0.321W_3$	68.68	$0.094(W_1 + 2.504W_5)$	87.32
44	$0.319W_3$	68.53	$0.092(W_1 + 2.543W_5)$	87.10
45	$0.317W_3$	68.37	$0.096(W_1 + 2.574W_6)$	86.97
46	$0.352W_4$	68.22	$0.094(W_1 + 2.615W_6)$	86.85
47	$0.349W_4$	68.24	$0.093(W_1 + 2.655W_6)$	86.73
48	$0.346W_4$	68.23	$0.091(W_1 + 2.695W_6)$	86.60
49	$0.344W_4$	68.22	$0.090(W_1 + 2.734W_6)$	86.47
50	$0.341W_4$	68.20	$0.088(W_1 + 2.772W_6)$	86.33
51	$0.339W_4$	68.17	$0.087(W_1 + 2.810W_6)$	86.19
52	$0.336W_4$	68.12	$0.086(W_1 + 2.847W_6)$	86.04
53	$0.334W_4$	68.07	$0.084(W_1 + 2.884W_6)$	85.89
54	$0.332W_4$	68.02	$0.083(W_1 + 2.921W_6)$	85.73
55	$0.330W_4$	67.95	$0.082(W_1 + 2.956W_6)$	85.57
56	$0.328W_4$	67.87	$0.137(W_2 + 1.682W_9)$	85.44
57	$0.326W_4$	67.80	$0.135(W_2 + 1.706W_9)$	85.46
58	$0.324W_4$	67.71	$0.134(W_2 + 1.730W_9)$	85.47
59	$0.322W_4$	67.62	$0.132(W_2 + 1.753W_9)$	85.48
60	$0.321W_4$	67.53	$0.130(W_2 + 1.777W_9)$	85.48
61	$0.347W_5$	67.52	$0.128(W_2 + 1.800W_9)$	85.47

Table 2.7.2. (*Cont.*)

k	Based on One Quasirange Estimate	Eff(%)	Based on a Linear Combination of Two Quasiranges Among Those with $i < i' \le 9$ Estimate	Eff(%)
62	$0.345W_5$	67.52	$0.127(W_2 + 1.823W_9)$	85.46
63	$0.343W_5$	67.52	$0.125(W_2 + 1.846W_9)$	85.44
64	$0.341W_5$	67.50	$0.123(W_2 + 1.868W_9)$	85.42
65	$0.339W_5$	67.49	$0.122(W_2 + 1.890W_9)$	85.40
66	$0.337W_5$	67.46	$0.121(W_2 + 1.912W_9)$	85.36
67	$0.335W_5$	67.44	$0.119(W_2 + 1.934W_9)$	85.33
68	$0.334W_5$	67.40	$0.118(W_2 + 1.955W_9)$	85.29
69	$0.332W_5$	67.36	$0.116(W_2 + 1.976W_9)$	85.24
70	$0.330W_5$	67.32	$0.115(W_2 + 1.997W_9)$	85.20
71	$0.329W_5$	67.27	$0.114(W_2 + 2.018W_9)$	85.14
72	$0.327W_5$	67.22	$0.113(W_2 + 2.038W_9)$	85.09
73	$0.326W_5$	67.16	$0.111(W_2 + 2.059W_9)$	85.04
74	$0.324W_5$	67.11	$0.110(W_2 + 2.079W_9)$	84.98
75	$0.346W_6$	67.07	$0.109(W_2 + 2.099W_9)$	84.91
76	$0.344W_6$	67.03	$0.108(W_2 + 2.118W_9)$	84.85
77	$0.342W_6$	67.07	$0.107(W_2 + 2.138W_9)$	84.78
78	$0.341W_6$	67.07	$0.106(W_2 + 2.157W_9)$	84.71
79	$0.339W_6$	67.06	$0.105(W_2 + 2.176W_9)$	84.63
80	$0.338W_6$	67.04	$0.104(W_2 + 2.195W_9)$	84.56
81	$0.336W_6$	67.03	$0.103(W_2 + 2.214W_9)$	84.48
82	$0.335W_6$	67.01	$0.102(W_2 + 2.232W_9)$	84.40
83	$0.334W_6$	66.98	$0.101(W_2 + 2.251W_9)$	84.32
84	$0.332W_6$	66.95	$0.100(W_2 + 2.269W_9)$	84.23
85	$0.331W_6$	66.92	$0.099(W_2 + 2.287W_9)$	84.15
86	$0.329W_6$	66.89	$0.099(W_2 + 2.305W_9)$	84.07
87	$0.328W_6$	66.85	$0.098(W_2 + 2.323W_9)$	83.97
88	$0.327W_6$	66.81	$0.097(W_2 + 2.340W_9)$	83.88
89	$0.345W_7$	66.77	$0.096(W_2 + 2.358W_9)$	83.79
90	$0.344W_7$	66.77	$0.095(W_2 + 2.375W_9)$	83.70
91	$0.342W_7$	66.77	$0.095(W_2 + 2.393W_9)$	83.61
92	$0.341W_7$	66.77	$0.094(W_2 + 2.409W_9)$	83.51
93	$0.340W_7$	66.76	$0.093(W_2 + 2.426W_9)$	83.42
94	$0.338W_7$	66.75	$0.092(W_2 + 2.443W_9)$	83.32
95	$0.337W_7$	66.74	$0.092(W_2 + 2.459W_9)$	83.22
96	$0.336W_7$	66.73	$0.091(W_2 + 2.476W_9)$	83.12
97	$0.335W_7$	66.71	$0.090(W_2 + 2.492W_9)$	83.02
98	$0.333W_7$	66.69	$0.090(W_2 + 2.508W_9)$	82.92
99	$0.332W_7$	66.67	$0.089(W_2 + 2.524W_9)$	82.82
100	$0.331W_7$	66.65	$0.088(W_2 + 2.540W_9)$	82.71

Source: Table 4 of Harter (1959). Harter's results have been truncated (not rounded) to three decimal places.

are with respect to the minimum variance unbiased estimator

$$\tilde{\sigma} = \frac{\Gamma[(k-1)/2]}{[2/(k-1)]^{1/2}\Gamma(k/2)}\left[\sum_{\alpha}^{k}(\hat{\theta}_{\alpha} - \hat{\bar{\theta}})^2/(k-1)\right]^{1/2}.$$

A final important estimator of σ is

$$\hat{\sigma}_5 = \frac{2\sqrt{\pi}}{k(k-1)}\sum_{\alpha=1}^{k}[\alpha - (k+1)/2]\hat{\theta}_{(\alpha)}. \tag{2.7.3}$$

Barnett *et al.* (1967) have found that $\hat{\sigma}_5$ is highly efficient (>97.8%) and more robust against outliers than $\hat{\sigma}_1$ and $\tilde{\sigma}$.

EXAMPLE 2.7.1. $k = 16$ random groups are to be used. Then, the estimators of σ take the form

$$\hat{\sigma}_1 = W/3.531$$

$$\hat{\sigma}_2 = (W + W_{(2)} + W_{(4)})/(3.531 + 2.569 + 1.526)$$

$$= (W + W_{(2)} + W_{(4)})/7.626$$

$$\hat{\sigma}_3 = W/3.531$$

$$= 0.283\,W$$

$$\hat{\sigma}_4 = 0.168(W + 1.216\,W_{(3)})$$

$$\hat{\sigma}_5 = \frac{2\sqrt{\pi}}{16(15)}\sum_{\alpha=1}^{16}(\alpha - 17/2)\hat{\theta}_{(\alpha)}.$$

The corresponding estimators of the standard error of $\hat{\bar{\theta}}$ are

$$\hat{\sigma}_1/4$$

$$\hat{\sigma}_2/4$$

$$\hat{\sigma}_3/4$$

$$\hat{\sigma}_4/4$$

$$\hat{\sigma}_5/4. \qquad \square$$

The methods discussed here could be criticized on grounds that the estimators are bound to be both inefficient and overly sensitive to the shape of the parent distribution of the $\hat{\theta}_{\alpha}$. There is evidence, however, that both criticisms may be misleading (cf. David (1970)). As we have already seen, the loss in efficiency is usually unimportant. Furthermore, the ratio $\mathscr{E}\{W/\sigma\}$ is remarkably stable for most reasonable departures from normality. Table 2.7.3 illustrates the stability well. The entries in the table are the percent bias in the efficient estimators $\hat{\sigma}_3$ and $\hat{\sigma}_4$ (constructed using normality assumptions) when the parent distribution is actually uniform or exponential. For the uniform distribution the percent bias is quite trivial for most

Table 2.7.3. Percent Bias of Estimators of σ that Assume Normality

k	When Population Is Uniform		When Population Is Exponential	
	One Quasirange	Two Quasiranges with $i < i' \leq 9$	One Quasirange	Two Quasiranges with $i < i' \leq 9$
2	2.33		−11.38	
3	2.33		−11.38	
4	0.96	1.98	−10.95	−11.27
5	−0.71	1.61	−10.43	−11.16
6	−2.37	1.13	−9.91	−11.01
7	−3.93	0.54	−9.41	−10.84
8	−5.37	−0.11	−8.93	−10.66
9	−6.69	−0.80	−8.49	−10.46
10	−7.90	−1.51	−8.08	−10.26
11	−9.02	−2.22	−7.69	−10.06
12	−10.04	−1.47	−7.32	−10.07
13	−10.99	−1.72	−6.98	−10.00
14	−11.87	−2.01	−6.65	−9.92
15	−12.69	−2.33	−6.34	−9.84
16	−13.46	−2.66	−6.05	−9.75
17	−14.18	−3.01	−5.77	−9.65
18	1.26	−3.37	−11.85	−9.56
19	0.41	−3.74	−11.61	−9.46
20	−0.39	−4.11	−11.37	−9.36
21	−1.15	−4.48	−11.14	−9.25
22	−1.87	−3.11	−10.92	−9.43
23	−2.56	−3.32	−10.71	−9.38
24	−3.22	−3.53	−10.51	−9.33
25	−3.85	−3.75	−10.31	−9.28
26	−4.45	−3.98	−10.12	−9.22
27	−5.03	−4.21	−9.93	−9.16
28	−5.58	−4.45	−9.75	−9.10
29	−6.11	−4.69	−9.58	−9.04
30	−6.63	−4.94	−9.41	−8.98
31	−7.12	−5.18	−9.24	−8.91
32	1.23	−5.43	−12.07	−8.85
33	0.71	−3.95	−11.92	−9.09
34	0.20	−4.11	−11.78	−9.05
35	−0.29	−4.28	−11.63	−9.02
36	−0.77	−4.45	−11.49	−8.98
37	−1.23	−4.63	−11.35	−8.93
38	−1.68	−4.81	−11.22	−8.89
39	−2.11	−4.99	−11.09	−8.85
40	−2.53	−5.17	−10.96	−8.81
41	−2.94	−5.36	−10.83	−8.76
42	−3.34	−5.54	−10.71	−8.72

Table 2.7.3. (*Cont.*)

	When Population Is Uniform		When Population Is Exponential	
k	One Quasirange	Two Quasiranges with $i < i' \leq 9$	One Quasirange	Two Quasiranges with $i < i' \leq 9$
43	−3.73	−5.72	−10.59	−8.67
44	−4.10	−5.91	−10.47	−8.62
45	−4.47	−4.45	−10.35	−8.89
46	1.24	−4.59	−12.18	−8.86
47	0.86	−4.73	−12.07	−8.83
48	0.49	−4.88	−11.96	−8.80
49	0.12	−5.02	−11.86	−8.76
50	−0.23	−5.17	−11.75	−8.73
51	−0.58	−5.32	−11.65	−8.70
52	−0.92	−5.46	−11.55	−8.66
53	−1.25	−5.61	−11.45	−8.63
54	−1.57	−5.76	−11.36	−8.59
55	−1.89	−5.91	−11.26	−8.55
56	−2.20	−0.63	−11.17	−10.80
57	−2.50	−0.76	−11.08	−10.78
58	−2.79	−0.89	−10.98	−10.75
59	−3.08	−1.02	−10.89	−10.71
60	−3.37	−1.15	−10.81	−10.67
61	0.95	−1.28	−12.16	−10.64
62	0.66	−1.41	−12.07	−10.60
63	0.37	−1.55	−11.99	−10.56
64	0.09	−1.67	−11.91	−10.53
65	−0.19	−1.81	−11.83	−10.49
66	−0.46	−1.94	−11.75	−10.46
67	−0.73	−2.07	−11.67	−10.42
68	−0.99	−2.20	−11.59	−10.38
69	−1.25	−2.34	−11.51	−10.35
70	−1.50	−2.47	−11.44	−10.31
71	−1.75	−2.59	−11.36	−10.27
72	−1.99	−2.73	−11.29	−10.24
73	−2.23	−2.85	−11.22	−10.20
74	−2.47	−2.98	−11.14	−10.17
75	1.01	−3.11	−12.21	−10.13
76	0.77	−3.24	−12.14	−10.09
77	0.53	−3.36	−12.07	−10.05
78	0.30	−3.49	−12.01	−10.02
79	0.07	−3.62	−11.94	−9.98
80	−0.16	−3.74	−11.87	−9.95
81	−0.38	−3.87	−11.81	−9.91
82	−0.60	−3.99	−11.74	−9.88
83	−0.82	−4.11	−11.68	−9.84

Table 2.7.3. (*Cont.*)

k	When Population Is Uniform		When Population is Exponential	
	One Quasirange	Two Quasiranges with $i < i' \leq 9$	One Quasirange	Two Quasiranges with $i < i' \leq 9$
84	−1.04	−4.24	−11.62	−9.81
85	−1.25	−4.36	−11.55	−9.77
86	−1.45	−4.48	−11.49	−9.74
87	−1.66	−4.60	−11.43	−9.70
88	−1.86	−4.72	−11.37	−9.67
89	1.06	−4.83	−12.25	−9.63
90	0.85	−4.95	−12.19	−9.60
91	0.65	−5.07	−12.13	−9.56
92	0.45	−5.19	−12.08	−9.53
93	0.25	−5.30	−12.02	−9.49
94	0.05	−5.42	−11.96	−9.46
95	−0.14	−5.53	−11.91	−9.43
96	−0.33	−5.64	−11.85	−9.39
97	−0.52	−5.76	−11.80	−9.36
98	−0.70	−5.87	−11.74	−9.32
99	−0.88	−5.98	−11.69	−9.29
100	−1.07	−6.09	−11.63	−9.26

Source: Table 6 of Harter (1959).

values of k between 2 and 100. The bias is somewhat more important for the skewed parent (i.e., the exponential distribution), but not alarmingly so. In almost all cases, the estimators tend towards an underestimate. Cox (1954) suggests that the ratio $\mathscr{E}\{W/\sigma\}$ does not depend on the skewness of the parent distribution, but only on the kurtosis. With approximate knowledge of the kurtosis (e.g., from a prior survey of a similar population) one may use Cox's tables to correct for the bias.

The stability of $\mathscr{E}\{W/\sigma\}$ is, of course, of central importance in applying these methods to samples from finite populations, because the random group estimators $\hat{\theta}_\alpha$ cannot, strictly speaking, be viewed as normally distributed. The fact that the ratio $\mathscr{E}\{W/\sigma\}$ is quite stable lends support to the use of these estimators in finite population sampling. Further support is derived from the various central limit theorems given in Appendix B. In many cases, when the size m of the random groups is large and the number of groups k is fixed, the $\hat{\theta}_\alpha$ will behave, roughly speaking, as a random sample from a normal distribution and the methods presented in this section will be appropriate.

2.8. Deviations from Strict Principles

The fundamental principles of the random group method were presented in Sections 2.2 and 2.4 for the independent and nonindependent cases, respectively. In practice, however, there are often computational or other advantages to some deviation from these principles. We may suggest a modified random group procedure if it results in both a substantial cost savings and gives essentially the same results as the unmodified procedure. In this section we discuss briefly two such modifications.

The first concerns the "weights" used in preparing the survey estimates. To be precise, we consider a survey estimator

$$\hat{\theta} = \sum_{i \in s} \omega_i Y_i \qquad (2.8.1)$$

with weights $\{\omega_i\}$. In a typical survey, the ω_i may be a product of several components, including the reciprocal of the inclusion probability (or the basic weight); an adjustment for nonresponse, undercoverage, and post-stratification; and possibly a seasonal adjustment. Strict adherence to random group principles would dictate that the adjustments of the basic weights be computed separately within each random group. That is, the sample is divided into, say, k groups and from each group $s(\alpha)$ an estimator

$$\hat{\theta}_\alpha = \sum_{i \in s(\alpha)} \omega_{\alpha i} y_i \qquad (2.8.2)$$

is developed of the same functional form as the parent estimator $\hat{\theta}$. The sum in (2.8.2) is only over units in the α-th random group, and the weights $\{\omega_{\alpha i}\}$ are developed from the inclusion probabilities with adjustments derived from information in the α-th group only. The reader will immediately recognize the computational cost and complexities associated with this procedure. In effect the weight adjustments must be computed $k + 1$ times: once for the full sample estimator and once for each of the k random group estimators. For extremely large samples, the cost associated with these multiple weight adjustments may be significant. A simpler procedure that sometimes gives satisfactory results is to compute random group estimates, say $\tilde{\theta}_\alpha$, using the weight adjustments appropriate to the full sample, rather than to the relevant random group. That is, $\tilde{\theta}_\alpha$ is defined as in (2.8.2) and the $\omega_{\alpha i}$ are now developed from the inclusion probabilities for the α-th random group with adjustments derived from all of the information in the full parent sample. In this way computational advantages are gained because only one set of weight adjustments is necessary.

Intuitively, we expect that this shortcut procedure may tend to under-represent the component of variability associated with random error in the weight adjustments, and thus underestimate the total variance of $\hat{\theta}$. In some cases, however, Taylor series approximations (see Chapter 6) suggest that this problem is not a serious one. To illustrate, consider a simple situation where the sample is classified into L poststrata. The poststratified estimator is in the form of (2.8.1) with $\omega_i = \pi_i^{-1}(N_h/\hat{N}_h)$, where π_i is the inclusion probability associated with the i-th unit, h is the stratum to which the i-th unit was classified, N_h is the known number of units in the h-th stratum, $\hat{N}_h = \sum_{j \in s_h} \pi_j^{-1}$ is the full-sample estimator of N_h, and s_h denotes the set of selected units that were classified into the h-th stratum. The α-th random group estimator is in the form of (2.8.2) with

$$\omega_{\alpha i} = k\pi_i^{-1}(N_h/\hat{N}_{h(\alpha)}),$$

$$\hat{N}_{h(\alpha)} = \sum_{j \in s_h(\alpha)} k\pi_j^{-1},$$

and $s_h(\alpha)$ denotes the set of units in the α-th random group that were classified into the h-th stratum. Note that the basic weight associated with the full sample is π_i^{-1}, and with the α-th random group is $k\pi_i^{-1}$. Then from first principles, a random group estimator of variance is

$$v_2(\hat{\theta}) = \frac{1}{k(k-1)} \sum_{\alpha}^{k} (\hat{\theta}_\alpha - \hat{\theta})^2.$$

Viewing this estimator as a function of $(\hat{N}_{1(1)}, \ldots, \hat{N}_{L(1)}, \ldots, \hat{N}_{L(k)})$ and expanding in a Taylor series about the point $(\hat{N}_1, \ldots, \hat{N}_L, \ldots, \hat{N}_L)$ gives

$$v_2(\hat{\theta}) \doteq \frac{1}{k(k-1)} \sum_{\alpha}^{k} (\tilde{\theta}_\alpha - \hat{\theta})^2$$

where

$$\tilde{\theta}_\alpha = \sum_i \{k\pi_i^{-1}(N_h/\hat{N}_h)\} y_i$$

is the α-th random group estimator derived using the full sample weight adjustment $N_h/\hat{N}_h$. This shows that the strict random group procedure $\hat{\theta}_\alpha$ and the modified shortcut procedure $\tilde{\theta}_\alpha$ should give similar results, at least to a local approximation.

Simmons and Baird (1968) and Bean (1975) have compared empirically the shortcut procedure to the standard procedure using the National Center for Health Statistic's Health Examination Survey. They found that the shortcut procedure gave slightly worse variance estimates, but saved greatly in computational costs. We are hesitant to recommend this procedure for all surveys, however, on the basis of these few studies to date. More work is needed, both theoretical and empirical work, before any general recommendations can be made.

These comments about weight adjustments also apply to the donor pool used in making imputations for missing data. Strict RG principles suggest that the α-th random group should serve as the donor pool for imputing for missing data within the α-th random group, for $\alpha = 1, \ldots, k$. An alternative procedure that may give satisfactory results in some circumstances is to let the entire parent sample serve as the donor pool for imputing for missing data within any of the random groups.

The second modification to the random group method concerns the manner in which the random groups are formed in the nonindependent case. To illustrate the modification, we suppose the population is divided into $L \geq 1$ strata and two or more primaries are selected within each stratum using some without replacement sampling scheme. Strict adherence to random group principles would dictate that random groups be formed by randomly selecting one or more of the ultimate clusters from each stratum. In this manner, each random group would have the same design features as the parent sample. However, if there are only a small number of selected primaries in some or all of the strata, the number of random groups, k, will be small and the resulting variance of the variance estimator large. For this situation we may seek a modified random group procedure which is biased but leads to greater stability through use of a larger number of random groups. The following modification may be acceptable:

(i) The ultimate clusters are ordered on the basis of the stratum from which they were selected. Within a stratum, the ultimate clusters are taken in a natural or random order.

(ii) The ultimate clusters are then systematically assigned to k (acceptably large) random groups. For example, the first ultimate cluster may be assigned random group α^* (a random integer between 1 and k), the second to group $\alpha^* + 1$, and so forth in a modulo k fashion.

The heuristic motivation for the modification is quite simple: the bias of the variance estimator should not be large since the systematic assignment procedure reflects approximately the stratification in the sample, while use of increased k should reduce the variance of the variance estimator. Unfortunately, we know of no theory to substantiate this claim, but a small empirical study by Isaki and Pinciaro (1977) is supportive. An example involving an establishment survey is reported in Section 2.10.

2.9. On the Condition $\hat{\bar{\theta}} = \hat{\theta}$ for Linear Estimators

At various points in this chapter, we have stated that the mean of the random group estimators is equal to the parent sample estimator, i.e., $\hat{\bar{\theta}} = \hat{\theta}$, whenever the estimator is linear. We shall make similar statements in Chapters 3 and 4 as we talk about balanced half-samples and the jackknife. We shall now demonstrate the meaning of this statement in the context of Definition 1.1 in Section 1.5. This work clarifies the distinction between $v(\hat{\bar{\theta}})$, $v_1(\hat{\theta})$, and $v_2(\hat{\theta})$, and suggests when the parent sample estimator $\hat{\theta}$ may be reproduced as the mean of the $\hat{\theta}_\alpha$. This later point is of interest from a computational point of view, because it is important to know when

$\hat{\theta}$ may be computed as a by-product of the $\hat{\theta}_\alpha$ calculations and when a separate calculation of $\hat{\theta}$ is required.

Using the notation of Section 1.5, we note that it is sufficient to work with (1.02) because the estimator in Definition 1.1 satisfies

$$\hat{\bar{\theta}} = \frac{1}{k}\sum_\alpha^k \hat{\theta}_\alpha$$
$$= \gamma_0 + \gamma_1\hat{\bar{\theta}}(1) + \ldots + \gamma_p\hat{\bar{\theta}}(p)$$
$$= \gamma_0 + \gamma_1\hat{\theta}(1) + \ldots + \gamma_p\hat{\theta}(p)$$
$$= \hat{\theta}$$

if and only if

$$\hat{\bar{\theta}}(j) = \hat{\theta}(j), \qquad \text{for } j = 1, \ldots, p,$$

where

$$\hat{\theta}_\alpha = \gamma_0 + \gamma_1\hat{\theta}_\alpha(1) + \ldots + \hat{\theta}_\alpha(p)$$

and $\hat{\theta}_\alpha(j)$ denotes the estimator for the j-th characteristic based on the α-th random group. As a consequence, we condense the notation, letting $\hat{\theta}$ denote an estimator of the form (1.02).

It is easy to establish that not all linear estimators in this form satisfy the property $\hat{\bar{\theta}} = \hat{\theta}$. A simple example is the classical ratio estimator for srs wor. For this case

$$\hat{\theta} = (\bar{y}/\bar{x})\bar{X},$$
$$\hat{\theta}_\alpha = (\bar{y}_\alpha/\bar{x}_\alpha)\bar{X},$$

and

$$\hat{\bar{\theta}} = \left(k^{-1}\sum_\alpha^k \bar{y}_\alpha/\bar{x}_\alpha\right)\bar{X}$$
$$\neq \hat{\theta}.$$

Unfortunately, it is difficult to specify precisely the class of linear estimators for which the property does hold. The best we can do is illustrate some cases where it does and does not hold.

(i) Suppose that a single-stage sample of fixed size $n = mk$ is selected without replacement and then divided into k random groups according to the principles given in Section 2.3. The Horvitz–Thompson (H–T) estimator of the population total satisfies $\hat{\bar{\theta}} = \hat{\theta}$, since

$$\hat{\theta} = \sum_i^n y_i/\pi_i$$

and

$$\hat{\theta}_\alpha = \sum_i^m y_i/(\pi_i k^{-1}).$$

(ii) Suppose k independent random groups are selected, each of size m, using pps wr sampling. The customary estimators of the population total based on the parent sample and on the α-th random group are

$$\hat{\theta} = (1/n) \sum_i^n y_i/p_i$$

and

$$\hat{\theta}_\alpha = (1/m) \sum_i^m y_i/p_i,$$

respectively. Clearly, $\hat{\bar{\theta}} = \hat{\theta}$.

(iii) If k independent random groups, each a srs wr of size m, are selected, the Horvitz–Thompson estimators are

$$\hat{\theta} = \sum_{i \in d(s)} Y_i/\{1 - (1 - 1/N)^n\}$$

and

$$\hat{\theta}_\alpha = \sum_{i \in d(s(\alpha))} Y_i/\{1 - (1 - 1/N)^m\},$$

where the summations are over the distinct units in the full sample s and in the α-th random group $s(\alpha)$, respectively. For this problem, it is easy to see that $\hat{\bar{\theta}} \neq \hat{\theta}$.

(iv) Let the sampling scheme be the same as in (iii) and let $\hat{\theta}$ and $\hat{\theta}_\alpha$ denote the sample means of the distinct units in the full sample and in the α-th random group, respectively. Once again we have $\hat{\bar{\theta}} \neq \hat{\theta}$.

These examples show that the kinds of sampling strategies which satisfy the condition $\hat{\bar{\theta}} = \hat{\theta}$ are quite varied and cut across all classes of linear estimators discussed in Section 1.5. To complicate matters, even some nonlinear estimators satisfy the condition, such as

$$\hat{\theta} = (N/n) \sum_{i \in s} \hat{Y}_i$$

$$\hat{\theta}_\alpha = (N/m) \sum_{i \in s(\alpha)} \hat{Y}_i,$$

where a srs wor of $n = km$ primaries is divided into k random groups, and $\hat{Y}_i$ denotes some nonlinear estimator of the total in the i-th primary. Thus, the statements that we have made in this chapter about $\hat{\bar{\theta}} = \hat{\theta}$ are somewhat imprecise without clarification of the meaning of the term *linear estimator*. This is equally true of our statements in Chapters 3 and 4. The reader should interpret all such statements in light of the exceptions described above. It should be observed, however, that for the sampling strategies used most commonly in practice, e.g., without replacement sampling and the Horvitz–Thompson estimator, the condition $\hat{\bar{\theta}} = \hat{\theta}$ does hold.

2.10. Example: The Retail Trade Survey

The Census Bureau's retail trade survey is a large, complex survey conducted monthly to obtain information about retail sales in the United States. In this section we discuss the problems of variance estimation for this survey. The case of nonindependent random groups is illustrated.

The target population for the retail trade survey consists of all business establishments in the United States that are primarily engaged in retail trade. A given month's sample consists of two principal components, each selected from a different sampling frame. The first component is a sample of approximately 12,000 units selected from a list of retail firms that have employees and that make Social Security payments for their employees. This is by far the larger component of the survey, contributing about 94% of the monthly estimates of total retail sales. A combination of several types of sampling units are used in this component, though a complete description of the various types is not required for present purposes. For purposes of this example, we will treat the company (or firm) as the sampling unit, this being only a slight oversimplification.

The second principal component of the retail trade survey is a multistage sample of land segments. All retail stores located in selected segments and not represented on the list frame are included in this component. Typically, such stores either do not have employees or have employees but only recently hired them. This component contributes only about 6% of the monthly estimates of total retail sales.

Due to its overriding importance, this example will only treat the problems of estimation for the list sample component. Before considering the estimators of variance, however, we discuss briefly the sampling design and estimators of total sales.

Important aspects of the sampling design for the list sample component include the following:

(i) The population of firms was stratified by kind of business (KB) and within KB by size of firm. A firm's measure of size was based on both the firm's annual payroll as reported to the Internal Revenue Service (IRS) and the firm's sales in the most recent Census of Retail Trade.

(ii) The highest size stratum within each KB was designated a *certainty* stratum. All firms in the certainty stratum were selected with probability one.

(iii) The remaining size strata within a KB were designated *noncertainty*. A simple random sample without replacement (srs wor) was selected independently within each noncertainty stratum.

(iv) A cutoff point was established for subsampling individual establishments within selected firms. The cutoff point was 25 for certainty firms and 10 for noncertainty firms. Within those selected firms having a "large" number of establishments (i.e., more establishments than the cutoff point), an establishment subsample was selected. The subsample was selected independently within each such firm using unequal probability systematic sampling. In this operation, an establishment's conditional inclusion probability (i.e., the probability of selection given that the firm was selected) was based on the same size measure as employed in stratification. Within those selected firms having a "small" number of establishments (i.e., fewer establishments than the cutoff point), all establishments were selected. Thus, the company (or firm) was the primary sampling unit and the establishment, the second stage unit.

(v) Each month lists of birth establishments are obtained from the previously selected companies. Additionally, lists of birth companies are obtained from administrative sources approximately once every third month. The birth establishments of previously selected companies are sampled using the sampling scheme described in (iv). Birth companies are subjected to a double sampling scheme. A large first phase sample is enumerated by mail, obtaining information on both sales size and the kind of business. Using this information, the second phase sample is selected from KB by sales size strata. Because the births represent a relatively small portion of the total survey, we shall not describe the birth sampling in any greater detail here. For more information, see Wolter et al. (1976).

(vi) From the noncertainty strata, two additional samples (or panels) were selected according to this sampling plan without replacement. The first panel was designated to report in the first, fourth, seventh, and tenth months of each calendar year; the second in the second, fifth, eighth, and eleventh months; and the third in the third, sixth, ninth, and twelfth months. Cases selected from the certainty stratum were designated to be enumerated every month.

For any given month, the firms selected for that month's sample are mailed a report form asking for total company sales and sales of the selected establishments in that month and in the immediately preceding month. Call backs are made to delinquent cases by telephone.

The principal parameters of retail trade which are estimated from the survey include total monthly sales, month-to-month trend in sales, and month-to-same-month-a-year-ago trend in sales. The estimates are computed for individual KBs, individual geographic areas, and across all KBs and geographic areas. In this example we shall focus on the estimation of total monthly sales.

To estimate the variability of the survey estimators, the random group method is employed in its nonindependent mode. Sixteen random groups are used. Important aspects of the assignment of firms and establishments to the random groups include the following:

(i) Strict application of the random group principles articulated in Section 2.4.1 would require at least 16 selected units in each noncertainty stratum, i.e., at least one unit per random group. This requirement was not met in the retail trade survey, and, in fact, in many of the KB by size strata only three units were selected. This meant that, at most, only three random groups could be formed and that the estimator of variance would itself have a variance that is much larger than desired. It was therefore decided to deviate somewhat from strict principles and to use a method of forming random groups that would create a larger number of groups. The method chosen accepts a small bias in the resulting variance estimator in exchange for a much reduced variance relative to what would occur if only three random groups were used.

(ii) To form 16 random groups such that as much of the stratification as possible was reflected in the formation of the random groups, the selected units in noncertainty strata were ordered by KB and within KB by size stratum. The order within a size stratum was by the units' identification numbers, an essentially random ordering. Then, a random integer, say α^*, between 1 and 16 was generated, and the first unit in the ordering was assigned to random group α^*, the second to group $\alpha^* + 1$ and so forth in a modulo 16 fashion. Thus, the random groups were formed systematically instead of in a stratified manner. The effect of stratification in the parent sample, however, was captured to a large extent by the ordering that was specified.

(iii) Within firms selected in noncertainty strata, all selected establishments were assigned the same random group number as was the firm. Thus, the ultimate cluster principle for multistage sampling designs was employed.

(iv) In the certainty stratum, although the component of variability due to the sampling of companies was zero, it was necessary to account for the component of variability due to subsampling within the companies. This was accomplished by ordering the selected establishments by KB, and within KB by size stratum. Within a size stratum ordering was by identification number. Then, the establishments were systematically assigned to random groups in the manner described in (ii). Of course, establishments associated with certainty firms, all of whose establishments were selected, do not contribute to the sampling variance and were not assigned to one of the 16 random groups.

(v) The selected birth establishments were also assigned to the 16 random groups. Again, for brevity we shall not describe the assignment process here. For details, see Wolter et al. (1976).

The basic estimator of total sales used in the retail trade survey is the Horvitz–Thompson estimator

$$\hat{Y} = \sum_{\alpha=0}^{16} \sum_{j} y_{\alpha j} / \pi_{\alpha j}, \qquad (2.10.1)$$

where $y_{\alpha j}$ is the sales of the j-th establishment in the α-th random group and $\pi_{\alpha j}$ is the associated inclusion probability. The subscript $\alpha = 0$ is reserved for the establishments of certainty companies that have not been subsampled, and for such cases $\pi_{0j} = 1$. For subsampled establishments of certainty companies and for establishments of noncertainty companies, the inclusion probability is less than one ($\pi_{\alpha j} < 1$). The inclusion probabilities used in (2.10.1) refer to the complete sample; the probability of inclusion in any given random group is $\pi_{\alpha j}/16$. Consequently, an alternative expression for $\hat{Y}$ is

$$\hat{Y} = \hat{Y}_0 + \sum_{\alpha=1}^{16} \hat{Y}_\alpha/16, \qquad\qquad (2.10.2)$$

where

$$\hat{Y}_0 = \sum_j y_{0j},$$

$$\hat{Y}_\alpha = \sum_j y_{\alpha j}(16/\pi_{\alpha j})$$

for $\alpha = 1, \ldots, 16$. In (2.10.2), $\hat{Y}_\alpha$ is the Horvitz–Thompson estimator from the α-th random group of the total sales due to the noncertainty portion of the population, and $\hat{Y}_0$ is the total of the certainty establishments. Since $\hat{Y}_0$ is fixed, it does not contribute to the sampling variance of $\hat{Y}$ and

$$\text{Var}\{\hat{Y}\} = \text{Var}\left\{\sum_{\alpha=1}^{16} \hat{Y}_\alpha/16\right\}.$$

The imputation of sales for nonresponding establishments is a complicated process in the retail trade survey. Here, we shall present a simplified version of the imputation process, but one that contains all of the salient features of the actual process. The imputed value $\tilde{y}_{\alpha j}$ of a nonresponding unit (α, j) is essentially the value of the unit at some previous point in time multiplied by a measure of change between the previous and current times. Specifically,

$$\tilde{y}_{\alpha j} = \tilde{\delta} x_{\alpha j},$$

where $x_{\alpha j}$ is the value of unit (α, j) at a previous time,

$$\tilde{\delta} = \frac{\sum^+ y_{\beta i}/\pi_{\beta i}}{\sum^+ x_{\beta i}/\pi_{\beta i}}$$

is a ratio measure of change, and the summations $\sum^+$ are over all similar units (e.g., in the same kind of business) that responded in both the present and previous time periods. The ratio is computed from data in all random groups, not just from the random group of the nonrespondent (α, j). Usually, the previous time period is three months ago for a noncertainty establishment and one month ago for a certainty establishment (i.e., the last time the establishment's panel was enumerated). To simplify notation, the "$.$" is deleted from the imputed values in (2.10.1), (2.10.2), and (2.10.3), although it should be understood there that the $y_{\alpha j}$ is the reported or imputed value depending upon whether the unit responded or not, respectively.

The random group estimator of $\mathrm{Var}\{\hat{Y}\}$ is then

$$v(\hat{Y}) = \{1/16(15)\} \sum_{\alpha=1}^{16} \left(\hat{Y}_\alpha - \sum_{\beta=1}^{16} \hat{Y}_\beta/16 \right)^2 \qquad (2.10.3)$$

and the estimator of the coefficient of variation (CV) is $\{v(\hat{Y})\}^{1/2}/\hat{Y}$. The reader will note that (2.10.2) and (2.10.3) are entirely equivalent to letting

$$\hat{\theta}_\alpha = \hat{Y}_0 + \hat{Y}_\alpha,$$

$$\hat{\bar{\theta}} = \sum_{\alpha=1}^{16} \hat{\theta}_\alpha/16$$

$$= \hat{Y},$$

and

$$v(\hat{\bar{\theta}}) = \{1/16(15)\} \sum_{\alpha=1}^{16} (\hat{\theta}_\alpha - \hat{\bar{\theta}})^2$$

$$= v(\hat{Y}).$$

In this notation, $\hat{\theta}_\alpha$ is an estimator of total sales, including both certainty and noncertainty portions of the population. Presented in this form, the estimators of both total sales and variance have the form in which they were originally presented in Section 2.4.

It is worth noting at this point that strict principles were violated when the change measure $\tilde{\delta}$ used in the imputation process was computed from all random groups combined, rather than computed individually within each random group. The overall $\tilde{\delta}$ has obvious computational advantages. This procedure probably does not seriously bias the variance estimator, although a rigorous proof has not been given (recall the discussion Section 2.8).

To illustrate the computations that are required, we consider the case of total August 1977 grocery store sales. The random group totals are given in Table 2.10.1. Computations associated with $\hat{Y}$ and $v(\hat{Y})$ are presented in Table 2.10.2. Some estimators of $\mathrm{Var}\{\hat{Y}\}$ based on the order statistics are computed in Table 2.10.3.

In the retail trade survey itself, the published statistics result from a "composite estimation" formula. The estimates presented here are not published but are the inputs to the composite formula. We shall return to this example in Chapter 6, and at that time discuss the composite estimator and estimators of the month-to-month trend in sales.

Table 2.10.1. Random Group
Totals $\hat{Y}_\alpha$ for August 1977 Grocery
Store Sales

Random Group α	$\hat{Y}_\alpha$($1000)
0	7,154,943
1	4,502,016
2	4,604,992
3	4,851,792
4	4,739,456
5	3,417,344
6	4,317,312
7	4,278,128
8	4,909,072
9	3,618,672
10	5,152,624
11	5,405,424
12	3,791,136
13	4,743,968
14	3,969,008
15	4,814,944
16	4,267,808

Table 2.10.2. Computation of $\hat{Y}$ and $v(\hat{Y})$ for
August 1977 Grocery Store Sales

By definition, we have

$$\hat{Y} = \hat{Y}_0 + \sum_{\alpha=1}^{16} \hat{Y}_\alpha/16$$

$$= 7,154,943 + 4,461,481$$

$$= 11,616,424$$

where the unit is $1000.
Also

$$v(\hat{Y}) = \{1/16(15)\} \sum_{\alpha=1}^{16} \left(\hat{Y}_\alpha - \sum_{\beta=1}^{16} \hat{Y}_\beta/16 \right)^2$$

$$= 19,208,267,520.$$

Thus, the estimated coefficient of variation is

$$cv(\hat{Y}) = \{v(\hat{Y})\}^{1/2}/\hat{Y}$$

$$= 138,594/11,616,424$$

$$= 0.012.$$

Table 2.10.3. Computations Associated with Estimates of Var$\{\hat{Y}\}$ Based on the Ordered $\hat{Y}_\alpha$

Corresponding to $\hat{\sigma}_1$, $\hat{\sigma}_2$, $\hat{\sigma}_3$, $\hat{\sigma}_4$, and $\hat{\sigma}_5$, of Section 2.7, we have the following estimates of Var$\{\hat{Y}\}$:

$$(\hat{\sigma}_1/4)^2 = \{(W/3.531)/4\}^2$$
$$= 140{,}759^2$$
$$(\hat{\sigma}_2/4)^2 = \{(W + W_{(2)} + W_{(4)}/7.626)/4\}^2$$
$$= 144{,}401^2$$
$$(\hat{\sigma}_3/4)^2 = (\hat{\sigma}_1/4)^2$$
$$= 140{,}759^2$$
$$(\hat{\sigma}_4/4)^2 = \{0.168(W + 1.216 W_{(3)})/4\}^2$$
$$= 140{,}595^2$$

and

$$(\hat{\sigma}_5/4)^2 = \left\{\left(\frac{2\sqrt{\pi}}{16(15)} \sum_{\alpha=1}^{16} (\alpha - 17/2)\hat{Y}_{(\alpha)}\right)/4\right\}^2$$
$$= 143{,}150^2.$$

The corresponding estimates of the coefficient of variation are

$$cv_1(\hat{Y}) = \frac{140{,}657}{11{,}616{,}424} = 0.012,$$
$$cv_2(\hat{Y}) = \frac{144{,}368}{11{,}616{,}424} = 0.012,$$
$$cv_3(\hat{Y}) = cv_1(\hat{Y}) = 0.012,$$
$$cv_4(\hat{Y}) = \frac{140{,}595}{11{,}616{,}424} = 0.012$$
$$cv_5(\hat{Y}) = \frac{143{,}150}{11{,}616{,}424} = 0.012.$$

Clearly, these "quick" estimators are very similar to the random group estimator for these data.

2.11. Example: The 1972–73 Consumer Expenditure Survey

The U.S. Bureau of Labor Statistics has sponsored eight major surveys of consumer expenditures, savings, and income since 1888. The 1972–73 survey, which is the main focus of this example, was undertaken principally to revise the weights and associated pricing samples for the Consumer Price Index and to provide timely, detailed, and accurate information on how American families spend their income.

The 1972–73 Consumer Expenditure Survey (CES) consisted of two main components, each using a separate probability sample and questionnaire. The first component, known as the quarterly survey, was a panel survey in which each consumer unit[7] in a given panel was visited by an interviewer every 3 months over a 15 month period. Repondents were asked primarily about their expenditures on major items, e.g., clothing, utilities, appliances, motor vehicles, real estate, and insurance. The second component of CES was the diary survey, in which diaries were completed at home by the respondents. This survey was intended to obtain expenditure data on food, household supplies, personal care products, nonprescription drugs, and other small items not included in the quarterly survey.

To simplify the presentation, this example will be concerned only with the quarterly survey. The sampling design, estimation procedure, and variance estimation procedure for the diary survey were similar to those of the quarterly survey.

The quarterly survey employed a multistage, self-weighting sampling design. Its principal features included the following:

1. The 1,924 primary sampling units (PSU) defined for the Census Bureau's Current Population Survey (see Hanson (1978)) were grouped into 216 strata on the basis of percent nonwhite and degree of urbanization. Fifty-four of these strata contained only one PSU (thus designated self-representing PSUs), while the remaining 162 strata contained two or more PSUs (thus designated nonself-representing PSUs).
2. From each of the 162 strata, one primary was selected using a controlled selection scheme. This scheme controlled on the number of SMSAs (Standard Metropolitan Statistical Area) from each of two size classes, and on the expected number of nonself-representing PSUs in each State.
3. Within each selected PSU a self-weighting sample of three types of units was selected:
    a. housing units which existed at the time of the 1970 Census;
    b. certain types of group quarters;
    c. building permits representing new construction since 1970.
    For simplicity this example will only describe the sampling of types *a* and *b* units. The sampling frame for types *a* and *b* units was the 20% sample of households in the 1970 Decennial Census that received the census long form.
4. Subsampling of types *a* and *b* units was performed independently in each nonself-representing PSU. Existing housing units were assigned a sampling code between 1 and 54 according to Table 2.11.1. Each group

---

[7] A consumer unit is a single financially independent consumer or a family of two or more persons living together, pooling incomes, and drawing from a common fund for major expenditures.

Table 2.11.1. Sampling Codes for Housing Units for Within PSU Sampling

Rent or Value		Owner Family Size					Renter Family Size				
Rent	Value	1	2	3	4	5 +	1	2	3	4	5 +
$0–$49	$0–$9,999	1	4	5	8	9	2	3	6	7	10
$50–$69	$10,000–$14,999	20	17	16	13	12	19	18	15	14	11
$70–$99	$15,000–$19,999	21	24	25	28	29	22	23	26	27	30
$100–$149	$20,000–$24,999	40	37	36	33	32	39	38	35	34	31
$150 +	$25,000 +	41	44	45	48	49	42	43	46	47	50

51: Low Value Vacants (Rent under $80 or value under $15,000)
52: Medium Value Vacants (Rent of $80–$119 or value of $15,000–$24,999)
53: High Value Vacants (Rent over $120 or value over $25,000)
54: Residual Vacants (Those not for sale or rent).

quarters persons was assigned sampling code 55. All types *a* and *b* units in each PSU were then arranged into the following order:
a. sampling code:
b. state;
c. county;
d. census enumeration district (ED).
For each PSU, a single random start was generated and a systematic sample of housing units (HU) and group quarters persons (GQP) was selected.
5. All consumer units in selected housing units were taken into the sample.
6. A panel number between 1 and 3 was systematically assigned to each selected HU and GQP in a modulo 3 fashion.
7. In self-representing primaries, the sampling of types *a* and *b* units occurred as in the nonself-representing primaries. In self-representing primaries, the selected units were then assigned to 15 random groups based on the order of the systematic selection. A random integer, α^*, between 1 and 15 was generated and the first unit selected was assigned to group α^*, the second to group $\alpha^* + 1$, and so forth in a modulo 15 fashion.
8. Finally, the sample was divided in half, and one half was enumerated between January, 1972 and March, 1973 and the second half between January, 1973 and March, 1974. Thirty of the original 54 self-representing PSUs were retained in sample for both years, but the subsample in each of these PSUs was randomly halved. The remaining 24 original self-representing PSUs and the original nonself-representing PSUs were paired according to stratum size. Then, one PSU from each pair was assigned to 1972 and the other to 1973. This step was not included in

the original specification of the CES sampling design. It was instituted later when budgetary limitations necessitated extending the sample over two fiscal years instead of over one year, as originally specified.

The estimator of total used in the quarterly survey was of the form

$$\hat{Y} = \sum_i w_i y_i \qquad (2.11.1)$$

where y_i is the value of the i-th consumer unit (CU) and w_i denotes the corresponding weight. The CU weights were generated by the following seven step process:

1. The *basic weight* was the reciprocal of the inclusion probability.
2. A so called *duplication control factor*[8] was applied to the basic weight of certain CUs selected in the new construction sample and existing CUs in two small PSUs.
3. The weight resulting from step 2 was multiplied by a *noninterview adjustment factor* which was calculated for each of 106 noninterview cells within each of four geographical regions.
4. *First stage ratio factors* were then applied to the weights of all CUs in the nonself-representing PSUs. This factor was computed for ten race-residence cells in each of the four geographic regions, and took the form

$$\frac{\sum_h (1970 \text{ Census total for stratum } h)}{\sum_h (\text{Sample PSU in stratum } h)/\pi_h}$$

where π_h is the inclusion probability associated with the selected primary in stratum h and the summation is taken over all strata in the region. Thus, the weight resulting from step 3 for a given CU was multiplied by the factor appropriate for the region and race-residence cell corresponding to the CU.
5. To adjust for noninterviewed CUs in multi-CU households, a *multi-CU factor* was applied to the weight resulting from step 4.
6. Next, a *second stage ratio factor* was applied to each person 14 years old and over in each CU. This was computed for 68 age-sex-race cells and took the form

$$\frac{\text{Independent population count of an age-sex-race cell}}{\text{Sample estimate of the population of an age-sex-race cell}},$$

where the independent population counts were obtained from the Census Bureau.
7. Until the assignment of the second stage ratio factor, all persons in a CU had the same weight. But after the second stage factor was applied,

---

[8] In a few cases, CUs were subsampled in large housing units. The duplication control factor simply adjusted the weight to include the conditional probability due to subsampling.

unequal weights were possible. To assign a final weight to a CU, a so-called *principal person procedure*[9] was employed.

For a comprehensive discussion of the CES and its sample design and estimation schemes, see U.S. Department of Labor (1978). Here we have only attempted to provide the minimal detail needed to understand the CES variance estimators. The description provided above suggests a complex sampling design and estimation scheme. The ensuing development shows the considerable simplification in variance estimation that results from the random group method.

In the CES, estimated totals and their associated variance estimates were computed separately for the self-representing (SR) and nonself-representing (NSR) PSUs. We begin by discussing the estimation for the SR PSUs. Subsequently, we discuss the estimation for the NSR PSUs and the combined estimates over both SR and NSR PSUs.

The variance due to subsampling within the SR PSUs was estimated in the following fashion:

1. The 30 SR PSUs were grouped into 15 clusters of one or more PSUs for purposes of variance estimation.
2. For each cluster, 15 random group totals were computed according to the relation

$$\hat{Y}_{1c\alpha} = \sum_i 15 w_{c\alpha i} y_{c\alpha i},$$

   where $y_{c\alpha i}$ denotes the value of the i-th CU in the α-th random group and c-th cluster and $w_{c\alpha i}$ denotes the corresponding weight as determined by the seven step procedure defined above. $\hat{Y}_{1c\alpha}$ is the estimator in (2.11.1) where the summation has been taken only over units in the α-th random group and c-th cluster. The $w_{c\alpha i}$ are the parent sample weights and the 15 $w_{c\alpha i}$ are the appropriate weights for a given random group.
3. Cluster totals and variances were estimated by

$$\hat{Y}_{1c} = \sum_{\alpha=1}^{15} \hat{Y}_{1c\alpha}/15$$

   and

$$v(\hat{Y}_{1c}) = \frac{1}{15(14)} \sum_{\alpha=1}^{15} (\hat{Y}_{1c\alpha} - \hat{Y}_{1c})^2,$$

   respectively.
4. Totals and variances over all SR PSUs were estimated by

$$\hat{Y}_1 = \sum_c^{15} \hat{Y}_{1c}$$

[9] Expenditure data was tabulated in the CES by designating a "principal person" and assigning that person's weight to the CU. This was done because the second stage ratio factor applied to persons, not CUs. Since expenditure data was to be based on CUs, there was a need to assign each CU a unique weight.

and

$$v(\hat{Y}_1) = \sum_c^{15} v(\hat{Y}_{1c}),$$

respectively.

The total variance (between plus within components) due to the sampling of and within the NSR primaries was estimated using the random group and collapsed stratum techniques.

1. The 93 strata were collapsed into 43 groups, 36 of which contained two strata and seven of which contained three strata.
2. For each NSR PSU, a weighted total was computed according to the relation

$$\hat{Y}_{2gh} = \sum_i w_{ghi} y_{ghi}$$

where y_{ghi} denotes the value of the i-th CU in the h-th PSU in group g and w_{ghi} denotes the corresponding weight as determined by the seven step procedure defined earlier. $\hat{Y}_{2gh}$ is the estimator in (2.11.1) where the summation has been taken only over units in the h-th PSU in the g-th group. The w_{ghi} are the full sample weights. Thus $\hat{Y}_{2gh}$ is an estimator of the total in the (g, h)-th stratum.
3. Totals and variances over all NSR PSUs were estimated by

$$\hat{Y}_2 = \sum_g^{43} \hat{Y}_{2g} = \sum_g^{43} \sum_h^{L_g} \hat{Y}_{2gh}$$

and

$$v(\hat{Y}_2) = \sum_{g=1}^{36} 2 \sum_{h=1}^{2} (\hat{Y}_{2gh} - P_{gh} \hat{Y}_{2g})$$

$$+ \sum_{g=37}^{43} \frac{3}{2} \sum_{h=1}^{3} (\hat{Y}_{2gh} - P_{gh} \hat{Y}_{2g})^2,$$

where P_{gh} is the proportion of the population in the g-th group living in the h-th stratum. Observe that $v(\hat{Y}_2)$ is a collapsed stratum estimator. The factors P_{gh} are analogous to the A_{gh}/A_g in equation (2.5.2). Population is an appropriate factor here because it should be well correlated with the expenditure items of interest in the CES.

Finally, totals and variances over both SR and NSR PSUs were estimated by

$$\hat{Y} = \hat{Y}_1 + \hat{Y}_2 \qquad (2.11.2)$$

and

$$v(\hat{Y}) = v(\hat{Y}_1) + v(\hat{Y}_2),$$

respectively. The variance of $\hat{Y}$ is the sum of the variances of $\hat{Y}_1$ and $\hat{Y}_2$ because sampling was performed independently in the SR and NSR strata.

Before presenting some specific variance estimates, two aspects of the estimation procedure require special discussion. First, the application of the random group method did not adhere strictly to the principles discussed in Sections 2.2 and 2.4 of this chapter. Recall that the weights attached to the sample units included nonresponse adjustments and first and second stage ratio adjustments. All of these adjustment factors were computed from the entire sample, whereas strict random group principles would suggest computing these factors individually for each random group. The adopted procedure is clearly preferable from a computational standpoint, and it also can be justified in some cases by Taylor series arguments. Second, the collapsed stratum feature of the variance estimation procedure probably tended to overstate the actual variance. As was shown in Section 2.5, such overstatement tends to occur when one unit is selected independently within each stratum. In the CES, however, primaries were sampled by a controlled selection procedure, and to the extent that this resulted in lower true variance than an independent selection procedure, the overestimation of variance was probably aggravated.

The principal estimates derived from the quarterly survey included the total number of CUs and the mean annual expenditure per CU for various expenditure categories. The estimator of mean annual expenditure was of the ratio form, and its variance was estimated using a combination of random group and Taylor series methodologies. For that reason, the discussion of variance estimation for mean annual expenditures will be deferred until Chapter 6 (see Section 6.8).

Estimation of the total number of CUs is described in Tables 2.11.2–2.11.5. Random group totals for the SR and NSR PSUs are presented in Tables 2.11.2 and 2.11.3, respectively. The factors P_{gh}, are given in Table 2.11.4. Both the SR and NSR variances are computed in Table 2.11.5.

Table 2.11.2. Random Group Totals $\hat{Y}_{1c\alpha}$ for 15 SR PSU Clusters for the Characteristic "Number of Consumer Units"

Cluster	Random Group														
	1	2	3	4	5	6	7	8	9	10	11	12	13	14	15
1	408,639.0	357,396.0	406,582.0	463,005.0	427,265.0	433,718.0	316,821.0	367,599.0	465,279.0	368,789.0	312,066.0	379,808.0	382,066.0	308,008.0	310,034.0
2	84,475.5	65,930.7	109,240.0	115,305.0	119,158.0	54,822.7	108,168.0	131,387.0	113,461.0	147,637.0	71,462.7	111,232.0	113,085.0	113,632.0	116,211.0
3	90,731.1	78,854.1	101,145.0	99,786.3	69,822.7	85,448.8	119,923.0	90,064.0	83,736.4	99,358.4	68,473.6	96,295.3	114,048.0	88,357.6	92,100.0
4	65,456.7	82,499.1	64,632.7	55,332.4	62,720.2	56,897.9	75,327.1	46,551.6	73,008.5	59,297.7	56,198.8	72,127.1	61,664.2	56,334.8	65,764.3
5	83,525.3	81,785.1	73,253.1	77,608.7	58,887.5	92,749.9	59,008.9	72,472.1	93,013.3	82,558.8	64,400.1	88,424.0	73,746.0	85,262.3	95,249.9
6	87,031.8	77,788.0	108,043.0	90,720.5	63,497.7	85,113.3	70,378.6	99,848.2	74,704.6	108,067.0	51,308.8	83,727.0	81,678.4	87,635.5	91,742.5
7	65,741.1	83,396.6	57,453.9	77,180.5	72,451.9	71,307.4	85,104.4	71,513.8	84,460.7	92,231.5	80,008.2	82,344.5	74,616.0	95,972.3	68,959.0
8	185,623.0	156,194.0	164,003.0	174,197.0	141,762.0	154,799.0	175,702.0	122,718.0	176,838.0	200,017.0	124,952.0	192,507.0	158,457.0	181,052.0	161,947.0
9	75,639.1	64,196.3	93,421.4	108,321.0	100,972.0	93,013.9	95,722.4	86,739.6	136,033.0	69,893.2	91,167.2	100,265.0	96,195.1	108,930.0	100,151.0
10	101,372.0	90,663.4	91,030.5	79,964.8	112,677.0	86,183.4	78,265.5	93,575.0	77,851.6	78,362.5	84,926.5	121,252.0	83,196.4	79,638.5	77,902.5
11	90,187.9	99,528.5	81,693.6	94,278.5	113,166.0	67,375.5	91,108.6	109,077.0	61,284.6	85,516.7	78,263.5	78,887.1	101,257.0	93,691.7	99,227.3
12	126,003.0	108,540.0	134,745.0	142,045.0	156,887.0	121,070.0	91,035.9	102,899.0	107,933.0	135,442.0	103,747.0	121,209.0	137,179.0	126,890.0	98,291.6
13	187,673.0	172,838.0	182,223.0	164,629.0	157,816.0	186,880.0	188,280.0	199,172.0	164,640.0	188,261.0	171,694.0	196,747.0	186,855.0	171,844.0	215,760.0
14	128,911.0	98,133.0	133,032.0	116,259.0	149,563.0	101,264.0	112,746.0	112,180.0	137,857.0	104,227.0	116,848.0	114,150.0	93,083.5	113,723.0	101,238.0
15	123,450.0	155,278.0	129,759.0	193,347.0	144,412.0	181,260.0	146,856.0	123,346.0	178,617.0	138,641.0	124,411.0	117,424.0	151,347.0	145,666.0	125,546.0

Source: C. Dippo, Personal Communication (1977).
Note: All entries in this table should be multiplied by 15.

Table 2.11.3. Estimated Totals $\hat{Y}_{2gh}$ for 43 Collapsed Strata for the Characteristic "Number of Consumer Units"

| | Stratum (h) | | |
Group (g)	1	2	3
1	361,336	434,324	
2	413,727	479,269	
3	446,968	408,370	
4	520,243	598,114	
5	375,400	467,515	
6	477,180	464,484	
7	494,074	496,722	
8	437,668	456,515	
9	387,651	430,562	
10	450,008	467,255	
11	485,998	502,247	
12	464,604	393,965	
13	415,047	472,583	
14	444,814	481,008	
15	375,815	442,793	
16	438,436	474,527	
17	451,239	382,624	
18	460,168	311,482	
19	462,894	470,407	
20	493,373	540,379	
21	469,461	394,530	
22	426,485	546,285	
23	515,182	974,332	
24	436,378	410,247	
25	436,449	362,472	
26	383,687	431,037	
27	387,268	419,426	
28	302,383	441,139	
29	432,195	454,737	
30	432,159	426,645	
31	440,998	374,043	
32	367,096	528,503	
33	428,326	549,871	
34	395,286	456,075	
35	410,925	220,040	
36	465,199	475,912	
37	449,720	387,772	471,023
38	441,744	437,025	640,130
39	651,431	364,652	638,782
40	441,244	420,171	362,705
41	489,315	463,869	384,602
42	443,885	476,963	397,502
43	821,244	692,441	431,657

Table 2.11.4. Factors P_{gh} Used in Computing the
Variance Due to Sampling in NSR Strata

Group (g)	Stratum (h)		
	1	2	3
1	0.486509	0.513491	
2	0.485455	0.514545	
3	0.496213	0.503787	
4	0.438131	0.561869	
5	0.493592	0.506408	
6	0.505083	0.494917	
7	0.503599	0.496401	
8	0.499901	0.500099	
9	0.501436	0.498564	
10	0.507520	0.492480	
11	0.503276	0.496724	
12	0.500381	0.499619	
13	0.501817	0.498183	
14	0.501071	0.498929	
15	0.500474	0.499526	
16	0.489670	0.510330	
17	0.495551	0.504449	
18	0.500350	0.499650	
19	0.496320	0.503680	
20	0.497922	0.502078	
21	0.498676	0.501324	
22	0.475579	0.524421	
23	0.459717	0.540283	
24	0.495257	0.505743	
25	0.499968	0.500032	
26	0.499178	0.500822	
27	0.498887	0.501113	
28	0.497341	0.502659	
29	0.499085	0.500915	
30	0.499191	0.500809	
31	0.498829	0.501171	
32	0.484498	0.515502	
33	0.510007	0.489993	
34	0.532363	0.467637	
35	0.471407	0.528593	
36	0.486899	0.513101	
37	0.318160	0.338869	0.342971
38	0.313766	0.323784	0.362450
39	0.409747	0.312005	0.278248
40	0.337979	0.333625	0.328396
41	0.325368	0.333347	0.341285
42	0.331581	0.334400	0.334019
43	0.414992	0.332518	0.252491

Table 2.11.5. Estimated Totals and Variances for Both SR and NSR
PSUs

$\hat{Y}_{1c\alpha}$ is the element in the c-th row, α-th column of Table 2.11.2. Thus, the
estimated total and variance for the SR PSUs are

$$\hat{Y}_1 = \sum_c^{15} \hat{Y}_{1c} = \sum_c^{15} \sum_\alpha^{15} \hat{Y}_{1c\alpha}/15 = 28.2549 \cdot 10^6$$

and

$$v(\hat{Y}_1) = \sum_c^{15} v(\hat{Y}_{1c}) = \sum_c^{15} \frac{1}{15(14)} \sum_\alpha^{15} (\hat{Y}_{1c\alpha} - \hat{Y}_{1c})^2 = 10.2793 \cdot 10^{10},$$

respectively.

$\hat{Y}_{2gh}$ and P_{gh} are the elements in the g-th row, h-th column of Tables 2.11.3 and
2.11.4, respectively. Thus, the estimated total and variance for the NSR PSUs are

$$\hat{Y}_2 = \sum_g^{43} \hat{Y}_{2g} = \sum_g^{43} \sum_h^{L_g} \hat{Y}_{2gh} = 42.5344 \cdot 10^6$$

and

$$v(\hat{Y}_2) = \sum_g^{36} 2 \sum_{h=1}^{2} (\hat{Y}_{2gh} - P_{gh}\hat{Y}_{2g})^2$$

$$+ \sum_g^{43} (3/2) \sum_{h=1}^{3} (\hat{Y}_{2gh} - P_{gh}\hat{Y}_{2g})^2$$

$$= 46.5747 \cdot 10^{10},$$

respectively.

The estimated total and variance over both SR and NSR PSUs are

$$\hat{Y} = \hat{Y}_1 + \hat{Y}_2 = 70.7893 \cdot 10^6$$

and

$$v(\hat{Y}) = v(\hat{Y}_1) + v(\hat{Y}_2) = 56.8540 \cdot 10^{10}.$$

2.12. Example: The 1972 Commodity Transportation Survey

The 1972 Commodity Transportation Survey was a part of the 1972 U.S.
Census of Transportation. Prime objectives of the survey included the
measurement of the transportation and geographic distribution of commodities shipped beyond the local area by manufacturing establishments
in the U.S.

This example is limited to the Shipper Survey, which was the major
component of the Commodity Transportation Survey program. The frame

for the Shipper Survey was derived from the 1972 U.S. Census of Manufacturers, and consisted of all manufacturing establishments in the U.S. with twenty or more employees, there being approximately 100,000 in number. The Mail Summary Data Survey, a minor component of the Commodity Transportation Survey program, covered the manufacturing plants with less than 20 employees.

Key features of the sampling design and estimation procedure for the Shipper Survey included the following:

(i) Using the Federal Reserve Board's Index of Industrial Production, manufacturing plants were divided into 85 *shipper classes.* Each class was composed of similar SIC (Standard Industrial Classification) codes.

(ii) Each shipper class was then assigned to one of nine *tonnage divisions* based on total tons shipped. Each tonnage division comprised a separate sampling stratum.

(iii) Within a tonnage division manufacturing plants were ordered by shipper class, by state, by SIC code.

(iv) Each plant was assigned an *expected tonnage rating* on the basis of the plant's total employment size.

(v) Based on the expected tonnage rating, a certainty cutoff was specified within each tonnage division. All plants with a rating greater than the cutoff were included in the sample with probability one.

(vi) An unequal probability, single-start systematic sample of plants was selected independently from within the noncertainty portion of each tonnage division. The selection was with probability proportional to expected tonnage rating. The sample sizes are given in Table 2.12.1.

Table 2.12.1. Allocation of the Sample for the Shipper Survey

Tonnage Division	Shipper Classes	Number of Selected Plants		
		Certainty	Noncertainty	Total
1	1	81	40	121
2	5	459	696	1,155
3	5	354	578	932
4	4	339	541	880
5	13	1,242	1,045	2,287
6	12	771	866	1,637
7	15	880	1,318	2,198
8	13	675	1,102	1,777
9	17	585	1,348	1,933
Total	85	5,386	7,534	12,920

Source: Wright (1973).

(vii) Within each selected manufacturing plant, an equal probability sample of bills of lading was selected. The filing system for shipping documents varies from plant to plant, but often the papers are filed by a serial number. If so, a single-start systematic sample of bills of lading was selected (cf. Table 2.12.2 for subsampling rates). If not, then a slightly different sampling procedure was employed. The alternative subsampling procedures are not of critical importance for this example, and thus are not described here.

(viii) Population totals were estimated using the Horvitz–Thompson estimator

$$\hat{Y} = \hat{Y}_0 + \hat{Y}_1 + \hat{Y}_2$$

$$= \sum_i \sum_j Y_{0ij} + \sum_i \sum_j Y_{1ij}/\pi_{1ij} + \sum_i \sum_j Y_{2ij}/\pi_{2ij}$$

where Y_{cij} denotes the value of the j-th document in the i-th plant and π_{cij} denotes the associated inclusion probability. The c subscript denotes

$c = 0$ document selected at the rate $1/1$ from within a certainty plant

$c = 1$ document selected at a rate $<1/1$ from within a certainty plant

$c = 2$ document selected from a noncertainty plant.

The variance of $\hat{Y}$ was estimated using the random group method. Plants were assigned to $k = 20$ random groups in the following fashion:

(ix) All noncertainty plants were placed in the following order:

     tonnage division
     shipper class
     state
     plant ID.

(x) A random integer between 1 and 20 was generated, say α^*. The first plant was then assigned to random group (RG)α^*, the second to RG $\alpha^* + 1$, and so forth in a modulo 20 fashion.

(xi) All selected second stage units (i.e., bills of lading) within a selected noncertainty plant were assigned to the same RG as the plant.

(xii) All second stage units selected at the rate $1/1$ within certainty plants were excluded from the 20 RGs.

(xiii) Second stage units selected at a rate $<1/1$ within certainty plants were placed in the following order:

     tonnage division
     shipper class
     state
     plant ID.

Table 2.12.2. Subsampling Rates for
Shipping Documents Filed in Serial
Number Order

Number of Documents in File	Sampling Rate
0–199	1/1
200–399	1/2
400–999	1/4
1,000–1,999	1/10
2,000–3,999	1/20
4,000–9,999	1/40
10,000–19,999	1/100
20,000–39,999	1/200
40,000–79,999	1/400
80,000–99,999	1/500

(xiv) The second stage units in (xiii) were assigned to the $k = 20$ random groups in the systematic fashion described in (x).

The random group estimator of $\text{Var}\{\hat{Y}\}$ is prepared by estimating the variance for certainty plants and noncertainty plants separately. The two estimates are then summed to give the estimate of the total sampling variance. The random group estimator for either the certainty $(c = 1)$ or noncertainty plants $(c = 2)$ is defined by

$$v(\hat{Y}_c) = \frac{1}{20(19)} \sum_{\alpha=1}^{20} (\hat{Y}_{c\alpha} - \hat{Y}_c)^2 \qquad (2.12.1)$$

where $c = 1, 2$,

$$\hat{Y}_{c\alpha} = \sum_{(c,i,j)\in s(\alpha)} Y_{cij}(20/\pi_{cij}),$$

$\sum_{(c,i,j)\in s(\alpha)}$ denotes a sum over units in the α-th random group, and the $\pi_{cij}/20$ are the inclusion probabilities associated with the individual random groups.

Table 2.12.3 presents some typical estimates and their estimated coefficients of variation (CV) from the Shipper Survey. These estimates include the contribution from both certainty and noncertainty plants.

To illustrate the variance computations, Table 2.12.4 gives the random group totals for the characteristic "U.S. total shipments over all commodities." The estimate of the total tons shipped is

$$\hat{Y} = \hat{Y}_0 + \hat{Y}_1 + \hat{Y}_2$$
$$= \hat{Y}_0 + \sum_{\alpha=1}^{20} \hat{Y}_{1\alpha}/20 + \sum_{\alpha=1}^{20} \hat{Y}_{2\alpha}/20$$
$$= 42.662 \cdot 10^6 + 236.873 \cdot 10^6 + 517.464 \cdot 10^6$$
$$= 796.999 \cdot 10^6,$$

where $\hat{Y}_0$ is the total of certainty shipments associated with certainty plants.

Table 2.12.3. Estimates of Total Tons Shipped and Corresponding
Coefficients of Variation

Transportation Commodity Code	Commodity	Tons Shipped (1,000s)	Estimated CV (%)
29	Petroleum and coal products	344,422	6
291	Products of petroleum refining	310,197	6
2911	Petroleum refining products	300,397	7
29111	Gasoline and jet fuels	123,877	8
29112	Kerosene	6,734	37
29113	Distillate fuel oil	58,601	13
29114	Petroleum lubricating and similar oils	23,348	5
29115	Petroleum lubricating greases	553	17
29116	Asphalt pitches and tars from petroleum	21,406	19
29117	Petroleum residual fuel oils	36,689	12
29119	Petroleum refining products, NEC	24,190	42
2912	Liquified petroleum and coal gases	9,800	9
29121	Liquified petroleum and coal gases	9,800	9
295	Asphalt paving and roofing materials	21,273	10
2951	Asphalt paving blocks and mixtures	6,426	24
29511	Asphalt paving blocks and mixtures	6,426	24
2952	Asphalt felts and coatings	14,847	10
29521	Asphalt and tar saturated felts	2,032	13
29522	Asphalt and tar cements and coatings	4,875	18
29523	Asphalt sheathings, shingles, and sidings	7,817	13
29529	Asphalt felts and coatings, NEC	124	27
299	Miscellaneous petroleum and coal products	12,952	33
2991	Miscellaneous petroleum and coal products	12,952	33
29912	Lubricants and similar compounds, other than petroleum	760	16
29913	Petroleum coke, exc. briquettes	2,116	42
29914	Coke produced from coal, exc. briquettes	2,053	14
29919	Petroleum and coal products, NEC	1,851	50

Source: U.S. Bureau of the Census (1976).

From Equation 2.12.1, the estimate of $\mathrm{Var}\{\hat{Y}\}$ is

$$v(\hat{Y}) = v(\hat{Y}_1) + v(\hat{Y}_2)$$
$$= 12.72 \cdot 10^{12} + 18,422.78 \cdot 10^{12}$$
$$= 18,435.50 \cdot 50 \cdot 10^{12}.$$

The estimated CV is

$$\mathrm{cv}\{\hat{Y}\} = \frac{v(\hat{Y})^{1/2}}{\hat{Y}} = 0.17.$$

Table 2.12.4. Random Group Totals for the Characteristic "U.S. Total Shipments Over All Commodities"

Random Group	Total		Certainty		Noncertainty	
	Documents[a]	Estimated Tons Shipped	Documents	Estimated Tons Shipped	Documents	Estimated Tons Shipped
1	4777	1,293,232,120	1881	216,178,580	2896	1,077,053,540
2	4950	426,484,160	1892	274,275,900	3058	152,208,260
3	4516	3,142,517,880	1880	235,322,300	2636	2,907,195,580
4	4669	509,366,100	1890	241,408,260	2779	267,957,840
5	4624	489,690,320	1889	233,960,940	2735	255,729,400
6	4571	526,581,360	1878	228,875,380	2693	297,705,980
7	4740	646,957,040	1885	231,806,940	2855	415,150,100
8	5137	467,648,500	1881	212,205,220	3256	255,443,300
9	4669	773,257,400	1877	240,278,440	2792	532,978,960
10	5118	514,719,500	1880	224,676,140	3238	290,043,360
11	4637	569,137,600	1881	232,158,760	2756	336,978,840
12	4688	1,054,605,100	1877	214,550,740	2811	840,054,360
13	4719	679,291,880	1880	245,138,240	2839	434,153,640
14	4667	513,100,860	1890	261,394,840	2777	251,706,020
15	4614	525,385,740	1892	234,419,440	2722	290,966,300
16	5127	404,842,340	1903	249,839,540	3224	155,002,800
17	4786	508,047,300	1884	245,807,860	2902	262,239,420
18	4959	574,508,140	1896	220,536,240	3063	353,971,900
19	4827	869,575,520	1885	237,553,960	2942	632,021,560
20	4738	597,770,900	1890	257,069,860	2848	340,701,020
Certainty[b]	7410	42,661,791	7410	42,661,791	0	0

[a] Unweighted number of second stage units.
[b] Total of certainty second stage units associated with certainty plants.

CHAPTER 3

Variance Estimation Based on Balanced Half-Samples

3.1. Introduction

Efficiency considerations often lead the survey designer to stratify to the point where only two primary units are selected from each stratum. In such cases, only two independent random groups (or replicates or half-samples) will be available for the estimation of variance and confidence intervals for the population parameters of interest will necessarily be wider than desired. To overcome this problem, several techniques have been suggested. One obvious possibility is to apply the first version of rule (iv), Section 2.4.1, letting the random group method operate within the strata instead of across them. Variations on the collapsed stratum method offer the possibility of ignoring some of the stratification, thus increasing the number of available random groups. A bias is incurred, however, when the variance is estimated in this manner. Other proposed techniques, including jackknife and half-sample replication, aim to increase the precision of the variance estimator through some form of "pseudoreplication."

In this chapter we discuss various aspects of balanced half-sample replication as a variance estimating tool. The jackknife method, first introduced as a tool for reducing bias, is related to half-sample replication and will be discussed in the next chapter.

The basic ideas of half-sample replication first emerged at the U.S. Bureau of the Census through the work of W. N. Hurwitz, M. Gurney, and others. During the late 1950s and early 1960s, this method was used to estimate the variances of both unadjusted and seasonally adjusted estimates derived from the Current Population Survey. Following Plackett and Burman (1946), McCarthy (1966, 1969a, 1969b) introduced and developed the mathematics of balancing. The terms balanced half-samples, balanced fractional samples, pseudoreplication, and balanced repeated replication (BRR) have since come into common usage and all refer to McCarthy's method.

3.2. Description of Basic Techniques

Suppose it is desired to estimate a population mean $\bar{Y}$ from a stratified design with two units per stratum, where the selected units in each stratum comprise a simple random sample with replacement (srs wr). Let L denote the number of strata, N_h the number of units within the h-th stratum, and $N = \sum_{h=1}^{L} N_h$ the size of the entire population. Suppose y_{h1} and y_{h2} denote the observations from stratum h ($h = 1, \ldots, L.$) Then an unbiased estimator of $\bar{Y}$ is

$$\bar{y}_{st} = \sum_{h=1}^{L} W_h \bar{y}_h,$$

where

$$W_h = N_h / N$$

$$\bar{y}_h = (y_{h1} + y_{h2})/2.$$

The textbook estimator of $\text{Var}\{\bar{y}_{st}\}$ is given by

$$v(\bar{y}_{st}) = \sum_{h=1}^{L} W_h^2 s_h^2 / 2$$

$$= \sum_{h=1}^{L} W_h^2 d_h^2 / 4,$$

where

$$d_h = y_{h1} - y_{h2}.$$

For a complete discussion of the theory of estimation for stratified sampling, see Cochran (1977, Chapter 5).

For the given problem, only two independent random groups (or replicates or half-samples) are available: $(y_{11}, y_{21}, \ldots, y_{L1})$ and $(y_{12}, y_{22}, \ldots, y_{L2})$. The random group estimator of $\text{Var}\{\bar{y}_{st}\}$ is then

$$v_{RG}(\bar{y}_{st}) = [2(2-1)]^{-1} \sum_{\alpha=1}^{2} (\bar{y}_{st,\alpha} - \bar{y}_{st})^2$$

$$= (\bar{y}_{st,1} - \bar{y}_{st,2})^2 / 4,$$

where

$$\bar{y}_{st,1} = \sum_{h=1}^{L} W_h y_{h1},$$

$$\bar{y}_{st,2} = \sum_{h=1}^{L} W_h y_{h2},$$

and

$$\bar{y}_{st} = (\bar{y}_{st,1} + \bar{y}_{st,2})/2.$$

Because this estimator is based on only one degree of freedom, its stability (or variance) will be poor relative to the textbook estimator $v(\bar{y}_{st})$.

We seek a method of variance estimation with both the computational simplicity of $v_{RG}(\bar{y}_{st})$ and the stability of $v(\bar{y}_{st})$. Our approach will be to consider half-samples comprised of one unit from each of the strata. This work will differ fundamentally from the random group methodology in that we shall now allow different half-samples to contain some common units (and some different units) in a systematic manner. Because of the overlapping units, the half-samples will be correlated with one another. In this sense, the methods to be presented represent a form of "pseudoreplication," as opposed to pure replication.

To begin, suppose that a half-sample replicate is formed by selecting one unit from each stratum. There are 2^L such half-samples for a given sample, and the estimator of $\bar{Y}$ from the α-th half-sample is

$$\bar{y}_{st,\alpha} = \sum_{h=1}^{L} W_h(\delta_{h1\alpha}y_{h1} + \delta_{h2\alpha}y_{h2}),$$

$$\delta_{h1\alpha} = \begin{cases} 1, & \text{if unit } (h, 1) \text{ is selected for the } \alpha\text{-th half-sample} \\ 0, & \text{otherwise,} \end{cases}$$

and

$$\delta_{h2\alpha} = 1 - \delta_{h1\alpha}.$$

It is interesting to note that the mean of the 2^L estimators $\bar{y}_{st,\alpha}$ is equal to the parent sample estimator $\bar{y}_{st}$. This follows because each unit in the parent sample is a member of exactly one-half of the 2^L possible half-samples (or $2^L/2 = 2^{L-1}$ half-samples). Symbolically we have

$$\sum_{\alpha=1}^{2^L} \bar{y}_{st,\alpha}/2^L = \sum_{h=1}^{L} W_h(y_{h1} + y_{h2})(2^{L-1}/2^L)$$

$$= \bar{y}_{st}.$$

We shall construct a variance estimator in terms of the $\bar{y}_{st,\alpha}$. Define

$$\delta_h^{(\alpha)} = 2\delta_{h1\alpha} - 1$$

$$= \begin{cases} 1, & \text{if unit } (h, 1) \text{ is in the } \alpha\text{-th half-sample} \\ -1, & \text{if unit } (h, 2) \text{ is in the } \alpha\text{-th half-sample.} \end{cases}$$

Then it is possible to write

$$\bar{y}_{st,\alpha} - \bar{y}_{st} = \sum_{h=1}^{L} W_h \delta_h^{(\alpha)} d_h/2 \qquad (3.2.1)$$

and

$$(\bar{y}_{st,\alpha} - \bar{y}_{st})^2 = \sum_{h=1}^{L} W_h^2 d_h^2/4 + \sum_{h<h'}^{L} \delta_h^{(\alpha)} \delta_{h'}^{(\alpha)} W_h W_{h'} d_h d_{h'}/2, \quad (3.2.2)$$

since $\delta_h^{(\alpha)2} = 1$. Note that the right side of (3.2.2) contains both the textbook estimator $v(\bar{y}_{st})$ and a cross-stratum term. Notwithstanding this cross-stratum term, (3.2.2) provides an unbiased estimator of $\mathrm{Var}\{\bar{y}_{st}\}$.

Theorem 3.2.1. *The statistic* $(\bar{y}_{st,\alpha} - \bar{y}_{st})^2$ *is an unbiased estimator of* $\mathrm{Var}\{\bar{y}_{st}\}$.

PROOF. Because

$$\sum_{\alpha=1}^{2^L} \delta_h^{(\alpha)} \delta_{h'}^{(\alpha)} = 0,$$

it follows that

$$E\{(\bar{y}_{st,\alpha} - \bar{y}_{st})^2 | d_1, \ldots, d_L\} = \sum_{\alpha=1}^{2^L} (\bar{y}_{st,\alpha} - \bar{y}_{st})^2/2^L$$

$$= v(\bar{y}_{st}). \quad (3.2.3)$$

The expectation $E\{\cdot | d_1, \ldots, d_L\}$ holds fixed the selected units, and is with respect to the formation of the α-th half-sample. Thus,

$$E\{(\bar{y}_{st,\alpha} - \bar{y}_{st})^2\} = E\{v(\bar{y}_{st})\}$$

$$= \mathrm{Var}\{\bar{y}_{st}\}. \qquad \square$$

A more direct proof of this result that avoids the conditional expectation (3.2.3) relies on the fact that sampling is independent in the various strata. Thus, $E\{d_h d_{h'}\} = E\{d_h\}E\{d_{h'}\} = 0$ and the expectation of the cross-stratum term in (3.2.2) is zero. Although the direct proof is appealing, the conditioning argument in (3.2.3) shows not only that $(\bar{y}_{st,\alpha} - \bar{y}_{st})^2$ is an unbiased estimator of $\mathrm{Var}\{\bar{y}_{st}\}$, but also that the textbook estimator may be reproduced by taking the mean of the $(\bar{y}_{st,\alpha} - \bar{y}_{st})^2$ over the 2^L half-samples. Thus there is no loss of information if all 2^L replicates are used to estimate $\mathrm{Var}\{\bar{y}_{st}\}$.

When L is large, the computation of $v(\bar{y}_{st})$ as the mean of the $(\bar{y}_{st,\alpha} - \bar{y}_{st})^2$ over the 2^L half-samples is clearly not feasible. A natural shortcut is to compute the mean only over a small subset of the replicates. In so doing we may or may not reproduce the textbook estimator $v(\bar{y}_{st})$, but we certainly simplify the computational difficulties. As we shall see, however, by choosing the subset of half-samples judiciously we may, in fact, reproduce $v(\bar{y}_{st})$.

Unfortunately, if the subset is chosen at random, then the variance of the resulting variance estimator may be much larger than the variance of $v(\bar{y}_{st})$. Specifically, suppose a simple random sample without replacement of k half-samples is selected, and consider the corresponding variance estimator

$$v_k(\bar{y}_{st}) = \sum_{\alpha=1}^{k} (\bar{y}_{st,\alpha} - \bar{y}_{st})^2/k. \quad (3.2.4)$$

From Theorem 3.2.1, this is seen to be an unbiased estimator of $\text{Var}\{\bar{y}_{st}\}$. Furthermore, since the conditional expectation of $v_k(\bar{y}_{st})$ over the 2^L half-samples for a given sample is $v(\bar{y}_{st})$, we have

$$\text{Var}\{v_k(\bar{y}_{st})\} = \text{Var}_1 E_2\{v_k(\bar{y}_{st})\} + E_1 \text{Var}_2\{v_k(\bar{y}_{st})\}$$

$$= \text{Var}_1\{v(\bar{y}_{st})\} + E_1 \text{Var}_2\{v_k(\bar{y}_{st})\}$$

$$\geq \text{Var}_1\{v(\bar{y}_{st})\},$$

where the operators E_2 and Var_2 are with respect to the selection of the sample of half-samples given $d_1, d_2, \ldots, d_L$, and E_1 and Var_1 are with respect to the sampling design generating the parent sample. Consequently, $v(\bar{y}_{st})$ is at least as stable as $v_k(\bar{y}_{st})$, with the excess of $\text{Var}\{v_k(\bar{y}_{st})\}$ over $\text{Var}\{v(\bar{y}_{st})\}$ arising from the cross-stratum contribution to $v_k(\bar{y}_{st})$. Specifically, from (3.2.2) we see that

$$\text{Var}\{v_k(\bar{y}_{st})\} = \text{Var}\{v(\bar{y}_{st})\}$$

$$+ \sum_{h<h'}^{L}\sum \frac{2^L - k}{k(2^L - 1)} W_h^2 W_{h'}^2 \text{Var}\{d_h\}\text{Var}\{d_{h'}\}/4.$$

How then must we choose the subset of half-samples so that $v_k(\bar{y}_{st}) = v(\bar{y}_{st})$, thus guaranteeing that $\text{Var}\{v_k(\bar{y}_{st})\} = \text{Var}\{v(\bar{y}_{st})\}$? By (3.2.2) this equality will obtain whenever the k half-samples satisfy the property

$$\sum_{\alpha=1}^{k} \delta_h^{(\alpha)}\delta_{h'}^{(\alpha)} = 0 \tag{3.2.5}$$

for all $h < h' = 1, \ldots, L$. Plackett and Burman (1946) have given methods for constructing $k \times k$ orthogonal matrices, k a multiple of 4, whose columns satisfy (3.2.5). For example, an 8×8 orthogonal matrix is presented in Table 3.2.1. In the present context, strata are represented by the columns

Table 3.2.1. Definition of Balanced Half-Sample Replicates for 5, 6, 7, or 8 Strata

Replicate	Stratum (h)							
	1	2	3	4	5	6	7	8
$\delta_h^{(1)}$	+1	−1	−1	+1	−1	+1	+1	−1
$\delta_h^{(2)}$	+1	+1	−1	−1	+1	−1	+1	−1
$\delta_h^{(3)}$	+1	+1	+1	−1	−1	+1	−1	−1
$\delta_h^{(4)}$	−1	+1	+1	+1	−1	−1	+1	−1
$\delta_h^{(5)}$	+1	−1	+1	+1	+1	−1	−1	−1
$\delta_h^{(6)}$	−1	+1	−1	+1	+1	+1	−1	−1
$\delta_h^{(7)}$	−1	−1	+1	−1	+1	+1	+1	−1
$\delta_h^{(8)}$	−1	−1	−1	−1	−1	−1	−1	−1

of the table and half-samples by the rows. An entry of $+1$ in the (α, h)-th cell signifies that unit $(h, 1)$ is part of the α-th replicate, while an entry of -1 signifies that unit $(h, 2)$ is part of the α-th replicate. Any set of 5 columns for the $L = 5$ case; 6 columns for the $L = 6$ case; 7 columns for the $L = 7$ case; or all 8 columns for the $L = 8$ case defines a set of $k = 8$ replicates satisfying (3.2.5). Thus, defining half-samples in this manner leads to the equality relation

$$v_k(\bar{y}_{st}) = v(\bar{y}_{st}).$$

These k half-samples contain all of the information with respect to $\mathrm{Var}\{\bar{y}_{st}\}$ contained in all 2^L half-samples. The cross-stratum component of $v_k(\bar{y}_{st})$ has been eliminated! McCarthy (1966) has referred to a set of half-samples satisfying (3.2.5) as being *balanced.*

Half-sample balancing also leads to another desirable property. From (3.2.1) it can be seen that the average of the $\bar{y}_{st,\alpha}(\alpha = 1, \ldots, k)$ will equal $\bar{y}_{st}$ whenever

$$\sum_{\alpha=1}^{k} \delta_h^{(\alpha)} = 0 \tag{3.2.6}$$

for each $h = 1, \ldots, L$. This condition is satisfied by Plackett's and Burman's matrices except when $k = L$ (e.g., see column 8 of Table 3.2.1). When $k = L$, one of the two units from the L-th stratum is used in each of the half-samples, thus defeating condition (3.2.6). In the example, the second unit from the 8-th stratum is used in each half-sample. It is intuitively clear that the mean of the $\bar{y}_{st,\alpha}$ can not equal $\bar{y}_{st}$ in this case, because the latter includes both units from the L-th stratum in the computations, whereas the former does not.

When both (3.2.5) and (3.2.6) are satisfied, we shall refer to the set of replicates as being in *full orthogonal balance.* This will be the case whenever k is an integral multiple of 4 which is greater than L. Choosing k to be the smallest such value minimizes the number of computations. For example, in the $L = 8$ case, $k = 12$ half-sample replicates would be required to achieve full orthogonal balance, and any 8 columns of Plackett's and Burman's 12×12 orthogonal matrix may be used to derive the replicates. If $k = 8$ half-samples are used instead (that is, all eight columns of the 8×8 matrix in Table 3.2.1), then balance is achieved but not full orthogonal balance. If $k = 16$ or more half-samples are used, full orthogonal balance is also achieved, but more computations are required than for $k = 12$ half-samples.

If fewer than $k = L$ half-samples are used, then neither balance nor full orthogonal balance can be achieved.

The orthogonal matrices discussed by Plackett and Burman are known in mathematics as Hadamard matrices. Strictly speaking, Hadamard matrices are not known to exist for every multiple of 4, although constructions have been given for all orders through 200×200, thus covering most situations of practical importance in survey sampling. To assist the reader in implementing the methods developed here, we have prepared Hadamard matrices for all orders through 100×100; see Appendix A. Hadamard matrices are not unique, and thus balance or full-orthogonal balance may be achieved with alternative sets of half-samples. An easy way to see this is to note that if $\mathbf{H}$ is a Hadamard matrix, then $-\mathbf{H}$ is also.

3.3. Usage with Multistage Designs

In Section 3.2 we introduced the basic balanced half-sample methodology using simple random sampling with replacement within strata. The methodology, however, has more general application and we now consider the case of multistage sampling with possibly unequal probabilities of selection.

We assume that primary sampling units (PSUs) are selected pps with replacement within each of L strata. We shall consider the problem of estimating a population total Y via the unbiased estimator

$$
\begin{aligned}
\hat{Y} &= \sum_{h=1}^{L} \hat{Y}_h \\
&= \sum_{h=1}^{L} (\hat{Y}_{h1}/2p_{h1} + \hat{Y}_{h2}/2p_{h2}),
\end{aligned}
\tag{3.3.1}
$$

where $\hat{Y}_{hi}$ is an unbiased estimator of the (h, i)-th primary total, say Y_{hi}, based upon sampling at the second and successive stages, and p_{hi} is the per-draw selection probability for the (h, i)-th primary. As usual we must have both 1) $p_{hi} > 0$ for all h and i and 2) $\sum_i p_{hi} = 1$ for all h. The textbook estimator of variance for this problem is

$$
v(\hat{Y}) = \sum_{h=1}^{L} (\hat{Y}_{h1}/p_{h1} - \hat{Y}_{h2}/p_{h2})^2/4.
\tag{3.3.2}
$$

As in the case of srs wr within strata, there are 2^L possible half-samples. We shall select k of them using the balancing ideas presented in Section 3.2. Note that the issue of balancing the half-samples is separate and distinct from the issue of the sampling design used in selecting the parent sample. Thus, the specification of a set of balanced half-samples is performed the same for pps wr sampling as for srs wr as for any other two-per-stratum sampling design. For pps wr sampling, the α-th half-sample estimator of Y is

$$
\hat{Y}_\alpha = \sum_{h=1}^{L} (\delta_{h1\alpha} \hat{Y}_{h1}/p_{h1} + \delta_{h2\alpha} \hat{Y}_{h2}/p_{h2}),
\tag{3.3.3}
$$

where

$$\delta_{h1\alpha} \begin{cases} = 1, & \text{if the } (h, 1)\text{-st PSU is in the } \alpha\text{-th half-sample} \\ = 0, & \text{otherwise} \end{cases}$$

and

$$\delta_{h2\alpha} = 1 - \delta_{h1\alpha}.$$

If a set of k balanced half-samples is specified, then

$$v_k(\hat{Y}) = \sum_{\alpha=1}^{k} (\hat{Y}_\alpha - \hat{Y})^2/k \tag{3.3.4}$$

provides the full-information, unbiased estimator of $\text{Var}\{\hat{Y}\}$. That is, following the approach of Section 3.2, it may be shown that

$$v_k(\hat{Y}) = v(\hat{Y}). \tag{3.3.5}$$

Indeed we may anticipate the result (3.3.5) because the sampling design considered in Section 3.2 is a special case of that considered here. Furthermore, relying on the development in Section 3.2 we may conclude that

- When $k < 2^L$ *randomly* selected half-samples are used, $v_k(\hat{Y})$ is an unbiased but possibly inefficient estimator of $\text{Var}\{\hat{Y}\}$. It may or may not equal the textbook estimator $v(\hat{Y})$.
- When $k = 2^L$, then the equality $v_k(\hat{Y}) = v(\hat{Y})$ is guaranteed. Computational costs will be prohibitive, however, in all circumstances where L is moderate or large.
- If k balanced half-samples are used, then $\sum_{\alpha=1}^{k} \hat{Y}_\alpha/k = \hat{Y}$, except when $k = L$.

We offer two final thoughts without providing a formal development of them. First, the balanced half-sample methodology may be used for estimating the components of the variance of $\hat{Y}$. The estimator presented in (3.3.4) estimates the total variance of $\hat{Y}$, which may be partitioned as

$$\text{Var}\{\hat{Y}\} = \text{Var}_1 \text{E}_2\{Y\} + \text{E}_1 \text{Var}_2\{\hat{Y}\},$$

Total Variance	Between PSU Variance	Within PSU Variance

where E_2 and Var_2 condition on the selected PSUs. Application of the balanced half-sample methodology to the second stage sampling units, allows one to estimate both $\text{Var}_2\{\hat{Y}\}$ and the within PSU component $\text{E}_1 \text{Var}_2\{\hat{Y}\}$. By subtraction, an estimator of the between PSU component may be derived. For multiple-stage sampling, the within PSU component may be further partitioned, with the elements of the partition estimated directly by the balanced half-sample method or indirectly by subtraction. Second, the balanced half-sample methodology may be applied to without replacement sampling designs, even though the designs presented in the last two sections featured with replacement sampling. Some overestimation of the variance tends to occur in this case. We consider this point further in Section 3.5.

3.4. Usage with Nonlinear Estimators

The balanced half-sample technique was introduced in Sections 3.2 and 3.3 in the context of simple linear estimators, a context in which the textbook variance estimating formulas may be computationally satisfactory. These methods, however, suggest techniques for estimating the variance of non-linear estimators, where simple and unbiased estimators of variance are not available.

We shall continue to use the stratified pps wr sampling design set forth in Section 3.3. Now suppose that an estimator $\hat{\theta}$, not necessarily linear, is constructed from the entire sample for some general population parameter θ. For example, θ may be a ratio, a difference of ratios, a regression coefficient, a correlation coefficient, etc. Suppose further that k balanced half-sample replicates are specified as described in Section 3.2.

Let $\hat{\theta}_\alpha$ ($\alpha = 1, \ldots, k$) denote the estimator of θ computed from the α-th half-sample. These estimators should be of the same functional form as the parent sample estimator $\hat{\theta}$. Thus, if $\hat{\theta}$ is the "combined" ratio estimator

$$\hat{\theta} = \frac{\hat{Y}}{\hat{X}} X,$$

where $\hat{Y}$ and $\hat{X}$ are of the form (3.3.1), then $\hat{\theta}_\alpha$ is the "combined" ratio estimator

$$\hat{\theta}_\alpha = \frac{\hat{Y}_\alpha}{\hat{X}_\alpha} X.$$

By analogy with the linear problem, an estimator of Var$\{\hat{\theta}\}$ based on the k balanced replicates is

$$v_k(\hat{\theta}) = \sum_{\alpha=1}^{k} (\hat{\theta}_\alpha - \hat{\theta})^2 / k. \tag{3.4.1}$$

This estimator is intuitively satisfying because it mimics the estimator developed for the linear problem. In general, however, its exact theoretical properties are unknown. The moments of $v_k(\hat{\theta})$ may be approximated by first "linearizing" the estimators (cf. Chapter 6) and then applying the results of Sections 3.2 and 3.3 to the linear approximation. Appendix B discusses the asymptotic properties of $v_k(\hat{\theta})$.

The nonlinear problem discussed here differs in two important respects from the linear problem discussed in earlier sections. First, we see almost immediately that the mean of the half-sample estimators

$$\hat{\bar{\theta}} = \sum_{\alpha=1}^{k} \hat{\theta}_\alpha / k$$

is not necessarily equal to the parent sample estimator $\hat{\theta}$. For the linear problem, we had the equality $\hat{\bar{\theta}} = \hat{\theta}$ provided that the half-samples were balanced and $k > L$. Even for the linear problem the equality breaks down when the half-samples are unbalanced or $k = L$. For the nonlinear problem, $\hat{\bar{\theta}}$ and $\hat{\theta}$ are never equal except by rare chance. They should be quite close, however, in most survey applications; certainly within sampling error of one another. Moderate or large differences between them should serve as a warning that either computational errors have occurred or bias exists in the estimators due to their nonlinear form.

The second unique aspect of a nonlinear problem concerns the fact that alternative variance estimators are available. Corresponding to the estimator $\hat{\theta}_\alpha$, one can define an estimator $\hat{\theta}_\alpha^c$ based on the half-sample which is complementary to α. This defines a different estimator of $\text{Var}\{\hat{\theta}\}$:

$$v_k^c(\hat{\theta}) = \sum_{\alpha=1}^{k} (\hat{\theta}_\alpha^c - \hat{\theta})^2 / k. \tag{3.4.2}$$

To estimate the variance of $\hat{\theta}$, we might also use

$$\bar{v}_k(\hat{\theta}) = [v_k(\hat{\theta}) + v_k^c(\hat{\theta})]/2 \tag{3.4.3}$$

or

$$v_k^\dagger(\hat{\theta}) = \sum_{\alpha=1}^{k} (\hat{\theta}_\alpha - \hat{\theta}_\alpha^c)^2 / 4k. \tag{3.4.4}$$

If the estimator $\hat{\theta}$ is linear, then the variance estimators $v_k(\hat{\theta})$, $v_k^c(\hat{\theta})$, $\bar{v}_k(\hat{\theta})$, and $v_k^\dagger(\hat{\theta})$ are identical; are equal to the textbook estimator $v(\hat{\theta})$; and are unbiased for $\text{Var}\{\hat{\theta}\}$. This follows because $\hat{\theta}_\alpha$ and $\hat{\theta}_\alpha^c$ are independent and

$$\hat{\bar{\theta}}_\alpha \overset{(\text{defn})}{=} (\hat{\theta}_\alpha + \hat{\theta}_\alpha^c)/2 = \hat{\theta}$$

for each α, whenever the estimator $\hat{\theta}$ is linear.

Additional estimators of variance can be constructed by employing squared deviations from $\hat{\bar{\theta}}$ instead of from $\hat{\theta}$. Such estimators would also be identical to $v_k(\hat{\theta})$ whenever $\hat{\theta}$ is linear. When $\hat{\theta}$ is nonlinear the estimators of variance are, in general, unequal.

In the case of nonlinear estimators, $v_k(\hat{\theta})$, $v_k^c(\hat{\theta})$, and $\bar{v}_k(\hat{\theta})$ are sometimes regarded as estimators of the mean squared error $\text{MSE}\{\hat{\theta}\}$, while $v_k^\dagger(\hat{\theta})$ is regarded as an estimator of variance $\text{Var}\{\hat{\theta}\}$. This follows because $v_k^\dagger(\hat{\theta})$ is an unbiased estimator of $\text{Var}\{\hat{\theta}_\alpha\}$ for any given α, and the variance of $\hat{\bar{\theta}}_\alpha$

is thought to be close to that of $\hat{\theta}$. We also note that

$$\bar{v}_k(\hat{\theta}) = v_k^\dagger(\hat{\theta}) + \sum_{\alpha=1}^{k} (\hat{\bar{\theta}}_\alpha - \hat{\theta})^2/k$$

so that $\bar{v}_k(\hat{\theta})$ is guaranteed to be larger than $v_k^\dagger(\hat{\theta})$. Similarly, $v_k(\hat{\theta})$ and $v_k^c(\hat{\theta})$ also tend to be larger than $v_k^\dagger(\hat{\theta})$. By symmetry, we have

$$
\begin{aligned}
E\{\bar{v}_k(\hat{\theta})\} = E\{v_k(\hat{\theta})\} &= E\{v_k^c(\hat{\theta})\} \\
&= \text{Var}\{\hat{\bar{\theta}}_\alpha\} + E\{(\hat{\bar{\theta}}_\alpha - \hat{\theta})^2\} \\
&\geq \text{Var}\{\hat{\bar{\theta}}_\alpha\} + [E\{\hat{\bar{\theta}}_\alpha - \hat{\theta}\}]^2 \\
&\doteq \text{Var}\{\hat{\bar{\theta}}_\alpha\} + [\text{Bias}\{\hat{\theta}\}]^2 \\
&\doteq \text{Var}\{\hat{\theta}\} + [\text{Bias}\{\hat{\theta}\}]^2 \\
&= \text{MSE}\{\hat{\theta}\},
\end{aligned}
$$

the first approximate equality holding whenever the bias of $\hat{\theta}$ is proportional to the number of selected PSUs (i.e., $2L$) and the second approximate equality holding whenever $\hat{\bar{\theta}}_\alpha$ and $\hat{\theta}$ have the same variance.

As an illustration of the above methods, suppose it is desired to estimate the ratio

$$R = Y/X,$$

where Y and X denote two population totals. The estimator based on the entire sample is

$$\hat{R} = \hat{Y}/\hat{X},$$

where $\hat{Y}$ and $\hat{X}$ are defined according to (3.3.1). And the α-th half-sample estimator is

$$\hat{R}_\alpha = \hat{Y}_\alpha/\hat{X}_\alpha,$$

where $\hat{Y}_\alpha$ and $\hat{X}_\alpha$ are defined according to (3.3.3). The estimator corresponding to (3.4.1) is given by

$$v_k(\hat{R}) = \sum_{\alpha=1}^{k} (\hat{R}_\alpha - \hat{R})^2/k, \qquad (3.4.5)$$

and the estimators $v_k^c(\hat{R})$, $\bar{v}_k(\hat{R})$, and $v_k^\dagger(\hat{R})$ are defined similarly. Clearly,

$$\hat{\bar{R}} = \frac{1}{k} \sum_{\alpha=1}^{k} \hat{R}_\alpha$$

$$\neq \hat{R}.$$

The variance estimator suggested for this problem in most textbooks is

$$v(\hat{R}) = \hat{X}^{-2}\{v(\hat{Y}) - 2\hat{R}c(\hat{Y}, \hat{X}) + \hat{R}^2 v(\hat{X})\},$$

where $v(\hat{Y})$, $c(\hat{Y}, \hat{X})$, and $v(\hat{X})$ are the textbook estimators of $\text{Var}\{\hat{Y}\}$, $\text{Cov}\{\hat{Y}, \hat{X}\}$, and $\text{Var}\{\hat{X}\}$, respectively. Using the approximation

$$\hat{R}_\alpha - \hat{R} \doteq (\hat{Y}_\alpha - \hat{R}\hat{X}_\alpha)/\hat{X},$$

we see that

$$v_k(\hat{R}) \doteq \hat{X}^{-2} \sum_{\alpha=1}^{k} \{(\hat{Y}_\alpha - \hat{Y}) - \hat{R}(\hat{X}_\alpha - \hat{X})\}^2/k$$

which equals the textbook estimator $v(\hat{R})$ whenever the half-samples are balanced! Using the same approximation we also see that $\hat{\bar{R}} \doteq \hat{R}$ whenever the half-samples are balanced and $k > L$.

Approximate equalities of this kind can be established between balanced half-sample estimators and textbook estimators for a wide class of nonlinear statistics $\hat{\theta}$.

These are approximate results, however, and there is a dearth of exact theoretical results for finite sample sizes. One exception is Krewski and Rao's (1981) finite sample work on the variance of the ratio estimator. Although there are few theoretical results, there is a growing body of empirical evidence that suggests balanced half-sample estimators provide satisfactory estimates of the true variance (or MSE). This is confirmed in Frankel's (1971b) investigation of means, differences of means, regression coefficients, and correlation coefficients; McCarthy's (1969a) investigation of ratios, regression coefficients, and correlation coefficients; Levy's (1971) work on the combined ratio estimator; Kish and Frankel's (1970) study of regression coefficients; Bean's (1975) investigation of poststratified means; and Mulry and Wolter's (1981) work on the correlation coefficient.

It is clear that if computational ease is a primary consideration, then the estimators $v_k(\hat{\theta})$ and $v_k^c(\hat{\theta})$ must be preferred to $\bar{v}_k(\hat{\theta})$ and $v_k^\dagger(\hat{\theta})$.

3.5. Without Replacement Sampling

Consider the simple linear estimator $\bar{y}_{st}$ discussed in Section 3.2, only now let us suppose the two units in each stratum are selected without replacement. The textbook estimator of $\text{Var}\{\bar{y}_{st}\}$ is now

$$v(\bar{y}_{st}) = \sum_{h=1}^{L} W_h^2(1 - 2/N_h)s_h^2/2$$

$$= \sum_{h=1}^{L} W_h^2(1 - 2/N_h)d_h^2/4,$$

where $(1 - 2/N_h)$ is the finite population correction (fpc). The balanced-half sample estimator $v_k(\bar{y}_{st})$ shown in (3.2.4) is identical to

$$\sum_{h=1}^{L} W_h^2 d_h^2/4,$$

and thus incurs the upward bias

$$\text{Bias}\{v_k(\bar{y}_{st})\} = \sum_{h=1}^{L} W_h^2 \frac{1}{N_h} S_h^2$$

for the without replacement problem.

It is possible to modify the half-sample replication technique to accommodate unequal fpc's and provide an unbiased estimator of $\text{Var}\{\bar{y}_{st}\}$. McCarthy (1966) observed that the fpc's could be taken into account by working with W_h^* instead of W_h, where

$$W_h^* = W_h \sqrt{1 - 2/N_h}.$$

The α-th half-sample estimator is then defined by

$$\bar{y}_{st,\alpha}^* = \bar{y}_{st} + \sum_{h=1}^{L} W_h^*(\delta_{h1\alpha} y_{h1} + \delta_{h2\alpha} y_{h2} - \bar{y}_h), \qquad (3.5.1)$$

and the estimator of variance is

$$v_k^*(\bar{y}_{st}) = \frac{1}{k} \sum_{\alpha=1}^{k} (\bar{y}_{st,\alpha}^* - \bar{y}_{st})^2. \qquad (3.5.2)$$

If the replicates are in full orthogonal balance, then the desirable properties

(i) $v_k^*(\bar{y}_{st}) = v(\bar{y}_{st})$

(ii) $k^{-1} \sum_{\alpha=1}^{k} \bar{y}_{st,\alpha}^* = \bar{y}_{st}$

are guaranteed, and $v_k^*(\bar{y}_{st})$ is an unbiased estimator of $\text{Var}\{\bar{y}_{st}\}$.

For nonlinear estimators, the modification for without replacement sampling is straightforward. In the case of the combined ratio estimator

$$\bar{y}_{RC} = \bar{X} \bar{y}_{st} / \bar{x}_{st}$$

Lee (1972) suggests the half-sample estimators be defined by

$$\bar{y}_{RC,\alpha}^* = \bar{X} \bar{y}_{st,\alpha}^* / \bar{x}_{st,\alpha}^*,$$

where $\bar{y}_{st,\alpha}^*$ and $\bar{x}_{st,\alpha}^*$ are defined according to (3.5.1) for the y- and x-variables, respectively. The variance of $\bar{y}_{RC}$ is then estimated by

$$v_k^*(\bar{y}_{RC}) = \sum_{\alpha}^{k} (\bar{y}_{RC,\alpha}^* - \bar{y}_{RC})^2 / k.$$

More generally, let $\hat{\theta} = g(\bar{\mathbf{y}}_{st})$ denote a class of nonlinear estimators, where $g(\cdot)$ is some real-valued function with continuous second derivatives and $\bar{\mathbf{y}}_{st} = (\bar{y}_{st}(1), \ldots, \bar{y}_{st}(p))$ is a p-vector of stratified sampling means based upon p different variables. Most of the nonlinear estimators used in applied survey work are of this form. The α-th half-sample estimator is defined by

$$\theta_\alpha^* = g(\bar{\mathbf{y}}_{\text{st},\alpha}^*)$$

$$\bar{\mathbf{y}}_{\text{st},\alpha}^* = (\bar{y}_{\text{st},\alpha}^*(1), \ldots, \bar{y}_{\text{st},\alpha}^*(p))$$

with corresponding estimator of variance

$$v_k^*(\hat{\theta}) = \frac{1}{k} \sum_{\alpha=1}^{k} (\theta_\alpha^* - \hat{\theta})^2.$$

This choice of estimator can be justified using the theory of Taylor series approximations (cf. Chapter 6).

Thus far we have been assuming srs wor within strata. Now let us suppose a single stage sample is selected with unequal probabilities, without replacement, and with inclusion probabilities $\pi_{hj} = 2p_{hj}$ for all strata h and units j, where $p_{hj} > 0$ and $\sum_j p_{hj} = 1$. This is a πps sampling design. We consider the Horvitz–Thompson estimator

$$\hat{Y} = \sum_{h=1}^{L} \hat{Y}_h$$

$$= \sum_{h=1}^{L} \left(\frac{y_{h1}}{\pi_{h1}} + \frac{y_{h2}}{\pi_{h2}} \right)$$

of the population total Y, and the balanced half-sample estimator of variance as originally defined in (3.2.4)

$$v_k(\hat{Y}) = \frac{1}{k} \sum_{\alpha=1}^{k} (\hat{Y}_\alpha - \hat{Y})^2$$

$$\hat{Y}_\alpha = \sum_{h=1}^{L} \left(\delta_{h\alpha 1} \frac{2y_{h1}}{\pi_{h1}} + \delta_{h\alpha 2} \frac{2y_{h2}}{\pi_{h2}} \right).$$

For this problem $v_k(\hat{Y})$ is not an unbiased estimator of $\text{Var}\{\hat{Y}\}$. Typically, $v_k(\hat{Y})$ tends to be upward biased. The reason is that $v_k(\hat{Y})$ estimates the variance as if the sample were selected with replacement, even though without replacement sampling is actually used. This issue was treated at length in Section 2.4.5, and here we restate briefly those results as they relate to the balanced half-sample estimator. It can be shown that

$$v_k(\hat{Y}) = \sum_{\alpha=1}^{k} (\hat{Y}_\alpha - \hat{Y})^2 / k$$

$$= \sum_{h=1}^{L} (y_{h1}/p_{h1} - y_{h2}/p_{h2})^2 / 4$$

$$= v(\hat{Y}_{\text{wr}}),$$

which is the textbook estimator of variance for pps wr sampling. By Theorem 2.4.6 it follows that

$$E\{v_k(\hat{Y})\} = \text{Var}\{\hat{Y}\} + 2(\text{Var}\{\hat{Y}_{wr}\} - \text{Var}\{\hat{Y}\})$$

where $\text{Var}\{\hat{Y}_{wr}\}$ is the variance of $\hat{Y}_{wr} = \sum_{h=1}^{L} 2^{-1}(y_{h1}/p_{h1} + y_{h2}/p_{h2})$ in pps wr sampling. Thus, the bias in $v_k(\hat{Y})$ is twice the reduction (or increase) in variance due to the use of without replacement sampling. In the useful applications of πps sampling (i.e., applications where πps is more efficient than pps wr), the balanced half-sample estimator $v_k(\hat{Y})$ tends to be upward biased.

If the use of without replacement sampling results in an important reduction in variance and if it is desired to reflect this fact in the variance calculations, then define the modified half-sample estimators

$$\hat{Y}_\alpha^* = \hat{Y} + \sum_{h=1}^{L} \left(\frac{\pi_{h1}\pi_{h2} - \pi_{h12}}{\pi_{h12}} \right)^{1/2} \left(\delta_{h\alpha 1} \frac{2y_{h1}}{\pi_{h1}} + \delta_{h\alpha 2} \frac{2y_{h2}}{\pi_{h2}} - \hat{Y}_h \right),$$

where π_{h12} is the joint inclusion probability in the h-th stratum. The half-sample estimator of variance takes the usual form

$$v_k^*(\hat{Y}) = \frac{1}{k} \sum_{\alpha=1}^{k} (\hat{Y}_\alpha^* - \hat{Y})^2.$$

And if the half-samples are in full orthogonal balance, then

(i) $$v_k^*(\hat{Y}) = \sum_{h=1}^{L} \frac{\pi_{h1}\pi_{h2} - \pi_{h12}}{\pi_{h12}} \left(\frac{y_{h1}}{\pi_{h1}} - \frac{y_{h2}}{\pi_{h2}} \right)^2$$

(the Yates–Grundy estimator)

and

(ii) $$k^{-1} \sum_{\alpha=1}^{k} \hat{Y}_\alpha^* = \hat{Y}.$$

Thus, the modified half-sample methods reproduce the Yates–Grundy estimator, which is the textbook unbiased estimator of variance.

For multistage sampling the estimators of total are

$$\hat{Y} = \sum_{h=1}^{L} \hat{Y}_h$$

$$= \sum_{h=1}^{L} \left(\frac{\hat{Y}_{h1}}{\pi_{h1}} + \frac{\hat{Y}_{h2}}{\pi_{h2}} \right)$$

$$\hat{Y}_\alpha = \sum_{h=1}^{L} \left(\delta_{h\alpha 1} \frac{2\hat{Y}_{h1}}{\pi_{h1}} + \delta_{h\alpha 2} \frac{2\hat{Y}_{h2}}{\pi_{h2}} \right),$$

where $\hat{Y}_{hi}$ is an estimator of the total of the (h, i)-th selected PSU based on sampling at the second and subsequent stages. Once again, it can be shown that

$$v_k(\hat{Y}) = \sum_{\alpha=1}^{k} (\hat{Y}_\alpha - \hat{Y})^2 / k$$

$$= \sum_{h=1}^{L} (\hat{Y}_{h1}/p_{h1} - \hat{Y}_{h2}/p_{h2})^2 / 4$$

$$= v(\hat{Y}_{\text{wr}}),$$

which is the textbook estimator of variance when pps wr sampling is employed in the selection of PSUs. By (2.4.16) we see that

$$\text{Bias}\{v_k(\hat{Y})\} = 2(\text{Var}\{\hat{Y}_{\text{wr}}\} - \text{Var}\{\hat{Y}\}),$$

where $\text{Var}\{\hat{Y}_{\text{wr}}\}$ is the variance assuming with replacement sampling of PSUs, and that this bias occurs only in the between PSU component of variance. In applications where the between variance is a minor part of the total variance, this bias may be unimportant.

3.6. Partial Balancing

When the number of strata is large, the cost of processing $k \geq L$ fully balanced replicates may be unacceptably high. For this case, we may use a set of k partially balanced half-sample replicates.

A *G-order partially balanced design* is constructed by dividing the L strata into G groups with L/G strata in each group. We temporarily assume that L/G is an integer. A fully balanced set of k half-samples is then specified for the first group, and this design is repeated in each of the remaining G-1 groups.

To illustrate, we construct a $G = 2$-order partially balanced design for $L = 6$ strata. Table 3.6.1 gives a fully balanced design for $L/G = 3$ strata.

Table 3.6.1. A Fully Balanced
Design for $L = 3$ Strata

Half-Sample	Stratum (h)		
	1	2	3
$\delta_h^{(1)}$	+1	+1	+1
$\delta_h^{(2)}$	−1	+1	−1
$\delta_h^{(3)}$	−1	−1	+1
$\delta_h^{(4)}$	+1	−1	−1

This uses columns 2, 3, 4 from Table A.2. The partially balanced design for $L = 6$ strata is given by repeating the set of replicates in the second group of $L/G = 3$ strata. A demonstration of this is in Table 3.6.2. Recall that a fully balanced design for $L = 6$ strata requires $k = 8$ replicates. Computational costs are reduced by using only $k = 4$ replicates.

The method of construction of partially balanced half-samples leads to the observation that any two strata are orthogonal (or balanced) provided they belong to the same group, or belong to different groups but are not corresponding columns in the two groups. That is

$$\sum_{\alpha=1}^{k} \delta_h^{(\alpha)} \delta_{h'}^{(\alpha)} = 0 \tag{3.6.1}$$

whenever h and h' are not corresponding strata in different groups. For example, in Table 3.6.2 strata 1 and 4 are corresponding strata in different groups.

To investigate the efficiency of partially balanced designs, consider the simple linear estimator $\bar{y}_{st}$ discussed in Section 3.2. The appropriate variance estimator is

$$v_{k,pb}(\bar{y}_{st}) = \sum_{\alpha=1}^{k} (\bar{y}_{st,\alpha} - \bar{y}_{st})^2 / k,$$

where the subscript pb denotes partially balanced. By (3.2.2) and (3.6.1), the variance estimator may be written as

$$v_{k,pb}(\bar{y}_{st}) = v(\bar{y}_{st}) + \sum_{h,h'}^{\dagger} W_h W_{h'} d_h d_{h'}/2,$$

where the summation $\sum_{h,h'}^{\dagger}$ is over all pairs (h, h') of strata such that $h < h'$ and h and h' are corresponding strata in different groups. From this expression it is clear that the variance estimator $v_{k,pb}(\bar{y}_{st})$ is not identical to the textbook estimator $v(\bar{y}_{st})$ because of the presence of cross-stratum terms. Evidently, the number of such terms, $L(G - 1)/2$, increases with G.

Table 3.6.2. A 2-Order Partially Balanced Design for $L = 6$ Strata

	Stratum (h)					
Half-Sample	1	2	3	4	5	6
$\delta_h^{(1)}$	+1	+1	+1	+1	+1	+1
$\delta_h^{(2)}$	−1	+1	−1	−1	+1	−1
$\delta_h^{(3)}$	−1	−1	+1	−1	−1	+1
$\delta_h^{(4)}$	+1	−1	−1	+1	−1	−1

Thus, while partial balancing offers computational advantages over complete balancing, its variance estimator cannot reproduce the textbook variance estimator. The estimator $v_{k,\text{pb}}(\bar{y}_{\text{st}})$ is unbiased, however. Because sampling is performed independently in the various strata, the d_h's are independent random variables. Thus

$$E\{v_{k,\text{pb}}(\bar{y}_{\text{st}})\} = E\{v(\bar{y}_{\text{st}})\} = \text{Var}\{\bar{y}_{\text{st}}\},$$

and

$$\text{Var}\{v_{k,\text{pb}}(\bar{y}_{\text{st}})\} = \text{Var}\{v(\bar{y}_{\text{st}})\} + \sum_{h,h'}^{\dagger} W_h^2 W_{h'}^2 \sigma_h^2 \sigma_{h'}^2 \quad (3.6.2)$$

where

$$\sigma_h^2 = \sum_{j=1}^{N_h} (Y_{hj} - \bar{Y}_h)^2 / N_h$$

is the population variance within the h-th stratum ($h = 1, \ldots, L$). The variance estimator has increased because of the presence of the cross-stratum terms.

We have seen that $v_{k,\text{pb}}(\bar{y}_{\text{st}})$ is unbiased but less precise than $v_k(\bar{y}_{\text{st}})$, the estimator based on a fully balanced design. For a given G-order partially balanced design, the loss in precision depends on the magnitudes of the $W_h^2 \sigma_h^2$ and on the manner in which the pairs ($W_h^2 \sigma_h^2$, $W_{h'}^2 \sigma_{h'}^2$) are combined as cross-products in the summation $\sum_{h,h'}^{\dagger}$. To evaluate the loss in precision more closely, we assume the L strata are arranged randomly into G groups each of size L/G strata. Let T_L be the set of all permutations on $\{1, 2, \ldots, L\}$. Then, the variance $\text{Var}\{v_{k,\text{pb}}(\bar{y}_{\text{st}})\}$ is equal to the expectation of (3.6.2) with respect to the random formation of groups:

$$\text{Var}\{v_{k,\text{pb}}(\bar{y}_{\text{st}})\}$$

$$= \text{Var}\{v(\bar{y}_{\text{st}})\} + (1/L!) \sum_{T_L} \left(\sum_{h,h'}^{\dagger} W_h^2 W_{h'}^2 \sigma_h^2 \sigma_{h'}^2 \right)$$

$$= \text{Var}\{v(\bar{y}_{\text{st}})\} + \sum_{h<h'}^{L} W_h^2 W_{h'}^2 \sigma_h^2 \sigma_{h'}^2 L(G-1)(L-2)!/L!$$

$$= \text{Var}\{v(\bar{y}_{\text{st}})\} + [(G-1)/(L-1)] \sum_{h<h'}^{L} W_h^2 W_{h'}^2 \sigma_h^2 \sigma_{h'}^2.$$

For the special case $W_1^2 \sigma_1^2 = \ldots = W_L^2 \sigma_L^2$ and $\beta_1 = \ldots = \beta_L = 3$ (β_h is the measure of kurtosis in the h-th stratum and $\beta_h = 3$ is equivalent to a normality assumption), Lee (1972) shows that

$$\text{Var}\{v_{k,\text{pb}}(\bar{y}_{\text{st}})\} / \text{Var}\{v(\bar{y}_{\text{st}})\} = G.$$

Thus, the loss in precision relative to the fully balanced design may be substantial. The loss is minimized when $G = 2$, but this choice of G will result in larger computational costs than when $G > 2$.

In an effort to improve the precision of partially balanced designs, Lee (1972, 1973a) has investigated several nonrandom techniques for grouping strata. His investigations suggest the SAOA (semi-ascending order arrangement) procedure:

(1) Arrange the L strata in ascending order of magnitude of $a_h = W_h^2 \sigma_h^2$ (in practice, the a_h will need to be estimated from a prior survey).
(2) Rearrange the last $L/2$ (or $(L-1)/2$ if L is odd) strata in descending order of the a_h's.
(3) Divide the L strata arranged in this order into G groups, each of size L/G.

When G is even and the monotonic increasing sequence $\{a_h\}$ is either strictly convex (i.e., $0 \le a_{h-1} - 2a_h + a_{h+1}$) or strictly concave (i.e., $0 \ge a_{h-1} - 2a_h + a_{h+1}$), Lee shows that AAA (alternate ascending order arrangement) procedure fares better then SAOA procedure:

(1) Same as step 1 of the SAOA procedure.
(2) Split the L strata arranged in this order into G groups, each of size L/G.
(3) Reverse the order of the L/G strata in each of the second, fourth, sixth, ... groups.

Ernst (1979) has proposed the NESA (nearly equal sums arrangement) procedure for increasing the precision of partially balanced designs:

(1) Arrange the L strata in decreasing order of magnitude of $a_h = W_h^2 \sigma_h^2$ (in practice, the a_h will need to be estimated from a prior survey).
(2) Recursively define a one-to-one onto map

$$g: \{1, \ldots, L\} \to \{1, \ldots, r\} \times \{1, \ldots, G\}$$

where $r = L/G$.
(a) $g(1) = (1, 1)$
(b) Assume that $g(h)$ is defined for $h = 1, \ldots, l$ and denote $g_1(h) = i$ if $g(h) = (i, j)$. The function $g_1(h)$ gives the position of stratum h in the group j to which it has been assigned.
(c) Define $H(l, i) = \{h: g_1(h) = i, h = 1, \ldots, l\}$ and let $\eta(l, i)$ denote the number of elements in $H(l, i)$, $i = 1, \ldots, r$.
(d) Define $J(l) = \{i: \eta(l, i) < G\}$ and let s be the smallest member of $J(l)$ satisfying

$$\sum_{h \in H(l, s)} a_h = \inf \left\{ \sum_{h \in H(l, i)} a_h : i \in J(l) \right\}$$

(e) $g(l + 1) = (s, \eta(l, s) + 1)$.

(3) Stratum h is assigned the i-th position in the j-th group, where $g(h) = (i, j)$.

This method aims to equalize the position sums

$$t_i = \sum_{j=1}^{G} a_{g^{-1}(i,j)}$$

as much as possible. At the l-th step, the stratum associated with a_l is assigned a position and group number; $H(l, i)$ then denotes the set of strata for which the i-th position has been assigned, and $\eta(l, i)$ denotes the number of such strata. At the $(l+1)$-st step, the stratum associated with a_{l+1} is assigned the position, s, with the smallest sum of the a's at that point. The method is initialized by assigning the largest stratum, a_1, to the first position, first group. We are unaware of any empirical comparisons of NESA versus SAOA and AAA. Ernst established an upper bound on the variance given NESA that can be exceeded by the variance given SAOA and AAA.

Of course the way in which strata are formed in the first place must, necessarily, affect the values of the a_h, and thus effect the application of the SAOA, AAA, or NESA procedures. If stratum boundaries are optimized using the cumulative $\sqrt{f}$ rule of Dalenius (1957) and Dalenius and Hodges (1959), then the values $a_h = W_h^2 S_h^2$ are approximately equal. In this case the three procedures SAOA, AAA, NESA are identical, and the loss in precision relative to the fully balanced design is the same as that experienced with a random formation of groups.

The methods of partial balancing discussed in this section can also be used to estimate the variance of an arbitrary nonlinear estimator. To apply the SAOA, AAA, or NESA procedures, however, it is necessary to modify the definition of a_h. For the combined ratio estimator

$$\bar{y}_{RC} = \bar{X} \bar{y}_{st} / \bar{x}_{st},$$

where $\bar{X}$ is the known population mean of an auxiliary variable, the appropriate definition is

$$a_h = W_h^2 \sigma_{he}^2$$
$$\sigma_{he}^2 = \sigma_{hy}^2 + R^2 \sigma_{hx}^2 - 2R\sigma_{hxy}$$
$$R = \bar{Y}/\bar{X}.$$

For a general estimator, $\hat{\theta}$, of the form

$$\hat{\theta} = g(\bar{y}_{st}(1), \ldots, \bar{y}_{st}(p)),$$

where $\bar{y}_{st}(1), \ldots, \bar{y}_{st}(p)$ are stratified sampling means associated with p

Table 3.6.3. A 3-Order Partially Balanced Design for
$L = 7$ Strata

				Stratum (h)			
Half-Sample	1	2	3	4	5	6	7
$\delta_h^{(1)}$	+1	+1	+1	+1	+1	+1	+1
$\delta_h^{(2)}$	−1	+1	−1	−1	+1	−1	−1
$\delta_h^{(3)}$	−1	−1	+1	−1	−1	+1	−1
$\delta_h^{(4)}$	+1	−1	−1	+1	−1	−1	+1

different survey variables, the appropriate definition is

$$a_h = W_h^2 \sigma_{he}^2,$$

$$\sigma_{he}^2 = \mathbf{e}\Sigma_h\mathbf{e}',$$

$$\mathbf{e} = (e_1, \dots, e_p)$$

$$e_i = \left.\frac{\partial g(x_1, \dots, x_p)}{\partial x_i}\right|_{(x_1, \dots, x_p) = (\bar{Y}_1, \dots, \bar{Y}_p)},$$

and Σ_h is the covariance matrix for a single observation, $(y_{hi1}, \dots, y_{hip})$ from the h-th stratum.

Lee (1972, 1973a) discusses suitable modifications to the a_h when the sampling design features multiple stages and unequal selection probabilities.

In the application of partial balancing, no additional complexities are encountered when L/G is not an integer. The solution is merely to employ unequally sized groups. For example, the case $L = 7$, $G = 3$ is presented in Table 3.6.3. Note that the replicate pattern for strata 1, 2, 3 is repeated for strata 4, 5, 6, and that the pattern for stratum 1 is repeated a third time for stratum 7.

3.7. Extensions of Half-Sample Replication to the Case $n_h \neq 2$

In the case of multistage surveys of households, stratification is often carried to the point of selecting only one primary sampling unit (PSU) per stratum. The estimator of the population total may be expressed by

$$\hat{Y} = \sum_h^L \hat{Y}_h$$

where $\hat{Y}_h$ denotes the estimator of the total in stratum h. From (2.5.2) the simple collapsed stratum estimator of Var$\{\hat{Y}\}$ is

$$v_{cs}(\hat{Y}) = \sum_{g=1}^{G} (\hat{Y}_{g1} - \hat{Y}_{g2})^2,$$

where G denotes the number of groups and $\hat{Y}_{gj}$ denotes the estimator of the total in the j-th stratum of the g-th group. The collapsed stratum estimator $v_{cs}(\hat{Y})$ may be reproduced exactly by specifying k balanced half-sample replicates (as in Section 3.2) and using

$$v_k(\hat{Y}) = \sum_{\alpha=1}^{k} (\hat{Y}_\alpha - \hat{Y})^2 / k,$$

where

$$\hat{Y}_\alpha = \sum_{g=1}^{G} (\delta_{g1\alpha} 2 \hat{Y}_{g1} + \delta_{g2\alpha} 2 \hat{Y}_{g2})$$

$$\delta_{g1\alpha} = \begin{cases} 1, & \text{if stratum } (g, 1) \text{ is selected} \\ & \text{into the } \alpha\text{-th half-sample} \\ 0, & \text{otherwise} \end{cases}$$

$$\delta_{g2\alpha} = 1 - \delta_{g1\alpha}.$$

Furthermore, $\sum_{\alpha=1}^{k} \hat{Y}_\alpha / k = \hat{Y}$ except when $k = G$. In this problem, the groups are treated as the strata for purposes of defining the half-sample replication scheme. Of course, as was demonstrated in Section 2.5, both $v_{cs}(\hat{Y})$ and $v_k(\hat{Y})$ will tend to overestimate the variance of $\hat{Y}$.

Conversely, in sampling economic establishments from a list frame, one frequently selects more than $n_h = 2$ units from some or all of the L strata. Balanced half-sample replication can be modified to accommodate this situation also. We first consider some simple ad hoc procedures.

To simplify matters, let n_h be a multiple of 2 for $h = 1, \ldots, L$, i.e., $n_h = 2m_h$, where m_h is an integer. We consider once again the linear estimator $\bar{y}_{st} = \sum_{h=1}^{L} W_h \bar{y}_h$, where $\bar{y}_h = \sum_{j=1}^{n_h} y_{hj} / n_h$. A simple ad hoc procedure is to divide the units in each stratum into two random groups, letting $\bar{y}_{h1}$ and $\bar{y}_{h2}$ denote the sample means of the m_h units in the first and second groups, respectively. Then, we form k half-sample replicates by operating on the two groups within each stratum, instead of on the individual observations. The estimator for the α-th half-sample is given by

$$\bar{y}_{st,\alpha} = \sum_{h=1}^{L} W_h (\delta_{h1\alpha} \bar{y}_{h1} + \delta_{h2} \bar{y}_{h2}),$$

$$\delta_{h1\alpha} = \begin{cases} 1, & \text{if group } (h, 1) \text{ is selected for} \\ & \text{the } \alpha\text{-th half-sample} \\ 0, & \text{otherwise} \end{cases}$$

$$\delta_{h2\alpha} = 1 - \delta_{h1\alpha}.$$

An unbiased estimator of $\text{Var}\{\bar{y}_{st}\}$ is

$$v_k(\bar{y}_{st}) = \sum_{\alpha=1}^{k} (\bar{y}_{st,\alpha} - \bar{y}_{st})^2/k \tag{3.7.1}$$

and $\sum_{\alpha=1}^{k} \bar{y}_{st,\alpha}/k = \bar{y}_{st}$ except when $k = L$. Contrary to the case of $n_h = 2$, the estimator in (3.7.1) is not algebraically equivalent to the textbook estimator

$$v(\bar{y}_{st}) = \sum_{h=1}^{L} W_h^2 s_h^2/n_h$$

$$s_h^2 = \sum_{j=1}^{n_h} (y_{hj} - \bar{y}_h)^2/(n_h - 1) \tag{3.7.2}$$

for this problem. Rather, it can be shown that

$$v_k(\bar{y}_{st}) = \sum_{h=1}^{L} W_h^2(\bar{y}_{h1} - \bar{y}_{h2})^2/4.$$

This ad hoc procedure "forces" the problem into the basic two-per-stratum situation discussed in Section 3.2. The procedure is computationally convenient, but some information is lost relative to the textbook variance estimator.

Another simple ad hoc procedure is to subdivide the h-th stratum into m_h artificial strata, each of sample size two, for $h = 1, \ldots, L$. Now there are $H = \sum_{h=1}^{L} m_h$ artificial strata, and we specify a balanced set of half-samples for the expanded set of strata. Corresponding to the simple linear estimator $\bar{y}_{st}$, we have half-sample estimators

$$\bar{y}_{st,\alpha} = \sum_{h=1}^{L} \sum_{i=1}^{m_h} (W_h')(\delta_{hi1\alpha}y_{hi1} + \delta_{hi2\alpha}y_{hi2}),$$

where $W_h' = W_h/m_h$

$$\delta_{hi1\alpha} = \begin{cases} 1, & \text{if the first unit in the } (h, i)\text{-th artificial} \\ & \text{stratum is in the } \alpha\text{-th half-sample} \\ 0, & \text{otherwise} \end{cases}$$

$$\delta_{hi2\alpha} = 1 - \delta_{hi1\alpha}.$$

These estimators satisfy the desirable property $\sum_{\alpha=1}^{k} \bar{y}_{st,\alpha}/k = \bar{y}_{st}$, except when $k = H$. The half-sample estimator of variance for this problem is

$$v_k(\bar{y}_{st}) = \sum_{\alpha=1}^{k} (\bar{y}_{st,\alpha} - \bar{y}_{st})^2$$

and this estimator is unbiased for $\text{Var}\{\bar{y}_{st}\}$. Furthermore, it is identical to

$$\sum_{h=1}^{L} \sum_{i=1}^{m_h} (W_h')^2(y_{hi1} - y_{hi2})^2/4, \tag{3.7.3}$$

which would be the textbook estimator if the H strata were *real* instead of *artificial*. In fact, though, there are only L real strata and the textbook estimator is given by (3.7.2). The difference between (3.7.2) and (3.7.3) is that the former estimates the within-stratum mean square S_h^2 by the full-information estimator s_h^2 based on $2m_h - 1$ degrees of freedom, whereas the latter uses the estimator

$$\frac{1}{4m_h} \sum_{i=1}^{m_h} (y_{hi1} - y_{hi2})^2$$

based on only m_h degrees of freedom. As was the case for the first ad hoc procedure, we have "forced" the problem into the basic two-per-stratum context. The resulting procedure is computationally convenient, but some information is lost. For large sample sizes, m_h, the loss may be unimportant.

In certain cases it is possible to construct "balanced n^{-1}-sample" replication schemes that exactly reproduce the textbook variance estimators, resulting in no loss of information. The theory for such schemes was first developed by Borack (1971).

We consider the simple sampling design and estimator discussed in Section 3.2, except we now allow $n_h = n$ units per stratum, where n is a positive integer greater than 2. An n^{-1}-sample consists of one unit from each stratum, and there are n^L possible n^{-1}-samples. The estimator from the α-th replicate is

$$\bar{y}_{st,\alpha} = \sum_{h=1}^{L} W_h (\delta_{h1\alpha} y_{h1} + \delta_{h2\alpha} y_{h2} + \ldots + \delta_{hn\alpha} y_{hn}),$$

where

$$\delta_{hi\alpha} = \begin{cases} 1, & \text{if the } (h, i)\text{-th unit is selected into} \\ & \text{the } \alpha\text{-th replicate} \\ 0, & \text{otherwise.} \end{cases}$$

Using all n^L possible n^{-1}-samples, it can be shown that

$$\sum_{\alpha=1}^{n^L} \bar{y}_{st,\alpha} / n^L = \bar{y}_{st}$$

and

$$\frac{1}{n^L(n-1)} \sum_{\alpha=1}^{n^L} (\bar{y}_{st,\alpha} - \bar{y}_{st})^2 = v(\bar{y}_{st}),$$

thus reproducing the textbook estimators.

Our goal is to produce a small set, k, of the n^L n^{-1}-samples wherein reproducibility of the textbook estimators is maintained. Such a set will be said to be *balanced*. The problem of constructing balanced n^{-1}-samples is analogous to the problem of constructing orthogonal designs in the context of statistical experiments. We divide the problem into three cases.

Case 1. Let $L = p^\beta$ and $n = L$, where β is a positive integer and p is a positive prime integer. For this case $k = L^2$ replicates are needed for balancing and are specified by the cell subscripts of $L \times L$ Greco^{L-3}-Latin Square designs. A replication pattern constructed for this problem can also be employed with sampling designs involving $L^* \le L$ strata. This case is for unusual sampling designs such as $(n, L) = (3, 3), (3, 2), (4, 4), (4, 3), (4, 2), (5, 5), (5, 4), (5, 3), (5, 2), (7, 7), (7, 6), (7, 5)$, and so on. Because such designs are not usually found in practice, we omit the specific rules needed for constructing the replication patterns. For a comprehensive discussion of the rules with examples, the reader may see Borack (1971) and the references contained therein.

Case 2. Let $n > L \ge 2$ and $n = p_1^{\beta_1} p_2^{\beta_2} \ldots p_r^{\beta_r}$, where the β_i are positive integers and the p_i positive prime integers, $i = 1, \ldots, r$. Also let $\min_i(p_i^{\beta_i}) + 1 \ge L$, i.e., the minimum prime power of n is at least $L - 1$. For this case $k = n^2$ replicates are needed for balancing and are specified by the cell subscripts of $n \times n$ Greco^{L-3}-Latin Square designs. A replication pattern constructed for this problem can also be employed with sampling designs involving $L^* \le L$ strata. As for Case 1, this case includes sampling designs that are not usually found in practice; construction rules and examples may be found in Borack (1971).

Case 3. Let $n = p$, $L = (p^\beta - 1)/(p - 1)$, where β is a positive integer and p is a positive prime integer. For this case, $k = p^\beta$ replicates are needed for balancing and are defined by a $p^\beta \times L$ matrix whose elements take the values $0, 1, \ldots, p - 1$. A value of 0 in the (α, h)-th cell signifies that unit $(h, 1)$ is included in the α-replicate. Similarly, values $1, 2, \ldots, p - 1$ signify units $(h, 2), (h, 3), \ldots, (h, p)$, respectively, for the α-th replicate. The columns of the matrix are orthogonal modulo p. A set of balanced replicates may also be used for sampling designs involving $L^* \le L$ strata: one simply uses any L^* columns of the $p^\beta \times L$ matrix. If we assume that 200 replicates are an upper bound in practical survey applications (even 200 replicates may be computationally prohibitive), then the only important practical problems are $(p, L, p^\beta) = (3, 4, 9)$, $(3, 13, 27)$, $(3, 40, 81)$, $(5, 6, 25)$, $(5, 31, 125)$, $(7, 8, 49)$, $(11, 12, 121)$, and $(13, 14, 169)$. This covers all three-per-stratum designs with $L = 40$ or fewer strata; five-per-stratum designs with $L = 31$ or fewer strata; seven-per-stratum designs with $L = 8$ or fewer strata; eleven-per-stratum designs with $L = 12$ or fewer strata; and thirteen-per-stratum designs with $L = 14$ or fewer strata. All other permissible problems (p, L, p^β) involve more than 200 replicates. Generators for the eight practical problems are presented in Table 3.7.1. For a given problem, use the corresponding column in the table as the first column in the orthogonal matrix. The other $L - 1$ columns are generated cyclically according to the rules

$$m(i, j + 1) = m(i + 1, j), \qquad i < p^\beta - 1$$

$$m(p^\beta - 1, j + 1) = m(1, j),$$

where $m(i, j)$ is the (i, j)-th element of the orthogonal matrix. Finally, a row of zeros is added at the bottom to complete the $p^\beta \times L$ orthogonal matrix. To illustrate the construction process, Table 3.7.2 gives the entire orthogonal matrix for the problem $(p, L, p^\beta) = (3, 13, 27)$. Generators such as those presented in Table 3.7.1 are defined on the basis of the Galois Field $\mathrm{GF}(p^\beta)$. General discussion of such generators is given in Gurney and Jewett (1975), as is the example $\mathrm{GF}(3^5)$.

Table 3.7.1. Generators for Orthogonal Matrices of Order $p^\beta \times L$

$p = 3$ $\beta = 2$ $L = 4$	$p = 3$ $\beta = 3$ $L = 13$	$p = 3$ $\beta = 4$ $L = 40$	$p = 5$ $\beta = 2$ $L = 6$	$p = 5$ $\beta = 3$ $L = 31$	$p = 7$ $\beta = 2$ $L = 8$	$p = 11$ $\beta = 2$ $L = 12$	$p = 13$ $\beta = 2$ $L = 14$
0	0	0	0	0	0	3	1
1	0	1	4	2	1	6	12
2	1	1	1	2	2	4	10
2	0	1	1	2	6	0	12
0	1	1	2	1	2	1	5
2	2	2	1	0	2	1	7
1	1	0	0	4	1	4	10
1	1	1	3	1	6	7	9
	2	2	2	1	0	8	2
	0	1	2	4	5	7	10
	1	1	4	1	3	9	6
	1	2	2	3	2	8	12
	1	1	0	1	3	2	0
	0	2	1	3	3	4	2
	0	0	4	4	5	10	2
	2	2	4	1	2	0	11
	0	0	3	2	0	8	7
	2	2	4	0	4	8	11
	1	2	0	2	1	10	10
	2	1	2	1	3	1	1
	2	1	3	1	1	9	7
	1	0	3	0	1	1	5
	0	2	1	2	4	6	4
	2	0	3	4	3	9	7
	2	1		4	0	5	12
	2	1		3	6	10	11
		0		1	5	3	0

Table 3.7.1. (*Cont.*)

$p = 3$ $\beta = 2$ $L = 4$	$p = 3$ $\beta = 3$ $L = 13$	$p = 3$ $\beta = 4$ $L = 40$	$p = 5$ $\beta = 2$ $L = 6$	$p = 5$ $\beta = 3$ $L = 31$	$p = 7$ $\beta = 2$ $L = 8$	$p = 11$ $\beta = 2$ $L = 12$	$p = 13$ $\beta = 2$ $L = 14$
			0	4	1	0	4
			1	0	5	9	4
			2	2	5	9	9
			2	0	6	3	1
			2	0	1	8	9
			0	4	0	6	7
			2	4	2	8	2
			1	4	4	4	1
			0	2	5	6	10
			0	0	4	7	8
			2	3	4	3	1
			0	2	2	2	11
			0	2	5	0	9
			0	3	0	6	0
			2	2	3	6	8
			2	1	6	2	8
			2	2	4	9	5
			2	1	6	4	2
			1	3	6	9	5
			0	2	3	10	1
			2	4	4	4	4
			1	0		1	2
			2	4		2	7
			2	2		5	3
			1	2		0	2
			2	0		4	9
			1	4		4	5
			0	3		5	0
			1	3		6	3
			0	1		10	3
			1	2		6	10
			1	3		3	4
			2	0		10	10
			2	4		8	2
			0	0		5	8
			1	0		7	4
			0	3		0	1
			2	3		10	6
			2	3		10	4
			0	4		7	5

Table 3.7.1. (*Cont.*)

$p = 3$ $\beta = 2$ $L = 4$	$p = 3$ $\beta = 3$ $L = 13$	$p = 3$ $\beta = 4$ $L = 40$	$p = 5$ $\beta = 2$ $L = 6$	$p = 5$ $\beta = 3$ $L = 31$	$p = 7$ $\beta = 2$ $L = 8$	$p = 11$ $\beta = 2$ $L = 12$	$p = 13$ $\beta = 2$ $L = 14$
		0		0		4	10
		2		1		3	0
		1		4		4	6
		1		4		2	6
		1		1		3	7
		0		4		9	8
		1		2		7	7
		2		4		1	4
		0		2		0	3
		0		1		3	8
		1		4		3	2
		0		3		1	12
		0		0		10	8
				3		2	10
				4		10	7
				4		5	0
				0		2	12
				3		6	12
				1		1	1
				1		8	3
				2		0	1
				4		2	8
				1		2	6
				0		8	3
				3		3	4
				0		5	11
				0		3	3
				1		7	7
				1		5	1
				1		4	0
				3		8	11
				0		9	11
				2		0	2
				3		5	6
				3		5	2
				2		9	3
				3		2	12
				4		7	6
				3		2	8
				4		1	9

Table 3.7.1. (*Cont.*)

$p = 3$	$p = 3$	$p = 3$	$p = 5$	$p = 5$	$p = 7$	$p = 11$	$p = 13$
$\beta = 2$	$\beta = 3$	$\beta = 4$	$\beta = 2$	$\beta = 3$	$\beta = 2$	$\beta = 2$	$\beta = 2$
$L = 4$	$L = 13$	$L = 40$	$L = 6$	$L = 31$	$L = 8$	$L = 12$	$L = 14$
				2		7	6
				3		10	1
				1		9	2
				0		6	0
				1		0	9
				3		7	9
				3		7	4
				0		6	12
				1		5	4
				2		1	6
				2		5	11
				4		8	12
				3		1	3
				2			5
				0			12
				1			2
				0			4
							0
							5
							5
							8
							11
							8
							12
							9
							11
							6
							10
							11
							4
							8
							0
							10
							10
							3
							9
							3
							11
							5
							9

Table 3.7.1. (*Cont.*)

$p = 3$	$p = 3$	$p = 3$	$p = 5$	$p = 5$	$p = 7$	$p = 11$	$p = 13$
$\beta = 2$	$\beta = 3$	$\beta = 4$	$\beta = 2$	$\beta = 3$	$\beta = 2$	$\beta = 2$	$\beta = 2$
$L = 4$	$L = 13$	$L = 40$	$L = 6$	$L = 31$	$L = 8$	$L = 12$	$L = 14$
							12
							7
							9
							8
							3
							0
							7
							7
							6
							5
							6
							9
							10
							5
							11
							1
							5
							3
							6
							0
							1

Note: $L = (p^\beta - 1)/(p - 1)$.

The n^{-1}-sample replication estimator of $\text{Var}\{\bar{y}_{st}\}$ is given by

$$v_k(\bar{y}_{st}) = \sum_{\alpha=1}^{k} (\bar{y}_{st,\alpha} - \bar{y}_{st})^2 / k(n-1).$$

When the replicates are balanced, the two desirable properties

(1) $\dfrac{1}{k} \sum_{\alpha=1}^{k} \bar{y}_{sy,\alpha} = \bar{y}_{st}$

(2) $v_k(\bar{y}_{st}) = v(\bar{y}_{st})$

are guaranteed. In the case of an arbitrary nonlinear estimator $\hat{\theta}$ of some population parameter θ, the n^{-1}-sample replication estimator of $\text{Var}\{\hat{\theta}\}$ is

$$v_k(\hat{\theta}) = \sum_{\alpha}^{k} (\hat{\theta}_\alpha - \hat{\theta})^2 / k(n-1),$$

where $\hat{\theta}_\alpha$ is the estimator computed from the α-th replicate.

Table 3.7.2. Fully Balanced Replication Scheme for the Case $p = 3$, $L = 13$, $\beta = 3$

Replicate	Stratum												
	1	2	3	4	5	6	7	8	9	10	11	12	13
1	0	0	1	0	1	2	1	1	2	0	1	1	1
2	0	1	0	1	2	1	1	2	0	1	1	1	0
3	1	0	1	2	1	1	2	0	1	1	1	0	0
4	0	1	2	1	1	2	0	1	1	1	0	0	2
5	1	1	1	1	2	0	1	1	1	0	0	2	0
6	2	1	1	2	0	1	1	1	0	0	2	0	2
7	1	1	2	0	1	1	1	0	0	2	0	2	1
8	1	2	0	1	1	1	0	0	2	0	2	1	2
9	2	0	1	1	1	0	0	2	0	2	1	2	2
10	0	1	1	1	0	0	2	0	2	1	2	2	1
11	1	1	1	0	0	2	0	2	1	2	2	1	0
12	1	1	0	0	2	0	2	1	2	2	1	0	2
13	1	0	0	2	0	2	1	2	2	1	0	2	2
14	0	0	2	0	2	1	2	2	1	0	2	2	2
15	0	2	0	2	1	2	2	1	0	2	2	2	0
16	2	0	2	1	2	2	1	0	2	2	2	0	0
17	0	2	1	2	2	1	0	2	2	2	0	0	1
18	2	1	2	2	1	0	2	2	2	0	0	1	0
19	1	2	2	1	0	2	2	2	0	0	1	0	1
20	2	2	1	0	2	2	2	0	0	1	0	1	2
21	2	1	0	2	2	2	0	0	1	0	1	2	1
22	1	0	2	2	2	0	0	1	0	1	2	1	1
23	0	2	2	2	0	0	1	0	1	2	1	1	2
24	2	2	2	0	0	1	0	1	2	1	1	2	0
25	2	2	0	0	1	0	1	2	1	1	2	0	1
26	2	0	0	1	0	1	2	1	1	2	0	1	1
27	0	0	0	0	0	0	0	0	0	0	0	0	0

The methods of n^{-1}-sample replication discussed here are also compatible with multistage survey designs, unequal selection probabilities, and the concept of partial balancing. See Dippo, Fay, and Morgenstein (1984) for an example of 3^{-1}-sample replication applied to the U.S. Occupational Changes in a Generation Survey.

3.8. Miscellaneous Developments

Thus far we have discussed balanced half-sample (or n^{-1}-sample) replication as a method for variance estimation in the context of descriptive surveys. Such replication has also proved to be useful in many analytical surveys.

Though our focus in this book is not on analytical applications, it does seem worthwhile to briefly list a few such applications. Undoubtedly the list will grow as the methods are made accessible to larger number of survey practitioners.

Koch and Lemeshow (1972) describe an application of balanced half-sample replication in the comparison of domain means in the U.S. Health Examination Survey. In this work, domain means are assumed, at least approximately, to follow a multivariate normal law. Both univariate and multivariate tests are presented wherein a replication estimate of the covariance matrix is employed.

Freeman (1975) presents an empirical investigation of balanced half-sample replication estimates of covariance matrices. The effects of such estimates on the weighted least squares analysis of categorical data is studied.

Also, see Koch, Freeman, and Freeman (1975) for a discussion of replication methods in the context of univariate and multivariate comparisons among cross-classified domains.

Chapman (1966), and later Nathan (1973), presented an approximate test for independence in contingency tables wherein balanced half-sample replication estimates of the covariance matrices were employed.

Nonparametric uses of balanced half-sample replication were first suggested by McCarthy (1966, 1969a, 1969b). A sign test based on the quantities $\bar{y}_{st,\alpha}$ was presented, where the y variable represented the difference between two other variables, say x and z.

Bean (1975), using data from the 1969 U.S. Health Interview Survey (HIS), studied the empirical behavior of poststratified means. Balanced half-sample estimates of variance were used in defining standardized deviates. The results showed that such standardized means agree well with the normal distribution for a variety of HIS variables, and thus that replication estimates of variance can be used for making inferential statements.

Finally, in a recent monograph Efron (1982) studies the balanced half-sample estimators along with other "resampling" estimators in an essentially analytical context. Some examples and simulations involving small samples are given.

3.9. Example: Southern Railway System

This example, due to Tepping (1976), is concerned with a survey of freight shipments carried by the Southern Railway System (SRS). The main survey objective was to estimate various revenue–cost relationships. Such relationships were to be used in an Interstate Commerce Commission hearing wherein SRS was objecting to the accounting methods used to allocate revenues between SRS and Seaboard Coast Line Railroad (SCL). Apparently, total railroad revenues are divided between the various rail carriers involved with a particular shipment according to a prespecified allocation formula. In this case SRS was claiming that the allocation formulae involving SRS and SCL were out of balance, favoring SCL.

The sample was selected from a file containing 44,523 records, each record representing a shipment carried by SRS in 1975. The records were ordered by a car-type code, and within each code according to an approximate cost-to-revenue ratio for the shipment. The sequence was ascending and descending in alternate car-type classes. In this ordering of the file, each set of 100 successive cars was designated a sampling stratum. Because the file of 44,523 shipments involved 44,582 cars, 446 sampling strata resulted (18 "dummy" cars were added to the final stratum in order to provide 100 cars in that stratum). The difference $44,582 - 44,523 = 59$ apparently represents large shipments that required more than one car.

Within each stratum a simple random sample of two cars was selected. The basic data obtained for the selected cars were actual costs and actual revenues of various kinds. Secondary data items included actual ton-mileage.

Estimates of variance were computed for the most important survey estimates via the balanced half-sample replication technique. Since 448 (smallest multiple of 4 larger than $L = 446$) replicates would have been needed for a fully balanced design, a partially balanced scheme involving only 16 replicates was chosen. This scheme results in great cost and computational savings relative to the fully balanced scheme.

Table 3.9.1. Designation of Half-Sample Replicates

Replicate	Stratum Group 1	2	3	4	5	6	7	8	9	10	11	12	13	14
1	1	1	1	1	1	1	1	1	1	1	1	1	1	1
2	1	-1	1	-1	1	-1	1	-1	1	-1	1	-1	1	-1
3	1	1	-1	-1	1	1	-1	-1	1	1	-1	-1	1	1
4	1	-1	-1	1	1	-1	-1	1	1	-1	-1	1	1	-1
5	1	1	1	1	-1	-1	-1	-1	1	1	1	1	-1	-1
6	1	-1	1	-1	-1	1	-1	1	1	-1	1	-1	-1	1
7	1	1	-1	-1	-1	-1	1	1	1	1	-1	-1	-1	-1
8	1	-1	-1	1	-1	1	1	-1	1	-1	-1	1	-1	1
9	1	1	1	1	1	1	1	1	-1	-1	-1	-1	-1	-1
10	1	-1	1	-1	1	-1	1	-1	-1	1	-1	1	-1	1
11	1	1	-1	-1	1	1	-1	-1	-1	-1	1	1	-1	-1
12	1	-1	-1	1	1	-1	-1	1	-1	1	1	-1	-1	1
13	1	1	1	1	-1	-1	-1	-1	-1	-1	-1	-1	1	1
14	1	-1	1	-1	-1	1	-1	1	-1	1	1	-1	1	-1
15	1	1	-1	-1	-1	-1	1	1	-1	-1	1	1	1	1
16	1	-1	-1	1	-1	1	1	-1	-1	1	-1	1	1	-1

The 16 replicates were constructed by dividing the 446 strata into 14 groups of 32 strata each (the last group contained 30 strata). Within each group, one of the two sample cars in each stratum was selected at random, and designated as the first "unit" for the group; the remaining cars were

Table 3.9.2. Replicate Estimates of Cost and Revenue/Cost Ratios, 1975

Replicate No. (α)	Replicate Estimates	
	Total Cost	Revenue/Cost Ratio
a. SCL		
1	11,689,909	1.54
2	12,138,136	1.53
3	11,787,835	1.55
4	11,928,088	1.53
5	11,732,072	1.55
6	11,512,783	1.56
7	11,796,974	1.53
8	11,629,103	1.56
9	11,730,941	1.54
10	11,934,904	1.54
11	11,718,309	1.57
12	11,768,538	1.55
13	11,830,534	1.55
14	11,594,309	1.57
15	11,784,878	1.54
16	11,754,311	1.59
b. SRS		
1	11,366,520	1.07
2	11,694,053	1.06
3	11,589,783	1.07
4	11,596,152	1.06
5	11,712,123	1.07
6	11,533,638	1.06
7	11,628,764	1.05
8	11,334,279	1.08
9	11,675,569	1.07
10	11,648,330	1.08
11	11,925,708	1.07
12	11,758,457	1.07
13	11,579,382	1.09
14	11,724,209	1.07
15	11,522,899	1.08
16	11,732,878	1.07

Table 3.9.3. Overall Estimates of Cost, Revenue/Cost Ratios, Differences in Revenue/Cost Ratios, and Associated Variance Estimates, 1975

The estimated total cost and total revenue for SCL are

$$\hat{Y}_{SCL} = 11{,}758{,}070$$

and

$$\hat{X}_{SCL} = 18{,}266{,}375$$

respectively. The revenue/cost ratio for SCL is

$$\hat{R}_{SCL} = \hat{X}_{SCL}/\hat{Y}_{SCL} = 1.554.$$

The analogous figures for SRS are

$$\hat{Y}_{SRS} = 11{,}628{,}627$$
$$\hat{X}_{SRS} = 12{,}414{,}633$$
$$\hat{R}_{SRS} = 1.068,$$

and the difference in revenue/cost ratios is estimated by

$$\hat{D} = \hat{R}_{SCL} - \hat{R}_{SRS} = 0.486.$$

Associated standard errors are estimated by the half-sample replication method as follows:

$$se(\hat{Y}_{SCL}) = [v_k(\hat{Y}_{SCL})]^{1/2} = \left[\sum_{\alpha=1}^{16} (\hat{Y}_\alpha - \hat{Y}_{SCL})^2 \Big/ 16 \right]^{1/2}$$
$$= 142{,}385$$

$$se(\hat{R}_{SCL}) = [v_k(\hat{R}_{SCL})]^{1/2} = \left[\sum_{\alpha=1}^{16} (\hat{R}_\alpha - \hat{R}_{SCL})^2 \Big/ 16 \right]^{1/2}$$
$$= 0.016$$

$$se(\hat{D}) = [v_k(\hat{D})]^{1/2} = \left[\sum_{\alpha=1}^{16} (\hat{D}_\alpha - \hat{D})^2 \Big/ 16 \right]^{1/2}$$
$$= 0.017.$$

Observe that

$$\hat{\bar{Y}}_{SCL} = \sum_{\alpha=1}^{16} \hat{Y}_\alpha/16 = 11{,}770{,}726.5$$
$$\neq \hat{Y}_{SCL},$$

because the first column in the replication pattern contains all one's, i.e., $\sum_{\alpha=1}^{16} \delta_1^{(\alpha)} \neq 0$. This is also true of the other linear estimators (and of course for the nonlinear estimators as well). Equality of $\hat{\bar{Y}}_{SCL}$ and $\hat{Y}_{SCL}$, and of other linear estimators, could have been obtained by using any 14 of the last 15 columns in Table A.5.

designated as the second "unit." Then, a half-sample replicate consisted of one "unit" from each of the 14 groups. A balanced set of $k = 16$ such replicates was specified according to the pattern in Table 3.9.1. The reader will note that this method of grouping and balancing is equivalent to the method of partial balancing discussed in Section 3.6, i.e., repeating the 16×14 replication pattern 32 times and omitting the final two columns, where $G = 32$ and $L/G = 14$ or 13.

To illustrate the variance computations, we consider three different survey estimates: a total, a ratio, and a difference of ratios. Table 3.9.2 displays the replicate estimates of the total cost and of the revenue/cost ratio for SCL and SRS. The weighted or Horvitz–Thompson estimator, which is linear, was used in estimating both total cost and total revenue, while the revenue/cost ratio was estimated by the ratio of the total estimators. For example, letting the variable y denote cost, the estimators of total cost are

$$\hat{Y} = \sum_{h=1}^{446} 100(y_{h1} + y_{h2})/2$$

$$\hat{Y}_\alpha = \sum_{h=1}^{446} 100(\delta_{h1\alpha} y_{h1} + \delta_{h2\alpha} y_{h2}).$$

The computations associated with variance estimation are presented in Table 3.9.3.

3.10. Example: The Health Examination Survey, Cycle II

The Health Examination Survey (HES), Cycle II was a large, multistage survey conducted by the U.S. National Center for Health Statistics to obtain information about the health status of the civilian, noninstitutional population of the United States. It operated between July 1963 and December 1965 and was concerned with children ages 6–11 years inclusive. Through direct medical and dental examinations and various tests and measurements, the survey gathered data on various parameters of growth and development; heart disease; congenital abnormalities; ear, nose, and throat diseases; and neuro-musculo-skeletal abnormalities.

The sample for HES, Cycle II consisted of approximately 7417 children selected in three fundamental stages.

Stage 1. The first fundamental stage of sampling was accomplished in two steps. First, the 3103 counties and independent cities that comprise the total land area of the United States were combined into 1891 primary sampling units (PSUs). These were the same PSUs used by the U.S. Bureau of the Census in the Current Population Survey (CPS) and in the Health Interview

Survey (HIS). See Hanson (1978). The PSUs were then clustered into 357 so-called first stage units (FSUs), where a FSU was actually a complete stratum of PSUs in the HIS or CPS design. Now, the FSUs were divided into 40 strata, there being 10 population density strata within each of four geographic regions. Seven of the strata were designated self-representing. From each of the remaining 33 strata, one FSU was selected with probability proportional to 1960 census population. This was accomplished using a controlled selection sampling scheme (see Goodman and Kish (1950)), where the "control classes" consisted of four rate-of-population-change classes and several state groups. Second, from each of the 40 selected FSUs one PSU was selected with probability proportional to its 1960 census population. This was accomplished simply by taking the HIS PSU selected from the FSU (or HIS stratum) into the HES sample.

Stage 2. In the second fundamental stage of sampling, each selected PSU was divided into mutually exclusive segments, where, with few exceptions, a segment was to contain about 11 children in the target population. Several kinds of segments were used in this work. In the case of housing units that were listed in the 1960 census with a usable address, the segments were clusters of the corresponding addresses. Whereas in other cases, such as housing units built since the 1960 census in an area not issuing building permits, the segments were defined in terms of areas of land (such as a city block). In the sequel we make no distinction between the kinds of segments because they are treated identically for purposes of estimation. A sample of the segments was selected in what amounted to two separate stages. First, a sample of about 20 to 30 Enumeration Districts (1960 census definition) was selected within each PSU using unequal probability systematic sampling. The probabilities were proportional to the number of children aged 5 to 9 in the 1960 census (or aged 6 to 11 at the time of the survey). Second, a simple random sample of one segment was selected within each Enumeration District.

Stage 3. In the third fundamental stage, a list was prepared of all eligible children living within the selected segments of each selected PSU. The list was prepared by enumerating via personal visit each housing unit within the segments selected in Stage 2. Table 3.10.1 gives some results from these screening interviews, including the number of eligible children listed. Then, the list of eligible children was subsampled to give approximately 190 to 200 children per PSU for the HES examination. The subsampling scheme was equal probability systematic sampling.

The HES, Cycle II estimation procedure consisted of the following features: 1. The basic estimator of a population total was the Horvitz–Thompson estimator, where the "weight" attached to a sample child was the reciprocal of its inclusion probability (taking account of all three stages

Table 3.10.1. Numbers of Segments, Interviewed
Housing Units, and Eligible Children in the
Sample by PSU

PSU	Segments	Interviewed Housing Units	Eligible Children
1	28	630	200
2	25	475	246
3	26	638	248
4	23	602	218
5	25	600	230
6	25	459	206
7	31	505	240
8	26	451	240
9	22	410	248
10	20	727	147
11	24	777	201
12	24	694	138
13	24	546	246
14	23	459	196
15	22	539	193
16	22	882	220
17	23	689	195
18	23	395	241
19	24	727	226
20	24	423	252
21	21	379	218
22	21	495	234
23	37	690	301
24	23	451	160
25	20	434	221
26	25	408	188
27	22	338	186
28	22	267	179
29	25	528	239
30	23	421	149
31	24	450	216
32	24	506	250
33	25	650	260
34	20	422	239
35	23	680	231
36	24	492	218
37	22	596	222
38	26	616	228
39	22	545	163
40	21	397	156

Source: Bryant, Baird, and Miller (1973).

of sampling). 2. The basic weight discussed in 1 was then adjusted to account for nonrespondents. Weight adjustments were performed separately within 12 age-sex classes within each sample PSU. 3. Finally, a poststratified ratio adjustment was performed using independent population totals in 24 age-sex-race classes.

Thus, the estimator $\hat{Y}$ of a population total Y was of the form

$$\hat{Y} = \sum_{g=1}^{24} \hat{R}_g \sum_{h=1}^{40} \sum_{j=1}^{12} \sum_{k=1}^{s} \sum_{l=1}^{n_{ghijk}} W_{1 \cdot hi} W_{2 \cdot hik} W_{3 \cdot hikl} a_{hij} y_{ghijkl}, \qquad (3.10.1)$$

where

y_{ghijkl} = value of y-characteristic for l-th sample person in the k-th segment, j-th age-sex class, i-th PSU, h-th stratum, and g-th age-sex-race class

$W_{1 \cdot hi}$ = first stage weight (reciprocal of probability of selecting the (PSU)

$W_{2 \cdot hik}$ = second stage weight (reciprocal of probability of selecting the segment, given the PSU)

$W_{3 \cdot hikl}$ = third stage weight (reciprocal of probability of selecting the child, given the PSU and segment)

a_{hij} = weight adjustment factor in the j-th age-sex class, i-th PSU, and h-th stratum.

= $\Sigma^{(j)} W_{1 \cdot hi} W_{2 \cdot hik} W_{3 \cdot hikl} / \Sigma^{1(j)} W_{1 \cdot hi} W_{2 \cdot hik} W_{3 \cdot hikl}$, where $\Sigma^{(j)}$ denotes summation over all selected children in the (h, i, j)-th adjustment class and $\Sigma^{1(j)}$ denotes summation only over the respondents therein

$\hat{R}_g$ = ratio of total U.S. noninstitutional population in the g-th age-sex-race class according to 1964 independent population figures (produced by the U.S. Bureau of the Census) to the sample estimate of the same population.

In discussing the estimation of variance for HES, Cycle II, we will only consider the total design variance of an estimator, and not the between or within components of variance. Variances were estimated using the balanced half-sample methodology.

Since only one PSU was selected from each stratum, it was necessary to collapse the strata into 20 stratum pairs or *pseudostrata* and to employ the half-sample methodology as outlined in Section 3.7. Pairing was on the basis of several characteristics of the original strata including population density, geographic region, rate of growth, industry, and size. Both original strata and pseudostrata are displayed in Table 3.10.2.

To estimate the total variance, it was necessary to account for the variability due to subsampling within the self-representing PSUs. In HES, Cycle II this was accomplished by first pairing two self-representing strata, and then randomly assigning all selected segments in the pair to one of two

Table 3.10.2. Collapsing Pattern of Strata for Replication Purposes

PSU Location	PSU No.	Pseudo-Stratum No.	Region	1960 Census Population of Stratum
Boston, Mass.	5	01	NE	4,994,736
Newark, N.J.	37	01	NE	4,183,250
Jersey City, N.J.	38	02	NE	3,759,760
Allentown, Pa.	35	02	NE	3,768,466
Columbia–Dutchess, N.Y. (Poughkeepsie, N.Y.)	3	03	NE	4,271,826
Hartford–Tolland, Conn. (Manchester and Bristol, Conn.)	36	03	NE	4,843,253
Columbia, S.C.	40	04	S	3,776,544
Charleston, S.C.	9	04	S	3,961,447
Crittenden–Poinsett (Marked Tree, Ark.)	27	05	S	4,961,779
Sussex (Georgetown, Del.)	39	05	S	4,622,338
Bell–Knox–Whitley, Ky. (Barbourville)	25	06	S	4,973,857
Breathitt–Lee, Ky. (West Liberty and Beattyville)	34	06	S	4,415,267
Cleveland, Ohio	33	07	NC	3,856,698
Minneapolis–St. Paul, Minn.	20	07	NC	5,155,715
Lapeer–St. Clair, Mich. (Lapeer and Marysville)	32	08	NC	4,507,428
Ashtabula–Geauga, Ohio	2	08	NC	4,156,090
San Francisco, Calif.	14	09	W	3,890,572
Denver, Colo.	6	09	W	4,899,898
Prowers, Colo. (Lamar)	8	10	W	5,519,588
Mariposa, Calif.	16	10	W	5,115,227
Atlanta, Ga.	13	11	S	4,318,307
Houston, Tex.	29	11	S	3,587,125
Des Moines, Iowa	24	12	NC	4,895,507
Wichita, Kans.	26	12	NC	5,047,027
Birmingham, Ala.	30	13	S	3,472,118
Grand Rapids, Mich.	21	13	NC	4,799,314
Clark, Wis. (Neillsville)	22	14	NC	4,384,792
Grant, Wash. (Moses Lake)	18	14	W	5,207,020
Portland, Maine	1	15	NE	3,759,516
Mahaska–Wapello, Iowa (Ottumwa)	4	15	NC	4,570,419
De Soto–Sarasota, Fla. (Sarasota)	11	16	S	4,739,463
Brownsville, Tex. (Brownsville)	28	16	S	4,841,990

Table 3.10.2. (*Cont.*)

PSU Location	PSU No.	Pseudo-Stratum No.	Region	1960 Census Population of Stratum
Philadelphia, Pa., and	7	01A, 01B	NE	4,342,897
Baltimore, Md.	15	01A, 01B	S	3,728,920
Chicago, Ill., and	23	02A, 02B	NC	6,794,461
Detroit, Mich.	31	02A, 02B	NC	3,762,360
Los Angeles, Calif.	10	03A, 03B	W	6,742,696
	12	03A, 03B		
New York, N.Y.	17	04A, 04B	NE	10,694,633
	19	04A, 04B		

Source: Bryant, Baird, and Miller (1973).

random groups. Thus, a given random group includes a random part of each of two original self-representing strata. For example, Chicago and Detroit are paired together and the two resulting random groups, 02A and 02B, are each comprised of segments from each of the original two strata. The half-sample methodology to be discussed will treat the two random groups within a pair of self-representing strata as the two "units" within a stratum. This procedure properly includes the variability due to sampling within the self-representing strata, while not including (improperly) any variability between self-representing strata.

Before proceeding, two remarks are in order. First, the nonself-representing PSU of Baltimore was paired with the self-representing PSU of Philadelphia and two random groups were formed in the manner just described for self-representing PSUs. It is easy to show, using the development in Sections 3.2 and 3.3, that this procedure has the effect of omitting the between component of variance associated with the Baltimore PSU. Baltimore's within component of variance is properly included, as is the total variance associated with the Philadelphia PSU. Of course, Philadelphia does not have a between component of variance because it is self-representing. Thus the variance estimators presented here are downward biased by the omission of Baltimore's between component.

The second remark concerns a bias that acts in the opposite direction. Recall that the collapsed stratum technique customarily gives an overestimate of the total sampling variance for the case where one PSU is selected independently within each stratum. In HES, Cycle II, moreover, a dependency exists between the strata due to the controlled selection of PSUs. Although the statistical properties of the collapsed stratum estimator are not fully known in this situation, it is believed that an upward bias still results (this is based on the premise that a controlled selection of PSUs is at least as efficient as an independent selection of PSUs for estimating the principal characteristics of the survey). Thus certain components of bias act in the upward direction, while others act in the downward direction. The net bias of the variance estimators is an open question.

Having defined the 20 pseudostrata, a set of $k = 20$ balanced half-samples was specified, each consisting of one PSU from each pseudostratum. To estimate the variance of a statistic computed from the parent sample, the estimate was calculated individually for each of the 20 half-samples. In computing the half-sample estimates, the original principles of half-sample replication were adhered to in the sense that the poststratification ratios, $\hat{R}_g$, were computed separately for each half-sample. Also, the weighting adjustments for nonresponse were calculated separately for each half-sample.

To illustrate the variance computations, we consider the characteristic "number of upper-arch permanent teeth among 8-year-old boys in which the annual family income is between \$5000 and \$6999." The population mean of this characteristic as estimated from the parent sample was $\hat{\theta} = 5.17$; the 20 half-sample estimates of the population mean are presented in Table 3.10.3. The estimator $\hat{\theta}$ was the ratio of two estimators (3.10.1) of total: total number of upper arch permanent teeth among 8-year-old boys in which the annual family income is between \$5000 and \$6999 divided by the total number of such boys. The variance of $\hat{\theta}$ was estimated using the methods discussed in Section 3.4:

$$v_k(\hat{\theta}) = (1/20) \sum_{\alpha=1}^{20} (\hat{\theta}_\alpha - \hat{\theta})^2$$

$$= 0.008545.$$

Table 3.10.3. Half-Sample Replicate Estimates of Mean Number of Upper-Arch Permanent Teeth for 8-Year-Old Boys with Family Income of \$5000–\$6999

Replicate Number	$\hat{\theta}_\alpha$	Replicate Number	$\hat{\theta}_\alpha$
1	5.1029	11	5.1899
2	5.0685	12	5.0066
3	5.1964	13	5.2291
4	5.2701	14	5.2074
5	5.1602	15	5.0424
6	5.2353	16	5.0260
7	5.1779	17	5.2465
8	5.2547	18	5.3713
9	5.1619	19	5.1005
10	5.1116	20	5.0737

Source: Bryant, Baird, and Miller (1973).

Further examples of HES, Cycle II estimates and their estimated standard errors are given in Table 3.10.4. For example, consider the characteristic "systolic blood pressure of white females age 6–7 years living in an SMSA with an annual family income of $5000–$10,000." The mean of this characteristic as estimated from the parent sample was $\hat{\theta} = 107.5$. The estimated standard error of $\hat{\theta}$, as given by the half-sample method, was 0.74. There were 336 sample individuals in this particular color-age-sex-residence-income class.

Table 3.10.4. Average Systolic Blood Pressure of White Females by Age, Income, and Residence. Means, Standard Errors, and Sample Sizes

Age Class	SMSA			Non-SMSA			Total
	<$5K	$5K–$10K	>$10K	<$5K	$5K–$10K	>$10K	
Total	110.3	111.0	110.9	110.6	109.9	110.2	110.6
	0.77	0.54	0.82	0.80	0.37	0.80	0.35
	384	923	390	513	456	132	2798
6–7 yr. old	107.1	107.5	107.0	107.5	105.7	106.7	107.0
	0.82	0.74	1.19	0.61	0.50	0.89	0.42
	121	336	108	175	155	35	930
8–9 yr. old	110.4	111.0	111.4	110.6	110.4	109.4	110.7
	1.08	0.84	1.03	0.98	1.25	0.82	0.50
	140	289	137	165	154	53	938
10–11 yr. old	113.6	115.1	113.9	114.0	114.0	114.1	114.3
	1.05	0.59	0.88	1.01	0.73	1.72	0.39
	123	298	145	173	147	44	930

Source: Brock, D. B., Personal Communication, 1977.

CHAPTER 4
The Jackknife Method

4.1. Introduction

In Chapters 2 and 3, we discussed variance estimating techniques based on random groups and balanced half-samples. Both of these methods are members of the class of variance estimators which employ the ideas of subsample replication. Another subsample replication technique, called the jackknife, has also been suggested as a broadly useful method of variance estimation. As in the case of the two previous methods, the jackknife derives estimates of the parameter of interest from each of several subsamples of the parent sample, and then estimates the variance of the parent sample estimator from the variability between the subsample estimates.

Quenouille (1949) originally introduced the jackknife as a method of reducing the bias of an estimator of a serial correlation coefficient. In a 1956 paper, Quenouille generalized the technique and explored its general bias reduction properties in an infinite population context. In an abstract, Tukey (1958) suggested that the individual subsample estimators might reasonably be regarded as independent and identically distributed random variables, which in turn suggests a very simple estimator of variance and an approximate t statistic for testing and interval estimation. Use of the jackknife in finite population estimation appears to have been considered first by Durbin (1959), who studied its use in ratio estimation. In the ensuing years, a great number of investigations of the properties of the jackknife have been published. The reference list contains many of the important papers, but it is by no means complete. A comprehensive bibliography to 1974 is given by Miller (1974a). Extensive discussion of the jackknife method is given in Gray and Schucany (1972) and in a recent monograph by Efron (1982).

Research on the jackknife method has proceeded along two distinct lines:
1) its use in bias reduction and 2) its use for variance estimation. Much of
the work has dealt with estimation problems in the infinite population. In
this chapter we do not present a complete account of jackknife methodology.
Our primary focus will be on variance estimation problems in the finite
population. We shall, however, follow the historical development of the
jackknife method, and introduce the estimators using the infinite population
model.

4.2. Some Basic Infinite Population Methodology

In this section, we review briefly the jackknife method as it applies to the
infinite population model. For additional details the reader should see Gray
and Schucany (1972) or Efron (1982). Discussion of jackknife applications
to the finite population model is deferred until Section 4.3.

4.2.1. Definitions

In this section we consider Quenouille's original estimator and discuss some
of its properties. We also study the variance estimator and approximate t
statistic proposed by Tukey.

We let $Y_1, \ldots, Y_n$ be independent, identically distributed random variables with distribution function $F(y)$. An estimator $\hat{\theta}$ of some parameter
of interest θ is computed from the full sample. We partition the complete
sample into k groups of m observations each, assuming (for convenience)
that n, m and k are all integers and $n = mk$. Let $\hat{\theta}_{(\alpha)}$ be the estimator of
the same functional form as $\hat{\theta}$, but computed from the reduced sample of
size $m(k - 1)$ obtained by omitting the α-th group, and define

$$\hat{\theta}_\alpha = k\hat{\theta} - (k - 1)\hat{\theta}_{(\alpha)}. \tag{4.2.1}$$

Quenouille's estimator is the mean of the $\hat{\theta}_\alpha$,

$$\hat{\bar{\theta}} = \sum_{\alpha=1}^{k} \hat{\theta}_\alpha / k, \tag{4.2.2}$$

and the $\hat{\theta}_\alpha$ are called "pseudovalues."

Quenouille's estimator has the property that it removes the order $1/n$
term from a bias of the form

$$E\{\hat{\theta}\} = \theta + a_1(\theta)/n + a_2(\theta)/n^2 + \ldots,$$

where $a_1(\cdot)$, $a_2(\cdot), \ldots$ are functions of θ but not of n. This is easily seen by noting that

$$E\{\hat{\theta}_{(\alpha)}\} = \theta + a_1(\theta)/m(k-1) + a_2(\theta)/(m(k-1))^2 + \ldots$$

and that

$$E\{\hat{\bar{\theta}}\} = k[\theta + a_1(\theta)/mk + a_2(\theta)/(mk)^2 + \ldots]$$
$$- (k-1)[\theta + a_1(\theta)/m(k-1) + a_2(\theta)/(m(k-1))^2 + \ldots]$$
$$= \theta - a_2(\theta)/m^2 k(k-1) + \ldots .$$

In addition, the estimator $\hat{\bar{\theta}}$ annihilates the bias for estimators $\hat{\theta}$ that are quadratic functionals. Let $\theta = \theta(F)$ be a functional statistic and let $\hat{\theta} = \theta(\hat{F})$, where $\hat{F}$ is the empirical distribution function. If $\hat{\theta}$ is a quadratic functional

$$\hat{\theta} = \mu^{(n)} + \frac{1}{n}\sum_{i=1}^{n}\alpha^{(n)}(Y_i) + \frac{1}{n^2}\sum_{i<j}^{n}\beta(Y_i, Y_j)$$

(i.e., $\hat{\theta}$ can be expressed in a form that involves the Y_i zero, one, and two at a time only), then $\hat{\bar{\theta}}$ is an unbiased estimator of θ

$$E\{\hat{\bar{\theta}}\} = \theta.$$

See Efron and Stein (1981) and Efron (1982). We shall return to the bias reducing properties of the jackknife later in this section.

Following Tukey's suggestion, we may *hope* that the pseudovalues $\hat{\theta}_\alpha$ are approximately independent and identically distributed. Let $\hat{\theta}_{(\cdot)}$ denote the mean of the k values $\hat{\theta}_{(\alpha)}$. The jackknife estimator of variance is then

$$v_1(\hat{\bar{\theta}}) = \frac{1}{k(k-1)}\sum_{\alpha=1}^{k}(\hat{\theta}_\alpha - \hat{\bar{\theta}})^2 \qquad (4.2.3)$$

$$= \frac{(k-1)}{k}\sum_{\alpha=1}^{k}(\hat{\theta}_{(\alpha)} - \hat{\theta}_{(\cdot)})^2,$$

and the statistic

$$\hat{t} = \frac{\sqrt{k}(\hat{\bar{\theta}} - \theta)}{\left\{\dfrac{1}{k-1}\sum_{\alpha=1}^{k}(\hat{\theta}_\alpha - \hat{\bar{\theta}})^2\right\}^{1/2}} \qquad (4.2.4)$$

should be distributed approximately as Student's t with $k-1$ degrees of freedom.

In practice, $v_1(\hat{\bar{\theta}})$ has been used to estimate the variance not only of Quenouille's estimator $\hat{\bar{\theta}}$, but also of $\hat{\theta}$. Alternatively, we may use the estimator

$$v_2(\hat{\bar{\theta}}) = \frac{1}{k(k-1)}\sum_{\alpha=1}^{k}(\hat{\theta}_\alpha - \hat{\theta})^2. \qquad (4.2.5)$$

This latter form is considered a more conservative estimator since

$$v_2(\hat{\bar{\theta}}) = v_1(\hat{\bar{\theta}}) + (\hat{\theta} - \hat{\bar{\theta}})^2/(k-1),$$

and the last term on the right side is guaranteed nonnegative.

4.2.2. Some Properties of the Jackknife Method

A considerable body of theory is now available to substantiate Tukey's conjectures about the properties of jackknife methodology, and we now review some of the important results.

Many important parameters are expressible as $\theta = g(\mu)$, where μ denotes the common mean $E\{Y_i\} = \mu$. Although

$$\bar{Y} = n^{-1} \sum_{j=1}^{n} Y_j$$

is an unbiased estimator of μ, $\hat{\theta} = g(\bar{Y})$ is generally a biased estimator of $\theta = g(\mu)$. Quenouille's estimator for this problem is

$$\hat{\bar{\theta}} = kg(\bar{Y}) - (k-1)k^{-1} \sum_{\alpha=1}^{k} g(\bar{Y}_{(\alpha)}),$$

where $\bar{Y}_{(\alpha)}$ denotes the sample mean of the $m(k-1)$ observations after omitting the α-th group. Theorem 4.2.1 establishes some asymptotic properties for $\hat{\bar{\theta}}$.

Theorem 4.2.1. *Let $\{Y_j\}$ be a sequence of independent, identically distributed random variables with mean μ and variance $0 < \sigma^2 < \infty$. Let $g(\cdot)$ be a function defined on the real line which, in a neighborhood of μ, has bounded second derivatives. Then, as $k \to \infty$, $k^{1/2}(\hat{\bar{\theta}} - \theta)$ converges in distribution to normal random variable with mean zero and variance $\sigma^2(g'(\mu))^2$, where $g'(\mu)$ is the first derivative of $g(\cdot)$ evaluated at μ.*

PROOF. See Miller (1964). ☐

Theorem 4.2.2. *Let $\{Y_j\}$ be a sequence of independent, identically distributed random variables as in Theorem 4.2.1. Let $g(\cdot)$ be a real valued function with continuous first derivative in a neighborhood of μ. Then, as $k \to \infty$,*

$$kv_1(\hat{\bar{\theta}}) \xrightarrow{p} \sigma^2\{g'(\mu)\}^2.$$

PROOF. See Miller (1964). ☐

Taken together, Theorems 4.2.1 and 4.2.2 prove that the statistic $\hat{t}$ is asymptotically distributed as a standard normal random variable. Thus, jackknife methodology is correct, at least asymptotically, for parameters of the form $\theta = g(\mu)$.

These results do not apply immediately to statistics such as

$$s^2 = (n-1)^{-1} \sum_{j=1}^{n} (Y_j - \bar{Y})^2 \quad \text{or} \quad \log(s^2)$$

since they are not of the form $g(\bar{Y})$. In a second paper, Miller (1968) showed that when the observations have bounded fourth moments and $\theta = \log(s^2)$, $\hat{t}$ converges in distribution to a standard normal random variable as $k \to \infty$. In this case, $\hat{\bar{\theta}}$ is defined by

$$\hat{\bar{\theta}} = k \log(s^2) - (k-1)k^{-1} \sum_{\alpha=1}^{k} \log(s_{(\alpha)}^2)$$

and $s_{(\alpha)}^2$ is the sample variance after omitting the m observations in the α-th group.

Miller's results generalize to U-statistics and functions of vector U-statistics. Let $f(Y_1, Y_2, \ldots, Y_r)$ be a statistic symmetrically defined in its arguments with $r \le n$,

$$E\{f(Y_1, Y_2, \ldots, Y_r)\} = \eta,$$

and

$$E\{(f(Y_1, Y_2, \ldots, Y_r))^2\} < \infty.$$

Define the U-statistic

$$U_n = U(Y_1, \ldots, Y_n) = \frac{1}{\binom{n}{r}} \sum f(Y_{i_1}, \ldots, Y_{i_r}), \tag{4.2.6}$$

where the summation is over all combinations of r variables $Y_{i_1}, \ldots, Y_{i_r}$ out of the full sample of n. The following theorems demonstrate the applicability of jackknife methods to such statistics.

Theorem 4.2.3. *Let ϕ be a real valued function with bounded second derivative in a neighborhood of η; let $\theta = \phi(\eta)$; and let $\hat{\theta} = \phi(U_n)$. Then, as $k \to \infty$*

$$k^{1/2}(\hat{\bar{\theta}} - \theta) \xrightarrow{d} N(0, r^2 \xi_1^2 \{\phi'(\eta)\}^2),$$

where

$$\hat{\bar{\theta}} = k^{-1} \sum_{\alpha=1}^{k} \hat{\theta}_\alpha,$$

$$\hat{\theta}_\alpha = k\phi(U_n) - (k-1)\phi(U_{m(k-1),(\alpha)}),$$

$$U_{m(k-1),(\alpha)} = \frac{1}{\binom{m(k-1)}{r}} \sum f(Y_{i_1}, \ldots, Y_{i_r}),$$

$$\xi_1^2 = \text{Var}\{E\{f(Y_1, \ldots, Y_r) | Y_1\}\},$$

Σ *denotes summation over all combinations of r integers chosen from* $(1, 2, \ldots, (j-1)m, jm+1, \ldots, n)$, *and* $\phi'(\eta)$ *is the first derivative of* $\phi(\cdot)$ *evaluated at* η.

PROOF. See Arvesen (1969).   □

Theorem 4.2.4. *Let the conditions of Theorem* 4.2.3 *hold, except now adopt the weaker condition that* $\phi(\cdot)$ *has continuous first derivative in a neighborhood of* η. *Then, as* $k \to \infty$

$$kv_1(\hat{\bar{\theta}}) \xrightarrow{p} r^2 \xi_1^2 \{\phi'(\eta)\}^2.$$

PROOF. See Arvesen (1969).   □

Theorems 4.2.3 and 4.2.4 generalize to functions of vector U-statistics, e.g., $(U_n^1, U_n^2, \ldots, U_n^q)$. Again, the details are given by Arvesen (1969). These results are important because they encompass an extremely broad class of estimators. Important statistics that fall within this framework include ratios, differences of ratios, regression coefficients, correlation coefficients, and the t statistic itself. The theorems show that jackknife methodology is correct, at least asymptotically, for all such statistics.

The reader will note that for all of the statistics studied thus far, $\hat{t}$ converges to a standard normal random variable as $k \to \infty$. If $n \to \infty$ with k fixed, it can be shown that $\hat{t}$ converges to Student's t with $(k-1)$ degrees of freedom.

All of these results are concerned with the asymptotic behavior of the jackknife method, and we have seen that Tukey's conjectures are correct asymptotically. Now we turn to some properties of the estimators in the context of finite samples.

Let $v_1(\hat{\theta})$ denote the jackknife estimator (4.2.3) viewed as an estimator of the variance of $\hat{\theta} = \hat{\theta}(Y_1, \ldots, Y_n)$, i.e., the estimator of θ based upon the parent sample of size n. Important properties of the variance estimator can be established by viewing $v_1(\hat{\theta})$ as the result of a two-stage process: (1) a direct estimator of the variance of $\hat{\theta}(Y_1, \ldots, Y_{m(k-1)})$, i.e., the estimator of θ based upon a sample of size $m(k-1)$, (2) a modification to the variance estimator to go from sample size $m(k-1)$ to size $n = mk$. The direct estimator of $\text{Var}\{\hat{\theta}(Y_1, \ldots, Y_{m(k-1)})\}$ is

$$v_1^{(n)}(\hat{\theta}(Y_1, \ldots, Y_{m(k-1)})) = \sum_{\alpha=1}^{k} (\hat{\theta}_{(\alpha)} - \hat{\theta}_{(\cdot)})^2,$$

and the sample size modification is

$$v_1(\hat{\theta}(Y_1, \ldots, Y_n)) = \left(\frac{k-1}{k}\right) v_1^{(n)}(\hat{\theta}(Y_1, \ldots, Y_{m(k-1)})).$$

Applying an ANOVA decomposition to this two-step process, we find that the jackknife method tends to produce conservative estimators of variance.

Theorem 4.2.5. *Let* $Y_1, \ldots, Y_n$ *be independent and identically distributed random variables,* $\hat{\theta} = \hat{\theta}(Y_1, \ldots, Y_n)$ *be defined symmetrically in its arguments, and* $\mathrm{E}\{\hat{\theta}^2\} < \infty$. *The estimator* $v_1^{(n)}(\hat{\theta}(Y_1, \ldots, Y_{m(k-1)}))$ *is conservative in the sense that*

$$\mathrm{E}\{v_1^{(n)}(\hat{\theta}(Y_1, \ldots, Y_{m(k-1)}))\} - \mathrm{Var}\{\hat{\theta}(Y_1, \ldots, Y_{m(k-1)})\} = 0\left(\frac{1}{k^2}\right) \geq 0.$$

PROOF. The ANOVA decomposition of $\hat{\theta}$ is

$$\hat{\theta}(Y_1, \ldots, Y_n) = \mu + \frac{1}{n}\sum_i \alpha_i + \frac{1}{n^2}\sum\sum_{i<i'} \beta_{ii'} + \frac{1}{n^3}\sum\sum\sum_{i<i'<i''} \gamma_{ii'i''}$$

$$+ \ldots + \frac{1}{n^n}\eta_{1,2,3,\ldots,n}, \tag{4.2.7}$$

where all $2^n - 1$ random variables on the right side of (4.2.7) have zero mean and are mutually uncorrelated with one another. The quantities in the decomposition are

$$\mu = \mathrm{E}\{\hat{\theta}\}$$

grand mean;

$$\alpha_i = n[\mathrm{E}\{\hat{\theta}|Y_i = y_i\} - \mu]$$

i-th main effect;

$$\beta_{ii'} = n^2[\mathrm{E}\{\hat{\theta}|Y_i = y_i, Y_{i'} = y_{i'}\} - \mathrm{E}\{\hat{\theta}|Y_i = y_i\} - \mathrm{E}\{\hat{\theta}|Y_{i'} = y_{i'}\} + \mu]$$

(i, i')-th second order interaction;

$$\gamma_{ii'i''} = n^3[\mathrm{E}\{\hat{\theta}|Y_i = y_i, Y_{i'} = y_{i'}, Y_{i''} = y_{i''}\}$$

$$- \mathrm{E}\{\hat{\theta}|Y_i = y_i, Y_{i'} = y_{i'}\}$$

$$- \mathrm{E}\{\hat{\theta}|Y_i = y_i, Y_{i''} = y_{i''}\}$$

$$- \mathrm{E}\{\hat{\theta}|Y_{i'} = y_{i'}, Y_{i''} = y_{i''}\}$$

$$+ \mathrm{E}\{\hat{\theta}|Y_i = y_i\}$$

$$+ \mathrm{E}\{\hat{\theta}|Y_{i'} = y_{i'}\}$$

$$+ \mathrm{E}\{\hat{\theta}|Y_{i''} = y_{i''}\} - \mu]$$

(i, i', i'')-th third order interaction; and so forth. See Efron and Stein (1981) both for the derivation of the ANOVA decomposition and for the remainder of the present proof. $\qquad\square$

The statistic $v_1^{(n)}(\hat{\theta}(Y_1, \ldots, Y_{m(k-1)})$ is based upon a sample of size $n = mk$, but estimates the variance of a statistic $\hat{\theta}(Y_1, \ldots, Y_{m(k-1)})$ associated with the reduced sample size $m(k - 1)$. Theorem 4.2.5 shows that $v_1^{(n)}$ tends to overstate the true variance $\mathrm{Var}\{\hat{\theta}(Y_1, \ldots, Y_{m(k-1)})\}$ associated with the reduced sample size $m(k - 1)$.

The next theorem describes the behavior of the sample size modification.

Theorem 4.2.6. *Let the conditions of Theorem 4.2.5 hold. In addition, let* $\hat{\theta} = \hat{\theta}(Y_1, \ldots, Y_n)$ *be a U-statistic and let* $m(k-1) \ge r$. *Then*

$$E\{v_1(\hat{\theta})\} \ge \text{Var}\{\hat{\theta}\}.$$

PROOF. From Hoeffding (1948), we have

$$\frac{(k-1)}{k} \text{Var}\{\hat{\theta}(Y_1, \ldots, Y_{m(k-1)})\} \ge \text{Var}\{\hat{\theta}(Y_1, \ldots, Y_n)\}.$$

The result follows from Theorem 4.2.5.   □

Thus, for U-statistics the overstatement of variance initiated in Theorem 4.2.5 for statistics associated with the reduced sample size $m(k-1)$ is preserved by the sample size modification factor $(k-1)/k$. In general, however, it is not true that the jackknife variance estimator $v_1(\hat{\theta})$ is always nonnegatively biased for statistics associated with the full sample size $n = mk$. For quadratic functionals, Efron and Stein (1981) show sufficient conditions for $v_1(\hat{\theta})$ to be nonnegatively biased.

For linear functionals, however, the biases vanish.

Theorem 4.2.7. *Let the conditions of Theorem 4.2.5 hold. For linear functionals, i.e., statistics* $\hat{\theta}$ *such that the interactions* $\beta_{ii'}$, $\gamma_{ii'i''}$, *etc. are all zero, the estimator* $v_1^{(n)}(\hat{\theta}(Y_1, \ldots, Y_{m(k-1)}))$ *is unbiased for* $\text{Var}\{\hat{\theta}(Y_1, \ldots, Y_{m(k-1)})\}$ *and the estimator* $v_1(\hat{\theta})$ *is unbiased for* $\text{Var}\{\hat{\theta}\}$.

PROOF. See Efron and Stein (1981).   □

In summary, Theorems 4.2.5, 4.2.6, and 4.2.7 are finite sample results, whereas earlier theorems presented asymptotic results. In the earlier theorems, we saw that the jackknife variance estimator was correct asymptotically. In finite samples, however, it tends to incur an upward bias of order $1/k^2$. But for linear functionals, the jackknife variance estimator is unbiased.

4.2.3. Bias Reduction

We have observed that the jackknife method was originally introduced as a means of reducing bias. Although our main interest is in variance estimation, we shall briefly review some additional ideas of bias reduction in this section.

The reader will recall that $\hat{\bar{\theta}}$ removes the order $1/n$ term from the bias in $\hat{\theta}$, and annihilates the bias entirely when $\hat{\theta}$ is a quadratic functional.

Quenouille (1956) also gave a method for eliminating the order $1/n^2$ term from the bias. And it is possible to extend the ideas to third, fourth, and higher order bias terms, if desired.

Schucany, Gray, and Owen (1971) show how to generalize the bias reducing properties of the jackknife. Let $\hat{\theta}^1$ and $\hat{\theta}^2$ denote two estimators of θ whose biases factor as

$$E\{\hat{\theta}^1\} = \theta + f_1(n)a(\theta)$$

$$E\{\hat{\theta}^2\} = \theta + f_2(n)a(\theta),$$

with

$$\begin{vmatrix} 1 & 1 \\ f_1(n) & f_2(n) \end{vmatrix} \neq 0.$$

Then, the *generalized jackknife*

$$G(\hat{\theta}^1, \hat{\theta}^2) = \frac{\begin{vmatrix} \hat{\theta}^1 & \hat{\theta}^2 \\ f_1(n) & f_2(n) \end{vmatrix}}{\begin{vmatrix} 1 & 1 \\ f_1(n) & f_2(n) \end{vmatrix}}$$

is exactly unbiased for estimating θ. This is analogous to Quenouille's original estimator with the following identifications:

$$k = n$$

$$\hat{\theta}^1 = \hat{\theta}$$

$$\hat{\theta}^2 = \sum_{\alpha=1}^{n} \hat{\theta}_{(\alpha)}/n$$

$$f_1(n) = 1/n$$

$$f_2(n) = 1/(n-1).$$

Now suppose $p + 1$ estimators of θ are available and that their biases factor as

$$E\{\hat{\theta}^i\} = \theta + \sum_{j=1}^{\infty} f_{ji}(n)a_j(\theta) \tag{4.2.8}$$

for $i = 1, \ldots, p + 1$. If

$$\begin{vmatrix} 1 & \cdots & 1 \\ f_{11}(n) & \cdots & f_{1,p+1}(n) \\ \vdots & & \vdots \\ f_{p1}(n) & & f_{p,p+1}(n) \end{vmatrix} \neq 0, \tag{4.2.9}$$

then the generalized jackknife estimator

$$
G(\hat{\theta}^1,\ldots,\hat{\theta}^{p+1}) = \frac{\begin{vmatrix} \hat{\theta}^1 & & \hat{\theta}^{p+1} \\ f_{11}(n) & \cdots & f_{1,p+1}(n) \\ \vdots & & \vdots \\ f_{p1}(n) & \cdots & f_{p,p+1}(n) \end{vmatrix}}{\begin{vmatrix} 1 & \cdots & 1 \\ f_{11}(n) & \cdots & f_{1,p+1}(n) \\ \vdots & & \vdots \\ f_{p1}(n) & \cdots & f_{p,p+1}(n) \end{vmatrix}}
$$

eliminates the first p terms from the bias.

Theorem 4.2.8. *Let conditions* (4.2.8) *and* (4.2.9) *be satisfied. Then,*

$$
\mathrm{E}\{G(\hat{\theta}^1,\ldots,\hat{\theta}^{p+1})\} = \theta + B(n,p,\theta),
$$

where

$$
B(n,p,\theta) = \frac{\begin{vmatrix} B_1 & \cdots & B_{p+1} \\ f_{11}(n) & \cdots & f_{1,p+1}(n) \\ \vdots & & \vdots \\ f_{p1}(n) & \cdots & f_{p,p+1}(n) \end{vmatrix}}{\begin{vmatrix} 1 & \cdots & 1 \\ f_{11}(n) & \cdots & f_{1,p+1}(n) \\ \vdots & & \vdots \\ f_{p1}(n) & \cdots & f_{p,p+1}(n) \end{vmatrix}}
$$

and

$$
B_i = \sum_{j=p+1}^{\infty} f_{ji}(n) a_j(\theta)
$$

for $i = 1, \ldots, p+1$.

PROOF. See, e.g., Gray and Schucany (1972).   □

An example of the generalized jackknife $G(\hat{\theta}^1,\ldots,\hat{\theta}^{p+1})$ is where we extend Quenouille's estimator by letting $\hat{\theta}^1 = \hat{\theta}$ and letting $\hat{\theta}^2,\ldots,\hat{\theta}^{p+1}$ be the statistic $k^{-1}\sum_{\alpha=1}^k \hat{\theta}_{(\alpha)}$ with $m = 1, 2, 3, 4, \ldots, p$, respectively. If the bias in the parent sample estimator $\hat{\theta}^1 = \hat{\theta}$ is of the form

$$
\mathrm{E}\{\hat{\theta}\} = \theta + \sum_{j=1}^{\infty} a_j(\theta)/n^j,
$$

then

$$
f_{ji} = 1/(n - i + 1)^j \tag{4.2.10}
$$

and the bias in $G(\hat{\theta}^1, \ldots, \hat{\theta}^{p+1})$ is order $n^{-(p+1)}$. Hence, the generalized jackknife reduces the order of bias from order n^{-1} to order $n^{-(p+1)}$.

4.2.4. Counter Examples

The previous subsections demonstrate the considerable utility of the jackknife method. We have seen how the jackknife method and its generalizations eliminate bias, and also how Tukey's conjectures regarding variance estimation and the $\hat{t}$ statistic are asymptotically correct in a wide class of problems. One must not, however, make the mistake of believing that the jackknife method is omnipotent, to be applied to every conceivable problem.

   In fact, there are many estimation problems, particularly in the area of order statistics and nonfunctional statistics, where the jackknife does not work well, if at all. Miller (1974a) gives a partial list of counter examples. To illustrate, we consider the case $\hat{\theta} = Y_{(n)}$, the largest order statistic. Miller (1964) demonstrates that $\hat{t}$ with $k = n$ can be degenerate or nonnormal. Quenouille's estimator for this case is

$$\hat{\theta}_\alpha = Y_{(n)}, \qquad\qquad \text{if } \alpha \neq n$$
$$= nY_{(n)} - (n-1)Y_{(n-1)}, \qquad \text{if } \alpha = n,$$
$$\hat{\hat{\theta}} = n^{-1} \sum_{\alpha=1}^{n} \hat{\theta}_\alpha$$
$$= Y_{(n)} + [(n-1)/n](Y_{(n)} - Y_{(n-1)}).$$

The limiting distribution of $\hat{t}$ is nonnormal with all its mass below $+1$. When Y_1 is distributed uniformly on the interval $[0, \theta]$, $\hat{\hat{\theta}}$ does not depend solely on the sufficient statistic $Y_{(n)}$, and the jackknife cannot be optimal for convex loss functions.

   The jackknife method with $k = n$ also fails for $\hat{\theta}$ = sample median. These and other counter examples led to the development of the bootstrap method, a close relative of the jackknife. See Efron (1979 and 1982). The bootstrap method, while extremely promising for many problems, is not treated in this book because its utility for modern, complex sample surveys has yet to be demonstrated.

4.2.5. Choice of Number of Groups k

There are two primary considerations in the choice of the number of groups k: (1) computational costs and (2) the precision or accuracy of the resuting estimators. As regards computational costs, it is clear that the choice $(m, k) = (1, n)$ is most expensive and $(m, k) = ((n/2), 2)$ is least expensive. For large data sets some value of (m, k) between the extremes may be preferred. The grouping, however, introduces a degree of arbitrariness in the formation of groups, a problem not encountered when $k = n$.

As regards the precision of the estimators, we generally prefer the choice $(m, k) = (1, n)$, at least when the sample size n is small to moderate. This choice is supported by much of the research on ratio estimation, including papers by Rao (1965), Rao and Webster (1966), Chakrabarty and Rao (1968), and Rao and Rao (1971). For reasonable models of the form

$$Y_i = \beta_0 + \beta_1 X_i + e_i$$

$$\mathrm{E}\{e_i|X_i\} = 0$$

$$\mathrm{E}\{e_i^2|X_i\} = \sigma^2 X_i^t$$

$$t \geq 0$$

$$\mathrm{E}\{e_i e_j|X_i X_j\} = 0, \qquad i \neq j,$$

both the bias and variance of $\hat{\bar{\theta}}$ are decreasing functions of k, where $\hat{\bar{\theta}}$ is Quenouille's estimator based on the ratio $\hat{\theta} = \bar{y}/\bar{x}$. Further, the bias of the variance estimator $v_1(\hat{\bar{\theta}})$ is minimized by the choice $k = n$, whenever $\{X_i\}$ is a random sample from a gamma distribution.

In the sequel, we shall present the jackknife methods for general k. The optimum k necessarily involves a complex tradeoff between computational costs and the precision of the estimators. In order to make an informed decision about this tradeoff, careful modeling of the costs of the jackknife process is needed. The cost function will undoubtedly vary from survey to survey and according to the computer environment. Further research is probably needed on the precision issue (for alternative k) as well. The results for the ratio estimator suggest unequivocally that the choice $(m, k) = (1, n)$ maximizes precision. But how far the results can be trusted in other estimation problems is an open question.

4.3. Basic Applications to the Finite Population

Throughout the remainder of this chapter, we shall be concerned with jackknife variance estimation in the context of finite population sampling. In general, the procedure is (1) to divide the parent sample into random groups in the manner articulated in Sections 2.2 (independence case) and 2.4 (nonindependence case) and (2) to apply the jackknife formulas displayed in Section 4.2 to the random groups. In the latter step, the random group plays the role of the observation in Section 4.2. The asymptotic properties of these methods are discussed in Appendix B, and the possibility of transforming the data prior to using these methods is discussed in Appendix C.

We shall describe the jackknife process in some detail and begin by demonstrating how the methodology applies to some simple linear estimators and basic sampling designs. We let N denote a finite population of identifiable units. Attached to each unit in the population is the value of an estimation variable, say y. Thus, Y_i is the value of the i-th unit with $i = 1, \ldots, N$. The population total and mean are denoted by

$$Y = \sum_{i}^{N} Y_i$$

and

$$\bar{Y} = Y/N,$$

respectively. It is assumed that we wish to estimate Y or $\bar{Y}$.

4.3.1. Simple Random Sampling with Replacement (srs wr)

Suppose a srs wr of size n is selected from the population N. It is known that the sample mean

$$\bar{y} = \sum_{i=1}^{n} y_i/n$$

is an unbiased estimator of the population mean $\bar{Y}$ with variance

$$\text{Var}\{\bar{y}\} = \sigma^2/n,$$

where

$$\sigma^2 = \sum_{i=1}^{N} (Y_i - \bar{Y})^2/N.$$

The unbiased textbook estimator of variance is

$$v(\bar{y}) = s^2/n, \tag{4.3.1}$$

where

$$s^2 = \sum_{i=1}^{n} (y_i - \bar{y})^2/(n - 1).$$

By analogy with (4.2.1), let $\hat{\theta} = \bar{y}$, and let the sample be divided into k random groups each of size m, $n = mk$. Quenouille's estimator of the mean $\bar{Y}$ is then

$$\hat{\hat{\theta}} = \sum_{\alpha=1}^{k} \hat{\theta}_\alpha/k, \tag{4.3.2}$$

where the α-th pseudovalue is

$$\hat{\theta}_\alpha = k\bar{y} - (k - 1)\bar{y}_{(\alpha)}$$

and

$$\bar{y}_{(\alpha)} = \sum_{i=1}^{m(k-1)} y_i / m(k-1)$$

denotes the sample mean after omitting the α-th group of observations. The corresponding variance estimator is

$$v(\hat{\hat{\theta}}) = \sum_{\alpha=1}^{k} (\hat{\theta}_\alpha - \hat{\hat{\theta}})^2 / k(k-1). \tag{4.3.3}$$

To investigate the properties of the jackknife, it is useful to rewrite (4.3.2) as

$$\hat{\hat{\theta}} = k\bar{y} - (k-1)\bar{y}_{(\cdot)}, \tag{4.3.4}$$

where

$$\bar{y}_{(\cdot)} = \sum_{\alpha=1}^{k} \bar{y}_{(\alpha)} / k.$$

We then have:

Lemma 4.3.1. *Quenouille's estimator is identically equal to the sample mean*

$$\hat{\hat{\theta}} = \bar{y}.$$

PROOF. Follows immediately from (4.3.4) since any given y_i appears in exactly $(k-1)$ of the $\bar{y}_{(\alpha)}$. □

From Lemma 4.3.1, it follows that the jackknife estimator of variance is

$$v_1(\hat{\hat{\theta}}) = \frac{(k-1)}{k} \sum_{\alpha=1}^{k} (\bar{y}_{(\alpha)} - \bar{y})^2. \tag{4.3.5}$$

The reader will note that $v_1(\hat{\hat{\theta}})$ is not, in general, equal to the textbook estimator $v(\bar{y})$. For the special case $k = n$ and $m = 1$, we see that

$$\bar{y}_{(\alpha)} = (n\bar{y} - y_\alpha)/(n-1)$$

and by (4.3.5) that

$$v_1(\hat{\hat{\theta}}) = \frac{(n-1)}{n} \sum_{\alpha=1}^{n} [(y_\alpha - \bar{y})/(n-1)]^2$$

$$= v(\bar{y}),$$

the textbook estimator of variance. In any case, whether $k = n$ or not, we have the following:

Lemma 4.3.2. *Given the conditions of this section,*

$$E\{v_1(\hat{\hat{\theta}})\} = \text{Var}\{\hat{\hat{\theta}}\} = \text{Var}\{\bar{y}\}.$$

PROOF. Left to the reader. □

We conclude that for srs wr the jackknife method preserves the linear estimator $\bar{y}$, gives an unbiased estimator of its variance, and reproduces the textbook variance estimator when $k = n$.

4.3.2. Probability Proportional to Size Sampling with Replacement (pps wr)

Suppose now that a pps wr sample of size n is selected from N using probabilities $\{p_i\}_{i=1}^{N}$, with $\sum_i^N p_i = 1$ and $p_i > 0$ for $i = 1, \ldots, N$. The srs wr design treated in the last section is the special case where $p_i = N^{-1}$, $i = 1, \ldots, N$. The customary estimator of the population total Y and its variance are given by

$$\hat{Y} = \frac{1}{n} \sum_{i=1}^{n} y_i/p_i$$

and

$$\text{Var}\{\hat{Y}\} = \frac{1}{n} \sum_{i=1}^{N} p_i (Y_i/p_i - Y)^2,$$

respectively. The unbiased textbook estimator of the variance $\text{Var}\{\hat{Y}\}$ is

$$v(\hat{Y}) = \frac{1}{n(n-1)} \sum_{i=1}^{n} (y_i/p_i - \hat{Y})^2.$$

Let $\hat{\theta} = \hat{Y}$, and suppose that the parent sample is divided into k random groups of size m, $n = mk$. Quenouille's estimator of the total Y is then

$$\hat{\bar{\theta}} = k\hat{Y} - (k-1)k^{-1} \sum_{\alpha=1}^{k} \hat{Y}_{(\alpha)},$$

where

$$\hat{Y}_{(\alpha)} = \frac{1}{m(k-1)} \sum_{i=1}^{m(k-1)} y_i/p_i$$

is the estimator based on the sample after omitting the α-th group of observations. The jackknife estimator of variance is

$$v_1(\hat{\theta}) = \frac{1}{k(k-1)} \sum_{\alpha=1}^{k} (\hat{\theta}_\alpha - \hat{\bar{\theta}})^2,$$

where

$$\hat{\theta}_\alpha = k\hat{Y} - (k-1)\hat{Y}_{(\alpha)},$$

is the α-th pseudovalue. For pps wr sampling the moment properties of $\hat{\bar{\theta}}$ and $v_1(\hat{\bar{\theta}})$ are identical with those for srs wr, provided we replace y_i by y_i/p_i.

Lemma 4.3.3. *Given the conditions of this section,*

$$\hat{\hat{\theta}} = \hat{Y}$$

and

$$E\{v_1(\hat{\hat{\theta}})\} = \text{Var}\{\hat{\hat{\theta}}\} = \text{Var}\{\hat{Y}\}.$$

Further, if $k = n$, *then*

$$v_1(\hat{\hat{\theta}}) = v(\hat{Y}).$$

PROOF. Left to the reader.   □

4.3.3. Simple Random Sampling Without Replacement (srs wor)

If a srs wor of size n is selected, then the customary estimator of $\bar{Y}$, its variance, and the unbiased textbook estimator of variance are

$$\hat{\theta} = \bar{y} = \sum_{i=1}^{n} y_i/n,$$

$$\text{Var}\{\bar{y}\} = (1 - f)S^2/n,$$

and

$$v(\bar{y}) = (1 - f)s^2/n,$$

respectively, where $f = n/N$ and

$$S^2 = \sum_{i=1}^{N} (Y_i - \bar{Y})^2/(N - 1).$$

We suppose that the parent sample is divided into k random groups, each of size m, $n = mk$. Because without replacement sampling is used, the random groups are necessarily nonindependent. For this case, Quenouille's estimator, $\hat{\hat{\theta}}$, and the jackknife variance estimator, $v_1(\hat{\hat{\theta}})$, are algebraically identical with the estimators presented in Section 4.3.1 for srs wr. These appear in (4.3.2) and (4.3.3), respectively. By Lemma 4.3.1, it follows that $\hat{\hat{\theta}} = \bar{y}$, and thus $\hat{\hat{\theta}}$ is also an unbiased estimator of $\bar{Y}$ for srs wor. However, $v_1(\hat{\hat{\theta}})$ is no longer an unbiased estimator of variance; in fact, it can be shown that

$$E\{v_1(\hat{\hat{\theta}})\} = S^2/n.$$

Clearly, we may use the jackknife variance estimator with little concern for the bias whenever the sampling fraction $f = n/N$ is negligible. In any case, $v_1(\hat{\hat{\theta}})$ will be a conservative estimator, overestimating the true variance of $\hat{\hat{\theta}} = \bar{y}$ by the amount

$$\text{Bias}\{v_1(\hat{\hat{\theta}})\} = fS^2/n.$$

If the sampling fraction is not negligible, a very simple unbiased estimator of variance is

$$(1 - f)v_1(\hat{\bar{\theta}}).$$

Another method of "correcting" the bias of the jackknife estimator is to work with

$$\hat{\theta}^*_{(\alpha)} = \bar{y} + (1 - f)^{1/2}(\bar{y}_{(\alpha)} - \bar{y})$$

instead of

$$\hat{\theta}_{(\alpha)} = \bar{y}_{(\alpha)}.$$

This results in the following definitions:

$$\hat{\theta}^*_\alpha = k\hat{\theta} - (k - 1)\hat{\theta}^*_{(\alpha)}, \qquad \text{(pseudovalue)},$$

$$\hat{\bar{\theta}}^* = \sum_{\alpha=1}^{k} \hat{\theta}^*_\alpha / k, \qquad \text{(Quenouille's estimator)},$$

$$v_1(\hat{\bar{\theta}}^*) = \frac{1}{k(k-1)} \sum_{\alpha=1}^{k} (\hat{\theta}^*_\alpha - \hat{\bar{\theta}}^*)^2, \qquad \begin{array}{l}\text{(jackknife estimator of} \\ \text{variance)}.\end{array}$$

We state the properties of these modified jackknife statistics in the following Lemma.

Lemma 4.3.4. *For srs wor, we have*

$$\hat{\bar{\theta}}^* = \bar{y}$$

and

$$E\{v_1(\hat{\bar{\theta}}^*)\} = \text{Var}\{\bar{y}\} = (1 - f)S^2/n.$$

Further, when k = n

$$v_1(\hat{\bar{\theta}}^*) = v(\bar{y}).$$

PROOF. Left to the reader.   □

Thus, the jackknife variance estimator defined in terms of the modified pseudovalues $\hat{\theta}^*_\alpha$ takes into account the finite population correction $(1 - f)$ and gives an unbiased estimator of variance.

4.3.4. Unequal Probability Sampling Without Replacement

Little is known about the properties of the jackknife method for unequal probability, without replacement sampling schemes. To describe the problem, we suppose that a sample of size n is drawn from N using some unequal probability sampling scheme without replacement, and let π_i denote the inclusion probability associated with the i-th unit in the population, i.e.,

$$\pi_i = \mathscr{P}\{i \in s\}$$

where s denotes the sample. The Horvitz–Thompson estimator of the population total is then

$$\hat{\theta} = \hat{Y} = \sum_{i=1}^{n} y_i/\pi_i. \qquad (4.3.6)$$

Again, we suppose that the parent sample has been divided into k random groups (nonindependent) of size m, $n = mk$. Quenouille's estimator $\hat{\hat{\theta}}$ for this problem is defined by (4.3.2), where the pseudovalues take the form

$$\hat{\theta}_\alpha = k\hat{Y} - (k-1)\hat{Y}_{(\alpha)}, \qquad (4.3.7)$$

and

$$\hat{Y}_{(\alpha)} = \sum_{i=1}^{m(k-1)} y_i/[\pi_i m(k-1)/n]$$

is the Horvitz–Thompson estimator based on the sample after removing the α-th group of observations. As was the case for the three previous sampling methods, $\hat{\hat{\theta}}$ is algebraically equal to $\hat{Y}$. The jackknife thus preserves the unbiased character of the Horvitz–Thompson estimator of total.

To estimate the variance of $\hat{\hat{\theta}}$ we have the jackknife estimator

$$v_1(\hat{\hat{\theta}}) = \frac{1}{k(k-1)} \sum_{\alpha=1}^{k} (\hat{\theta}_\alpha - \hat{\hat{\theta}})^2,$$

where the $\hat{\theta}_\alpha$ are defined in (4.3.7). If $\pi_i = np_i$ for $i = 1, \ldots, n$ (i.e., a πps sampling scheme) and $k = n$, then it can be shown that

$$v_1(\hat{\hat{\theta}}) = \frac{1}{n(n-1)} \sum_{i=1}^{n} (y_i/p_i - \hat{Y})^2. \qquad (4.3.8)$$

The reader will recognize this as the textbook estimator of variance for pps wr sampling. More generally, when $k < n$ it can be shown that the equality in (4.3.8) does not hold algebraically, but does hold in expectation

$$E\{v(\hat{\hat{\theta}})\} = E\left\{\frac{1}{n(n-1)} \sum_{i=1}^{n} (y_i/p_i - \hat{Y})^2\right\},$$

where the expectations are with respect to the πps sampling design. We conclude that the jackknife estimator of variance acts as if the sample were selected with unequal probabilities *with* replacement, rather than *without* replacement! The bias of this procedure may be described as follows:

Lemma 4.3.5. *Let* $\mathrm{Var}\{\hat{Y}_{\pi\mathrm{ps}}\}$ *denote the variance of the Horvitz–Thompson estimator and let* $\mathrm{Var}\{\hat{Y}_{\mathrm{wr}}\}$ *denote the variance of* $\hat{Y}_{\mathrm{wr}} = n^{-1}\sum_{i=1}^{n} y_i/p_i$ *in* pps wr *sampling. Let* $\pi_i = np_i$ *for the without replacement scheme. If the true design features without replacement sampling, then*

$$\text{Bias}\{v_1(\hat{\bar{\theta}})\} = (\text{Var}\{\hat{Y}_{\text{wr}}\} - \text{Var}\{\hat{Y}_{\pi\text{ps}}\})n/(n-1).$$

That is, the bias of the jackknife estimator of variance is a factor $n/(n-1)$ *times the gain (or loss) in precision from use of without replacement sampling.*

PROOF. Follows from Durbin (1953). See Section 2.4, particularly Theorem 2.4.6.   □

We conclude that the jackknife estimator of variance is conservative (upward biased) in the useful applications of πps sampling (applications where πps beats pps wr).

Some practitioners may prefer to use an approximate finite population correction (fpc) to correct for the bias in $v_1(\hat{\bar{\theta}})$. One such approximate fpc is $(1 - \bar{\pi})$, with $\bar{\pi} = \sum_{i=1}^{n} \pi_i/n$. This may be incorporated in the jackknife calculations by working with

$$\hat{\theta}^*_{(\alpha)} = \hat{Y} + (1 - \bar{\pi})^{1/2}(\hat{Y}_{(\alpha)} - \hat{Y})$$

instead of $\hat{\theta}_{(\alpha)} = \hat{Y}_{(\alpha)}$.

4.4. Application to Nonlinear Estimators

In Section 4.3 we applied the various jackknifing techniques to linear estimators, an application in which the jackknife probably has no real utility. The reader will recall that the jackknife simply reproduces the textbook variance estimators in most cases. Further, no worthwhile computational advantages are to be gained by using the jackknife rather than traditional formulas. Our primary interest in the jackknife lies in variance estimation for nonlinear statistics, and this is the topic of the present section. At the outset we note that few finite sample, distributional results are available concerning use of the jackknife for nonlinear estimators. See Appendix B for the relevant asymptotic results. It is for this reason that we dealt at some length with linear estimators. In fact, the main justification for the jackknife in *nonlinear* problems is that it works well and its properties are known in *linear* problems. If a nonlinear statistic has a local linear quality, then, on the basis of the results presented in Section 4.3, the jackknife method should give reasonably good variance estimates.

To apply the jackknife to nonlinear survey statistics, we

(1) form k random groups and
(2) follow the jackknifing principles enumerated in Section 4.2 for the infinite population model.

No restrictions on the sampling design are needed for application of the jackknife method. Whatever the design might be in a particular application, one simply forms the random groups according to the rules set forth in Section 2.2 (for the independent case) and Section 2.4 (for the nonindependent case). Then, as usual, the jackknife operates by omitting random groups from the sample. The jackknifed version of a nonlinear estimator $\hat{\theta}$ of some population parameter θ is

$$\hat{\bar{\theta}} = \sum_{\alpha=1}^{k} \hat{\theta}_{\alpha}/k,$$

where the pseudovalues $\hat{\theta}_{\alpha}$ are defined in (4.2.1), and $\hat{\theta}_{(\alpha)}$ is the estimator of the same functional form as $\hat{\theta}$ obtained after omitting the α-th random group. For linear estimators, we found that the estimator $\hat{\bar{\theta}}$ is equal to the parent sample estimator $\hat{\theta}$. For nonlinear estimators, however, we generally have $\hat{\bar{\theta}} \neq \hat{\theta}$.

The jackknife variance estimator

$$v_1(\hat{\bar{\theta}}) = \frac{1}{k(k-1)} \sum_{\alpha=1}^{k} (\hat{\theta}_{\alpha} - \hat{\bar{\theta}})^2$$

was first given in (4.2.3). A conservative alternative, corresponding to (4.2.5), is

$$v_2(\hat{\bar{\theta}}) = \frac{1}{k(k-1)} \sum_{\alpha}^{k} (\hat{\theta}_{\alpha} - \hat{\theta})^2.$$

We may use either $v_1(\hat{\bar{\theta}})$ or $v_2(\hat{\bar{\theta}})$ to estimate the variance of either $\hat{\theta}$ or $\hat{\bar{\theta}}$.

Little else is known about the relative accuracy of these estimators in finite samples. Brillinger (1966) shows that both v_1 and v_2 give plausible estimates of the asymptotic variance. The result for v_2 requires that $\hat{\theta}$ and $\hat{\theta}_{(\alpha)}$ have small bias, while the result for v_1 does not, instead requiring that the asymptotic correlation between $\hat{\theta}_{(\alpha)}$ and $\hat{\theta}_{(\beta)}$ ($\alpha \neq \beta$) be of the form $(k-2)(k-1)^{-1}$. The latter condition will obtain in many applications because $\hat{\theta}_{(\alpha)}$ and $\hat{\theta}_{(\beta)}$ have $(k-2)$ random groups in common out of $(k-1)$. For additional asymptotic results see Appendix B.

We close this section by giving two examples.

4.4.1. Ratio Estimation

Suppose that it is desired to estimate

$$R = Y/X,$$

the ratio of two population totals. The usual estimator is

$$\hat{R} = \hat{Y}/\hat{X},$$

where $\hat{Y}$ and $\hat{X}$ are estimators of the population totals based on the particular sampling design. Quenouille's estimator is obtained by working with

$$\hat{R}_{(\alpha)} = \hat{Y}_{(\alpha)}/\hat{X}_{(\alpha)},$$

where $\hat{Y}_{(\alpha)}$ and $\hat{X}_{(\alpha)}$ are estimators of Y and X, respectively, after omitting the α-th random group from the sample. Then, we have the pseudovalues

$$\hat{R}_\alpha = k\hat{R} - (k-1)\hat{R}_{(\alpha)}, \tag{4.4.1}$$

and Quenouille's estimator

$$\hat{\hat{R}} = k^{-1} \sum_{\alpha=1}^{k} \hat{R}_\alpha. \tag{4.4.2}$$

To estimate the variance of either $\hat{R}$ or $\hat{\hat{R}}$, we have either

$$v_1(\hat{\hat{R}}) = \frac{1}{k(k-1)} \sum_{\alpha=1}^{k} (\hat{R}_\alpha - \hat{\hat{R}})^2 \tag{4.4.3}$$

or

$$v_2(\hat{\hat{R}}) = \frac{1}{k(k-1)} \sum_{\alpha=1}^{k} (\hat{R}_\alpha - \hat{R})^2. \tag{4.4.4}$$

Specifically, let us assume srs wor. Then,

$$\hat{Y} = N\bar{y},$$
$$\hat{X} = N\bar{x},$$
$$\hat{Y}_{(\alpha)} = N\bar{y}_{(\alpha)},$$
$$\hat{X}_{(\alpha)} = N\bar{x}_{(\alpha)},$$
$$\hat{R} = \hat{Y}/\hat{X},$$
$$\hat{R}_{(\alpha)} = \hat{Y}_{(\alpha)}/\hat{X}_{(\alpha)},$$
$$\hat{R}_\alpha = k\hat{R} - (k-1)\hat{R}_{(\alpha)},$$
$$\hat{\hat{R}} = k^{-1} \sum_{\alpha=1}^{k} \hat{R}_\alpha,$$
$$= k\hat{R} - (k-1)\hat{R}_{(\cdot)}.$$

If the sampling fraction $f = n/N$ is *not* negligible, then the modification

$$\hat{R}_{(\alpha)}^* = \hat{R} + (1-f)^{1/2}(\hat{R}_{(\alpha)} - \hat{R})$$

might usefully be applied in place of $\hat{R}_{(\alpha)}$.

4.4.2. A Regression Coefficient

A second illustrative example of the jackknife is given by the regression coefficient

$$\hat{\beta} = \frac{\sum\limits_{i=1}^{n} (x_i - \bar{x})(y_i - \bar{y})}{\sum\limits_{i=1}^{n} (x_i - \bar{x})^2}$$

based on a srs wor of size n. Quenouille's estimator for this problem is formed by working with

$$\hat{\beta}_{(\alpha)} = \frac{\sum\limits_{i=1}^{m(k-1)} (x_i - \bar{x}_{(\alpha)})(y_i - \bar{y}_{(\alpha)})}{\sum\limits_{i=1}^{m(k-1)} (x_i - \bar{x}_{(\alpha)})^2},$$

where the summations are over all units not in the α-th random group. This gives the pseudovalue

$$\hat{\beta}_\alpha = k\hat{\beta} - (k-1)\hat{\beta}_{(\alpha)}$$

and Quenouille's estimator

$$\hat{\hat{\beta}} = k^{-1} \sum_{\alpha=1}^{k} \hat{\beta}_\alpha.$$

To estimate the variance of either $\hat{\beta}$ or $\hat{\hat{\beta}}$, we have either

$$v_1(\hat{\hat{\beta}}) = \frac{1}{k(k-1)} \sum_{\alpha=1}^{k} (\hat{\beta}_\alpha - \hat{\hat{\beta}})^2$$

or

$$v_2(\hat{\hat{\beta}}) = \frac{1}{k(k-1)} \sum_{\alpha=1}^{k} (\hat{\beta}_\alpha - \hat{\beta})^2.$$

An fpc may be incorporated in the variance computations by working with

$$\hat{\beta}^*_{(\alpha)} = \hat{\beta} + (1-f)^{1/2}(\hat{\beta}_{(\alpha)} - \hat{\beta})$$

in place of $\hat{\beta}_{(\alpha)}$.

4.5. Usage in Stratified Sampling

The jackknife runs into some difficulty in the context of stratified sampling plans because the observations are no longer identically distributed. We shall describe some methods for handling this problem. The reader should be especially careful not to apply the classical jackknife estimators (cf. Sections 4.2, 4.3, 4.4) to stratified sampling problems.

We assume the population is divided into L strata, where N_h describes the size of the h-th stratum. Sampling is carried out independently in the various strata. Within the strata, simple random samples are selected, either with or without replacement, n_h denoting the sample size in the h-th stratum. The population is assumed to be p-variate, with

$$\mathbf{Y}_{hi} = (Y_{1hi}, Y_{2hi}, \ldots, Y_{phi})$$

denoting the value of the i-th unit in the h-th stratum. We let

$$\bar{\mathbf{Y}}_h = (\bar{Y}_{1h}, \bar{Y}_{2h}, \ldots, \bar{Y}_{ph})$$

denote the p-variate mean of the h-th stratum, $h = 1, \ldots, L$.

The problem we shall be addressing is that of estimating a population parameter of the form

$$\theta = g(\bar{\mathbf{Y}}_1, \ldots, \bar{\mathbf{Y}}_L),$$

where $g(\cdot)$ is a "smooth" function of the stratum means $\bar{Y}_{rh}$, for $h = 1, \ldots, L$ and $r = 1, \ldots, p$. The natural estimator of θ is

$$\hat{\theta} = g(\bar{\mathbf{y}}_1, \ldots, \bar{\mathbf{y}}_L), \tag{4.5.1}$$

i.e., the same function of the sample means $\bar{y}_{rh} = \sum_{i=1}^{n_h} y_{rhi}/n_h$. The class of functions satisfying these specifications is quite broad, including for example

$$\hat{\theta} = \hat{R} = \frac{\displaystyle\sum_{h=1}^{L} N_h \bar{y}_{1h}}{\displaystyle\sum_{h=1}^{L} N_h \bar{y}_{2h}}$$

the combined ratio estimator;

$$\hat{\theta} = \bar{y}_{11}/\bar{y}_{12}$$

the ratio of one stratum mean to another;

$$\hat{\theta} = \hat{\beta} = \frac{\displaystyle\sum_{h=1}^{L} N_h \bar{y}_{4h} - \left(\sum_{h}^{L} N_h \bar{y}_{1h}\right)\left(\sum_{h}^{L} N_h \bar{y}_{2h}\right)\Big/N}{\displaystyle\sum_{h=1}^{L} N_h \bar{y}_{3h} - \left(\sum_{h}^{L} N_h \bar{y}_{2h}\right)^2\Big/N}$$

the regression coefficient (where $Y_{3hi} = Y_{2hi}^2$ and $Y_{4hi} = Y_{1hi}Y_{2hi}$); and

$$\hat{\theta} = (\bar{y}_{11}/\bar{y}_{21}) - (\bar{y}_{12}/\bar{y}_{22})$$

the difference of ratios.

As in the case of the original jackknife, the methodology for stratified sampling works with estimators of θ obtained by removing observations from the full sample. Accordingly, let $\hat{\theta}_{(hi)}$ denote the estimator of the same functional form as $\hat{\theta}$ obtained after deleting the (h, i)-th observation from the sample. Let

$$\hat{\theta}_{(h\cdot)} = \sum_{i=1}^{n_h} \hat{\theta}_{(hi)}/n_h,$$

$$\hat{\theta}_{(\cdot\cdot)} = \sum_{h=1}^{L} \sum_{i=1}^{n_h} \hat{\theta}_{(hi)}/n,$$

$$n = \sum_{h=1}^{L} n_h,$$

and

$$\bar{\theta}_{(\cdot\cdot)} = \sum_{h=1}^{L} \hat{\theta}_{(h\cdot)}/L.$$

Then define the pseudovalues $\hat{\theta}_{hi}$ by setting

$$\hat{\theta}_{hi} = (Lw_h + 1)\hat{\theta} - Lw_h\hat{\theta}_{(hi)}$$

$$w_h = (n_h - 1)(1 - n_h/N_h), \quad \text{for without replacement sampling}$$

$$= (n_h - 1), \quad \quad \text{for with replacement sampling}$$

for $i = 1, \ldots, n_h$ and $h = 1, \ldots, L$.[1]

In comparison with earlier sections, we see that the quantity $(Lw_h + 1)$ plays the role of the sample size and Lw_h the sample size minus one, although this apparent analogy must be viewed as tenuous at best. The jackknife estimator of θ is now defined by

$$\hat{\theta}^1 = \sum_{h=1}^{L} \sum_{i=1}^{n_h} \hat{\theta}_{hi}/Ln_h$$

$$= \left(1 + \sum_{h=1}^{L} w_h\right)\hat{\theta} - \sum_{h=1}^{L} w_h\hat{\theta}_{(h\cdot)}, \quad\quad (4.5.2)$$

and its moments are described in the following theorem.

---

[1] In the case $L = 1$, contrast this definition with the special pseudovalues defined in Section 4.3. Here we have (dropping the 'h' subscript)

$$\hat{\theta}_i = \hat{\theta} - (n - 1)(1 - f)(\hat{\theta}_{(i)} - \hat{\theta}),$$

whereas in Section 4.3 we had the special pseudovalues

$$\hat{\theta}_i^* = \hat{\theta} - (n - 1)(1 - f)^{1/2}(\hat{\theta}_{(i)} - \hat{\theta}).$$

For linear estimators $\hat{\theta}$, both pseudovalues lead to the same unbiased estimator of θ. For nonlinear $\hat{\theta}$, the pseudovalue defined here removes both the order n^{-1} and the order N^{-1} (in the case of without replacement sampling) bias in the estimation of θ. The pseudovalue $\hat{\theta}_i^*$ attempts instead to include an fpc in the variance calculations. In this section fpc's are incorporated in the variance estimators, but not via the pseudovalues. See (4.5.3).

Theorem 4.5.1. *Let $\hat{\theta}$ and $\hat{\theta}^1$ be defined by (4.5.1) and (4.5.2), respectively. Let $g(\cdot)$ be a function of stratum means, which does not explicitly involve the sample sizes n_h, with continuous third derivatives in a neighborhood of $\bar{\mathbf{Y}} = (\bar{\mathbf{Y}}_1, \ldots, \bar{\mathbf{Y}}_L)$. Then, the expectations of $\hat{\theta}$ and $\hat{\theta}^1$ to second order moments of the $\bar{y}_{rh}$ are*

$$\mathrm{E}\{\hat{\theta}\} = \theta + \sum_{h=1}^{L} \left[\frac{N_h - n_h}{(N_h - 1)n_h} \right] c_h, \qquad \begin{array}{l} \textit{for without replacement} \\ \textit{sampling} \end{array}$$

$$= \theta + \sum_{h=1}^{L} c_h / n_h, \qquad \textit{for with replacement sampling}$$

$$\mathrm{E}\{\hat{\theta}^1\} = \theta,$$

where the c_h are constants that do not depend on the n_h. Further, their variances to third order moments are

$$\mathrm{Var}\{\hat{\theta}\} = \sum_{h=1}^{L} \left[\frac{N_h - n_h}{(N_h - 1)n_h} \right] d_{1h} + \sum_{h=1}^{L} \left[\frac{(N_h - n_h)(N_h - 2n_h)}{(N_h - 1)(N_h - 2)n_h^2} \right] d_{2h},$$

$$\textit{for without replacement sampling}$$

$$= \sum_{h=1}^{L} n_h^{-1} d_{1h} + \sum_{h=1}^{L} n_h^{-2} d_{2h}, \qquad \textit{for with replacement sampling}$$

$$\mathrm{Var}\{\hat{\theta}^1\} = \sum_{h=1}^{L} \left[\frac{N_h - n_h}{(N_h - 1)n_h} \right] d_{1h} - \sum_{h=1}^{L} \left[\frac{(N_h - n_h)}{(N_h - 1)(N_h - 2)n_h} \right] d_{2h},$$

$$\textit{for without replacement sampling}$$

$$= \sum_{h=1}^{L} n_h^{-1} d_{1h}, \qquad \textit{for with replacement sampling,}$$

where the d_{1h} and d_{2h} are constants, not dependent upon the n_h, that represent the contributions of the second and third order moments of the $\bar{y}_{rh}$.

PROOF. See Jones (1974) and Dippo (1981).   □

This theorem shows that the jackknife estimator $\hat{\theta}^1$ is approximately unbiased for θ. In fact, it is strictly unbiased whenever θ is a linear or quadratic function of the stratum means. This remark applies to estimators such as

$$\hat{\theta} = \sum_{h=1}^{L} N_h \bar{y}_{1h}$$

the estimator of total;

$$\hat{\theta} = \bar{y}_{11} - \bar{y}_{12}$$

the estimated difference between stratum means; and

$$\hat{\theta} = \left(\sum_{h=1}^{L} (N_h/N)\bar{y}_{1h} \right) \left(\sum_{h=1}^{L} (N_h/N)\bar{y}_{2h} \right) \bigg/ \sum_{h=1}^{L} (N_h/N)\bar{Y}_{2h}.$$

the combined product estimator.

As was the case for sampling without stratification (i.e., $L = 1$), the jackknife may be considered for its bias reduction properties. The estimator $\hat{\theta}^1$, called by Jones the first-order jackknife estimator, eliminates the order n_h^{-1} and order N_h^{-1} terms from the bias of $\hat{\theta}$ as an estimator of θ. This is the import of the first part of Theorem 4.5.1. Jones also gives a second order jackknife estimator, say $\hat{\theta}^2$, which is unbiased for θ through third order moments of the $\bar{y}_{rh}$:

$$\hat{\theta}^2 = \left(1 + \sum_{h=1}^{L} w_{(h)} - \sum_{h=1}^{L} w_{(hh)} \right) \hat{\theta} - \sum_{h=1}^{L} w_{(h)} \hat{\theta}_{(h\cdot)} + \sum_{h=1}^{L} w_{(hh)} \hat{\theta}_{(h\cdot)(h\cdot)}$$

$$w_{(h)} = a_h a_{(hh)} / \{ (a_{(h)} - a_h)(a_{(hh)} - a_{(h)}) \}$$

$$w_{(hh)} = a_h a_{(h)} / \{ (a_{(hh)} - a_h)(a_{(hh)} - a_{(h)}) \}$$

$$\hat{\theta}_{(h\cdot)} = \sum_{i=1}^{n_h} \hat{\theta}_{(hi)} / n_h$$

$$\hat{\theta}_{(h\cdot)(h\cdot)} = 2 \sum_{i<j}^{n_h} \hat{\theta}_{(hi)(hj)} / \{ n_h(n_h - 1) \}$$

$$\begin{aligned} a_h &= n_h^{-1} - N_h^{-1}, & &\text{for without replacement sampling} \\ &= n_h^{-1}, & &\text{for with replacement sampling} \\ a_{(h)} &= (n_h - 1)^{-1} - N_h^{-1}, & &\text{for without replacement sampling} \\ &= (n_h - 1)^{-1}, & &\text{for with replacement sampling} \\ a_{(hh)} &= (n_h - 2)^{-1} - N_h^{-1}, & &\text{for without replacement sampling} \\ &= (n_h - 2)^{-1}, & &\text{for with replacement sampling} \end{aligned}$$

where $\hat{\theta}_{(hi)(hj)}$ is the estimator of the same functional form as $\hat{\theta}$ based upon the sample after removing both the (h, i)-th and the (h, j)-th observations. The second order jackknife is strictly unbiased for estimators $\hat{\theta}$ that are cubic functions of the stratum means $\bar{y}_{rh}$. For linear functions, we have

$$\hat{\theta} = \hat{\theta}^1 = \hat{\theta}^2.$$

The jackknife estimator of variance for the stratified sampling problem is defined by

$$v_1(\hat{\theta}) = \sum_{h=1}^{L} \frac{w_h}{n_h} \sum_{i=1}^{n_h} (\hat{\theta}_{(hi)} - \hat{\theta}_{(h\cdot)})^2. \tag{4.5.3}$$

This estimator and the following theorem are also due to Jones.

Theorem 4.5.2. *Let the conditions of Theorem 4.5.1 hold. Then, to second order moments of $\bar{y}_{rh}$, $v_1(\hat{\theta})$ is an unbiased estimator of both $\mathrm{Var}\{\hat{\theta}\}$ and $\mathrm{Var}\{\hat{\theta}^1\}$. To third order moments, the expectation is*

$$E\{v_1(\hat{\theta})\} = \sum_{h=1}^{L}\left[\frac{N_h - n_h}{(N_h - 1)n_h}\right]d_{1h} + \sum_{h=1}^{L}\left[\frac{(N_h - n_h)(N_h - 2n_h + 2)}{(N_h - 1)(N_h - 2)n_h(n_h - 1)}\right]d_{2h},$$

for without replacement sampling

$$= \sum_{h=1}^{L} n_h^{-1}d_{1h} + \sum_{h=1}^{L} n_h^{-1}(n_h - 1)^{-1}d_{2h},$$

for without replacement sampling,

where d_{1h} and d_{2h} are defined in Theorem 4.5.1.

PROOF. See Jones (1974) and Dippo (1981). □

Thus, $v_1(\hat{\theta})$ is unbiased to second order moments of the $\bar{y}_{rh}$ as an estimator of both $\mathrm{Var}\{\hat{\theta}\}$ and $\mathrm{Var}\{\theta^1\}$. When third order moments are included, however, $v_1(\hat{\theta})$ is seen to be a biased estimator of variance. Jones gives further modifications to the variance estimator that correct for even these "lower-order" biases.

In addition to Jones' work, McCarthy (1966) and Lee (1973b) have studied the jackknife for the case $n_h = 2(h = 1, \ldots, L)$. McCarthy's jackknife estimator of variance is

$$v_M(\hat{\theta}) = \sum_{h=1}^{L} (1/2) \sum_{i=1}^{2}\left(\hat{\theta}_{(hi)} - \sum_{h'=1}^{L}\hat{\theta}_{(h'i)}/L\right)^2$$

and Lee's estimator is

$$v_L(\hat{\theta}) = \sum_{h=1}^{L} (1/2) \sum_{h=1}^{2}(\hat{\theta}_{(hi)} - \hat{\theta})^2.$$

For general sample size n_h, the natural extensions of these estimators are

$$v_2(\hat{\theta}) = \sum_{h=1}^{L} (w_h/n_h) \sum_{i=1}^{n_h}(\hat{\theta}_{(hi)} - \hat{\theta}_{(\cdot\cdot)})^2, \tag{4.5.4}$$

$$v_3(\hat{\theta}) = \sum_{h=1}^{L} (w_h/n_h) \sum_{i=1}^{n_h}(\hat{\theta}_{(hi)} - \bar{\theta}_{(\cdot\cdot)})^2, \tag{4.5.5}$$

and

$$v_4(\hat{\theta}) = \sum_{h=1}^{L} (w_h/n_h) \sum_{i=1}^{n_h}(\hat{\theta}_{(hi)} - \hat{\theta})^2. \tag{4.5.6}$$

In the following theorem, we show that these are unbiased estimators of variance to a first order approximation.

Theorem 4.5.3. *Given the conditions of this section, the expectations of the jackknife variance estimators, to second order moments of the $\bar{y}_{rh}$, are*

$$E\{v_2(\hat{\theta})\} = \sum_{h=1}^{L} \left[\frac{N_h - n_h}{(N_h - 1)n_h} \right] d_{1h}, \qquad \textit{for without replacement sampling}$$

$$= \sum_{h=1}^{L} n_h^{-1} d_{1h}, \qquad \textit{for with replacement sampling}$$

and

$$E\{v_3(\hat{\theta})\} = E\{v_4(\hat{\theta})\} = E\{v_2(\hat{\theta})\},$$

where the d_{1h} are as in Theorem 4.5.1.

PROOF. Left to the reader.   □

Theorems 4.5.1, 4.5.2, and 4.5.3 show that to second order moments v_1, v_2, v_3, and v_4 are unbiased estimators of the variance of both $\hat{\theta}$ and $\hat{\theta}^1$.

Some important relationships exist between the estimators both for with replacement sampling, and for without replacement sampling when the sampling fractions are negligible, $w_h \doteq n_h - 1$. Given either of these conditions, Jones' estimator is

$$v_1(\hat{\theta}) = \sum_{h=1}^{L} [(n_h - 1)/n_h] \sum_{i=1}^{n_h} (\hat{\theta}_{(hi)} - \hat{\theta}_{(h\cdot)})^2,$$

and we may partition the sum of squares in Lee's estimator as

$$
\begin{aligned}
v_4(\hat{\theta}) = {}& \sum_{h=1}^{L} [(n_h - 1)/n_h] \sum_{i=1}^{n_h} (\hat{\theta}_{(hi)} - \hat{\theta})^2 \\
= {}& \sum_{h=1}^{L} [(n_h - 1)/n_h] \sum_{i=1}^{n_h} (\hat{\theta}_{(hi)} - \hat{\theta}_{(h\cdot)})^2 \\
& + \sum_{h=1}^{L} (n_h - 1)(\hat{\theta}_{(h\cdot)} - \hat{\theta}_{(\cdot\cdot)})^2 + (n + L)(\hat{\theta}_{(\cdot\cdot)} - \bar{\theta}_{(\cdot\cdot)})^2 \\
& + (n - L)(\bar{\theta}_{(\cdot\cdot)} - \hat{\theta})^2 + 2n(\bar{\theta}_{(\cdot\cdot)} - \hat{\theta})(\hat{\theta}_{(\cdot\cdot)} - \bar{\theta}_{(\cdot\cdot)}).
\end{aligned}
$$

(4.5.7)

The first term on the right side of (4.5.7) is $v_1(\hat{\theta})$. The estimator $v_2(\hat{\theta})$ is equal to the first two terms on the right side, and $v_3(\hat{\theta})$ is equal to the first three terms. When the n_h are equal ($h = 1, \ldots, L$) the fifth term is zero. Thus, we make the following observations:

(i) $v_4(\hat{\theta}) \geq v_1(\hat{\theta})$,
(ii) $v_3(\hat{\theta}) \geq v_2(\hat{\theta}) \geq v_1(\hat{\theta})$,
(iii) $v_4(\hat{\theta}) \geq v_3(\hat{\theta}) = v_2(\hat{\theta}) \geq v_1(\hat{\theta})$, whenever the n_h are roughly equal.

These results hold algebraically irrespective of the particular sample selected. They are important in view of the result (cf. Theorem 4.5.3) that the four estimators have the same expectation to second order moments of the $\bar{y}_{rh}$. We may say that $v_2(\hat{\theta})$ and $v_3(\hat{\theta})$ are conservative estimators of variance relative to $v_1(\hat{\theta})$, and that $v_4(\hat{\theta})$ is very conservative. Although in large complex sample surveys, there may be little difference between the four estimators.

EXAMPLE. To illustrate the application of the above methods, we consider the combined ratio estimator $\hat{\theta} = \hat{R}$. The estimator obtained by deleting the (h, i)-th observation is

$$\hat{\theta}_{(hi)} = \hat{R}_{(hi)} = \frac{\displaystyle\sum_{h' \neq h}^{L} N_{h'}\bar{y}_{1h'} + N_h \sum_{j \neq i}^{n_h} y_{1hj}/(n_h - 1)}{\displaystyle\sum_{h' \neq h}^{L} N_{h'}\bar{y}_{2h'} + N_h \sum_{j \neq i}^{n_h} y_{2hj}/(n_h - 1)}.$$

The jackknife estimator of $\theta = R$ is

$$\hat{\theta}^1 = \hat{R}^1 = \left(1 + \sum_{h=1}^{L} w_h\right)\hat{R} - \sum_{h=1}^{L} w_h \hat{R}_{(h\cdot)},$$

where $\hat{R}_{(h\cdot)} = \sum_{i=1}^{n_h} \hat{R}_{(hi)}/n_h$. The corresponding variance estimators are

$$v_1(\hat{\theta}) = v_1(\hat{R}) = \sum_{h=1}^{L} \frac{w_h}{n_h} \sum_{i=1}^{n_h} (\hat{R}_{(hi)} - \hat{R}_{(h\cdot)})^2,$$

$$v_2(\hat{\theta}) = v_2(\hat{R}) = \sum_{h=1}^{L} \frac{w_h}{n_h} \sum_{i=1}^{n_h} (\hat{R}_{(hi)} - \hat{R}_{(\cdot\cdot)})^2,$$

$$v_3(\hat{\theta}) = v_3(\hat{R}) = \sum_{h=1}^{L} \frac{w_h}{n_h} \sum_{i=1}^{n_h} (\hat{R}_{(hi)} - \bar{R}_{(\cdot\cdot)})^2,$$

$$v_4(\hat{\theta}) = v_4(\hat{R}) = \sum_{h=1}^{L} \frac{w_n}{n_h} \sum_{i=1}^{n_h} (\hat{R}_{(hi)} - \hat{R})^2,$$

and are applicable to either $\hat{R}$ or $\hat{R}^1$. For $n_h = 2$ and $(1 - n_h/N_h) \doteq 1$, the estimators reduce to

$$v_1(\hat{R}) = \sum_{h=1}^{L} (1/2) \sum_{i=1}^{2} (\hat{R}_{(hi)} - \hat{R}_{(h\cdot)})^2 = \sum_{h=1}^{L} (\hat{R}_{(h1)} - \hat{R}_{(h2)})^2/4,$$

$$v_2(\hat{R}) = \sum_{h=1}^{L} (1/2) \sum_{i=1}^{2} (\hat{R}_{(hi)} - \hat{R}_{(\cdot\cdot)})^2,$$

$$v_3(\hat{R}) = \sum_{h=1}^{L} (1/2) \sum_{i=1}^{2} (\hat{R}_{(hi)} - \bar{R}_{(\cdot\cdot)})^2,$$

$$v_4(\hat{R}) = \sum_{h=1}^{L} (1/2) \sum_{i=1}^{2} (\hat{R}_{(hi)} - \hat{R})^2. \qquad \square$$

All of the results stated thus far have been for the case where the jackknife operates on estimators $\hat{\theta}_{(hi)}$ obtained by eliminating *single* observations from the full sample. Valid results may also be obtained if we divide the sample n_h into k random groups of size m_h and define $\hat{\theta}_{(hi)}$ to be the estimator obtained after deleting the m_h observations in the i-th *random group* from stratum h.

The results stated thus far have also been for the case of simple random sampling, either with or without replacement. Now suppose sampling is carried out pps with replacement within the L strata. The natural estimator

$$\hat{\theta} = g(\bar{x}_1, \ldots, \bar{x}_L) \tag{4.5.8}$$

is now defined in terms of

$$\bar{x}_{rh} = (1/n_h) \sum_{i=1}^{n_h} x_{rhi},$$

where

$$x_{rhi} = y_{rhi}/N_h p_{hi}$$

and p_{hi} denotes the probability associated with the (h, i)-th unit. As usual, $p_{hi} > 0$ for all h and i, and $\sum_i p_{hi} = 1$, for all h. Similarly, $\hat{\theta}_{(hi)}$ denotes the estimator of the same form as (4.5.8) obtained after deleting the (h, i)-th observation from the sample,

$$\hat{\theta}_{(h\cdot)} = \sum_{i=1}^{n_h} \hat{\theta}_{(hi)}/n_h, \ \hat{\theta}_{(\cdot\cdot)} = \sum_{h=1}^{L} \sum_{i=1}^{n_h} \hat{\theta}_{(hi)}/n,$$

and

$$\bar{\theta}_{(\cdot\cdot)} = \sum_{h=1}^{L} \hat{\theta}_{(h\cdot)}/L.$$

The first order jackknife estimator of θ is

$$\hat{\theta}^1 = \left(1 + \sum_{h=1}^{L} w_h\right)\hat{\theta} + \sum_{h=1}^{L} w_h\hat{\theta}_{(h\cdot)},$$

where $w_h = (n_h - 1)$. To estimate the variance of either $\hat{\theta}$ or $\hat{\theta}^1$, we may use

$$v_1(\hat{\theta}) = \sum_{h=1}^{L} \{(n_h - 1)/n_h\} \sum_{i=1}^{n_h} (\hat{\theta}_{(hi)} - \hat{\theta}_{(h\cdot)})^2,$$

$$v_2(\hat{\theta}) = \sum_{h=1}^{L} \{(n_h - 1)/n_h\} \sum_{i=1}^{n_h} (\hat{\theta}_{(hi)} - \hat{\theta}_{(\cdot\cdot)})^2,$$

$$v_3(\theta) = \sum_{h=1}^{L} \{(n_h - 1)/n_h\} \sum_{i=1}^{n_h} (\hat{\theta}_{(hi)} - \bar{\theta}_{(\cdot\cdot)})^2,$$

or

$$v_4(\hat{\theta}) = \sum_{h=1}^{L} \{(n_h - 1)n_h\} \sum_{i=1}^{n_h} (\hat{\theta}_{(hi)} - \bar{\theta})^2.$$

The reader should note that the earlier results for simple random sampling extend to the present problem.

If sampling is with unequal probability without replacement with inclusion probabilities

$$\pi_{hi} = \mathscr{P}\{(h, i)\text{-th unit in sample}\} = n_h p_{hi},$$

i.e., a πps sampling scheme, then we may use the jackknife methods just given for pps with replacement sampling. This is a conservative procedure in the sense that the resulting variance estimators will tend to overestimate the true variance. See Section 2.4.5.

EXAMPLE. If sampling is pps with replacement and $\hat{\theta} = \hat{R}$ is the combined ratio estimator, then the jackknife operates on

$$\hat{\theta}_{(hi)} = \hat{R}_{(hi)} = \frac{\displaystyle\sum_{h' \neq h}^{L} N_{h'} \bar{x}_{1h'} + N_h \sum_{j \neq i}^{n_h} x_{1hj}/(n_h - 1)}{\displaystyle\sum_{h' \neq h}^{L} N_{h'} \bar{x}_{2h'} + N_h \sum_{j \neq i}^{n_h} x_{2hj}/(n_h - 1)},$$

where $x_{rhj} = y_{rhj}/N_h p_{hj}$. □

Finally, there is another variant of the jackknife which is appropriate for stratified samples. This variant is analogous to the *second option* of rule (iv), Section 2.4, whereas previous variants of the jackknife have been analogous to the *first option* of that rule. Let the sample n_h (which may be srs, pps, or πps) be divided into k random groups of size m_h, for $h = 1, \ldots, L$, and let $\hat{\theta}_{(\alpha)}$ denote the estimator of θ obtained after removing the α-th group of observations from *each* stratum. Define the pseudovalues

$$\hat{\theta}_\alpha = k\hat{\theta} - (k - 1)\hat{\theta}_{(\alpha)}.$$

Then, the estimator of θ and the estimator of its variance are

$$\hat{\bar{\theta}} = \sum_{\alpha=1}^{k} \hat{\theta}_\alpha/k$$

$$v_1(\hat{\bar{\theta}}) = \frac{1}{k(k - 1)} \sum_{\alpha=1}^{k} (\hat{\theta}_\alpha - \hat{\bar{\theta}})^2. \tag{4.5.9}$$

Little is known about the relative merits of this method vis-à-vis earlier methods. It does seem that the estimators in (4.5.9) have computational advantages over earlier methods. Unless k is large, however, (4.5.9) may be subject to greater instability.

4.6. Application to Cluster Sampling

Throughout this chapter we have treated the case where the elementary units and sampling units are identical. We now assume that clusters of elementary units, comprising primary sampling units (PSUs), are selected, with possibly several stages of subsampling occurring within each selected PSU. We continue to let $\hat{\theta}$ be the parent sample estimator of an arbitrary parameter θ. Now, however, N and n denote the number of PSUs in the population and sample, respectively.

No new principals are involved in the application of jackknife methodology to clustered samples. When forming k random groups of m units each ($n = mk$), we simply work with the *ultimate clusters* rather than the elementary units. The reader will recall from Chapter 2 that the term ultimate cluster refers to the aggregate of elementary units, second stage units, third stage units, and so on from the same primary unit. See rule (iii), Section 2.4. The estimator $\hat{\theta}_{(\alpha)}$ is then computed from the parent sample after eliminating the α-th random group of ultimate clusters ($\alpha = 1, \ldots, k$). Pseudovalues $\hat{\theta}_{\alpha}$, Quenouille's estimator $\hat{\hat{\theta}}$, and the jackknife variance estimators $v_1(\hat{\hat{\theta}})$ and $v_2(\hat{\hat{\theta}})$ are defined in the usual way.

As an illustration, consider the estimator

$$\hat{\theta} = \hat{Y} = \frac{1}{n} \sum_{i=1}^{n} \hat{y}_i / p_i \qquad (4.6.1)$$

of the population total $\theta = Y$ based on a pps wr sample of n primaries, where $\hat{y}_i$ denotes an estimator of the total in the i-th selected primary based on sampling at the second and successive stages. For this problem, the $\hat{\theta}_{(\alpha)}$ are defined by

$$\hat{\theta}_{(\alpha)} = \hat{Y}_{(\alpha)} = \frac{1}{m(k-1)} \sum_{i=1}^{m(k-1)} \hat{y}_i / p_i, \qquad (4.6.2)$$

where the summation is taken over all selected PSUs not in the α-th group. Quenouille's estimator is

$$\hat{\hat{\theta}} = \hat{\hat{Y}} = \frac{1}{k} \sum_{\alpha=1}^{k} \hat{Y}_{(\alpha)},$$

$$\hat{\theta}_{\alpha} = \hat{Y}_{\alpha} = k\hat{Y} - (k-1)\hat{Y}_{(\alpha)},$$

and

$$v_1(\hat{\hat{\theta}}) = \frac{1}{k(k-1)} \sum_{\alpha=1}^{k} (\hat{Y}_{\alpha} - \hat{\hat{Y}})^2$$

is an unbiased estimator of the variance of $\hat{\hat{\theta}}$.

If the sample of PSUs is actually selected without replacement with inclusion probabilities $\pi_i = np_i$, then the estimators $\hat{\theta}$ and $\hat{\theta}_{(\alpha)}$ take the same form as indicated in (4.6.1) and (4.6.2) for with replacement sampling. In this case, the estimator $v_1(\hat{\hat{\theta}})$ will tend to overestimate the variance of $\hat{\theta}$. See Section 2.4.5 for some discussion of this point. Also, using a proof similar to Theorem 2.4.5 it can be shown that the bias in $v_1(\hat{\hat{\theta}})$ arises only in the between PSU component of variance.

For the example $\hat{\theta} = \hat{Y}$, regardless of whether the sample primaries are drawn with or without replacement, the reader will note that $\hat{\hat{\theta}} = \hat{\theta}$ and $v_1(\hat{\hat{\theta}}) = v_2(\hat{\theta})$, because the parent sample estimator $\hat{\theta}$ is linear in the $\hat{y}_i$. For nonlinear $\hat{\theta}$, the estimators $\hat{\hat{\theta}}$ and $\hat{\theta}$ are generally not equal, nor are the estimators of variance $v_1(\hat{\hat{\theta}})$ and $v_2(\hat{\theta})$. For the nonlinear case exact results about the moment properties of the estimators are not generally available. Approximations are possible using the theory of Taylor series expansions. See Chapter 6. Also see Appendix B for some discussion of the asymptotic properties of the estimators.

If the sample of primaries is selected independently within each of $L \geq 2$ strata, then we use the rule of ultimate clusters together with the techniques discussed in Section 4.5. The methods are now based upon the estimators $\hat{\theta}_{(hi)}$ formed by deleting the i-th ultimate cluster from the h-th stratum. Continuing the example $\hat{\theta} = \hat{Y}$, we have

$$\hat{\theta}^1 = \left(1 + \sum_{h=1}^{L} w_h\right)\hat{\theta} - \sum_{h=1}^{L} w_h \hat{\theta}_{(h\cdot)}$$

$$= \left(1 + \sum_{h=1}^{L} w_h\right)\hat{Y} - \sum_{h=1}^{L'} w_h \hat{Y}_{(h\cdot)},$$

the first order jackknife estimator of θ, and

$$v(\hat{\theta}^1) = \sum_{h=1}^{L} \frac{n_h - 1}{n_h} \sum_{i=1}^{n_h} (\hat{\theta}_{(hi)} - \hat{\theta}_{(h\cdot)})^2$$

$$= \sum_{h=1}^{L} \frac{n_h - 1}{n_h} \sum_{i=1}^{n_h} (\hat{Y}_{(hi)} - \hat{Y}_{(h\cdot)})^2$$

the first order jackknife estimator of the variance, where

$$\hat{\theta}_{(h\cdot)} = \hat{Y}_{(h\cdot)} = \frac{1}{n_h} \sum_{j=1}^{n_h} \hat{Y}_{(hj)}$$

is the mean of the $\hat{\theta}_{(hj)} = \hat{Y}_{(hj)}$ over all selected primaries in the h-th stratum.

The discussion in this section has dealt solely with the problems of estimating θ and of estimating the total variance of the estimator. In survey planning, however, we often face the problem of estimating the individual components of variance due to the various stages of selection. This is important in order to achieve an efficient allocation of the sample. See Folsom, Bayless, and Shah (1971) for a jackknife methodology for variance component problems.

4.7. Example: The Methods Test Panel

We illustrate the use of the jackknife for variance estimation using data from the Census Bureau's Methods Test Panel (MTP). The MTP was a survey research experiment designed to investigate the impact on survey data of alternative data collection methodologies. It was intended to provide insight into nonsampling errors, with the ultimate goal of improving the quality, reliability, and utility of survey data. The MTP was itself a monthly household survey that operated between May 1978 and November 1979. Data presented here are for the months June and July 1978. The questionnaire used for the MTP was essentially the one used in the U.S. Current Population Survey (CPS), and contained mainly labor force inquiries. The MTP data, however, were not used in preparing the monthly CPS estimates. The CPS itself is discussed further in Section 5.5.

The MTP sample was selected in four stages. The primary sampling units (PSUs) were the Los Angeles–Long Beach, California Standard Metropolitan Statistical Area (SMSA); the Chicago, Illinois SMSA; Lackawanna County, Pennsylvania; and Macon, Dooly, and Houston Counties, Georgia. The four PSUs were selected purposively to display different types of unemployment problems, a mix of urban and rural characteristics, a representation of Blacks and Hispanics, and a wide geographic distribution. Another consideration was the availability of field staff.

Within each PSU an unequal probability systematic sample of 32 1970 Census Enumeration Districts (EDs), or block groups, was chosen with probability proportional to 1970 housing counts. All EDs, or block groups, within a given PSU were sorted geographically prior to the systematic selection. In the Los Angeles and Chicago SMSAs two strata, central city and balance of SMSA, were used and the sample allocated equally to each. In the less urbanized areas, EDs were used, whereas block groups were selected in the more urbanized areas. The EDs, or block groups, are the second stage sampling units (SSUs) and each consisted of approximately 250 housing units.

The selected SSUs were then grouped into eight blocks of four using the order in which they were selected. See Table 4.7.1. Within each block, the four SSUs (including all housing units to be selected in the SSUs) were randomly assigned to one of four rotation groups. The rotation groups play the role of random groups in the analysis. Notice that rule (iii), Section 2.4 was employed in the definition of the rotation (random) groups. We employ the terminology "rotation" group here, because these groups will rotate over time in and out of the parent sample.

The selected SSUs were canvassed before the initial enumeration in May 1978, listing every housing unit in the area. Then, every fourth month during the span of the MTP, one cluster of approximately 20 housing units was selected at random from the SSU. The listings for these clusters or third stage units (TSUs) were updated the month before the clusters were to

Table 4.7.1. Assignment of
SSUs to Blocks Within a PSU

ED Number	Block Number
1	1
2	1
3	1
4	1
5	2
6	2
7	2
8	2
.	.
.	.
.	.
.	.
29	8
30	8
31	8
32	8

come into sample to identify units which no longer existed (e.g., demolished).

In the final stage of sampling, 12 housing units were subsampled from among the currently occupied units or those available for occupancy in each cluster, and it was these units that were enumerated. The housing units in a given rotation group remained in sample for four consecutive months, and at that time a new sample of 12 units rotated in to take its place. In any given month, one of the rotation groups was contacted for the first time, one for the second time, one for the third time, and one for the fourth and final time.

Table 4.7.2 describes a monthly MTP sample for one PSU. Since there were four PSUs, a total of $4 \times 384 = 1536$ housing units were contacted each month. One-fourth of the housing units (or 384, or 96 from each of four PSUs) were members of each of the four rotation groups.

Three experimental data collection methodologies (or treatments) were investigated in the MTP:

  (i) interviewer-assignment
 (ii) mode-of-interview
(iii) type-of-respondent.

These were at two, two, and three levels, respectively, giving a total of 12 treatment combinations. The 12 combinations were assigned at random to the 12 housing units selected within each third stage unit. The interviewer-

Table 4.7.2. Number of Selected Housing Units for One PSU for a Given Month

Block	R1	R2	R3	R4	No. of Units
	\multicolumn{4}{Rotation Group}			
Block 1	12	12	12	12	48
Block 2	12	12	12	12	48
Block 3	12	12	12	12	48
Block 4	12	12	12	12	48
Block 5	12	12	12	12	48
Block 6	12	12	12	12	48
Block 7	12	12	12	12	48
Block 8	12	12	12	12	48
No. of Units	96	96	96	96	384

assignment treatment was intended to investigate the possibility of interviewer conditioning. The first level of this treatment specified that the same interviewer be used in each month that a housing unit was to be contacted, while the second level specified a different interviewer each month. The levels of the mode-of-interview treatment were telephone interview and personal interview. The type-of-respondent treatment was intended to investigate respondent conditioning and the accuracy of reporting by proxy respondents. The first level of this treatment, called the household respondent rule, used any one responsible adult (usually the person who answers the door or telephone) to respond for all persons in the housing unit. The second level of the treatment, called the designated respondent rule, used one randomly selected adult to respond for all persons in the unit, and used a different random selection each month. The third level of this treatment was the self-respondent rule.

The reader will recognize the MTP as a split-plot experiment imbedded in a sample survey. The whole-plots are the SSUs (i.e., ED or block group) and the whole-plot treatment is time-in-sample at four levels. The split-plots are housing units, and the split-plot treatments are the combinations of a complete $2 \times 2 \times 3$ factorial experiment.

To illustrate the jackknife method of variance estimation, we consider the problem of estimating the unemployment rate. The universe (or population) for which estimates will be prepared is the union of the four PSUs.

Let $y_{ti\alpha j}$ denote the unemployment indicator (i.e., 1 if unemployed and 0 otherwise) for the j-th individual in the α-th rotation group, i-th treatment combination, and t-th month. Let $x_{ti\alpha j}$ denote the corresponding individuals' indicator of civilian labor force status (i.e., 1 if in the civilian labor force and 0 otherwise). Also let $w_{ti\alpha j}$ denote the reciprocal of the probability of inclusion in month t, treatment combination i, and rotation group α. The estimators of total unemployed and total civilian labor force from the (t, i, α) sample are then given by

$$\hat{Y}_{ti\alpha} = \sum_j w_{ti\alpha j} \, y_{ti\alpha j}$$

and

$$\hat{X}_{ti\alpha} = \sum_j w_{ti\alpha j} \, x_{ti\alpha j},$$

respectively. The estimators for month t and treatment combination i from the combined sample over all eight rotation groups are

$$\hat{Y}_{ti} = \frac{1}{8} \sum_\alpha \hat{Y}_{ti\alpha}$$

and

$$\hat{X}_{ti} = \frac{1}{8} \sum_\alpha \hat{X}_{ti\alpha}.$$

The reader will note that $8 \, w_{ti\alpha j}^{-1}$ is the inclusion probability associated with the parent sample.

Tables 4.7.3 and 4.7.4 present the estimates $\hat{Y}_{ti\alpha}$, $\hat{X}_{ti\alpha}$, $\hat{Y}_{ti}$, and $\hat{X}_{ti}$ for June and July 1978. Data are presented for two treatment combinations i. They are

$i = 1$,   same interviewer, personal interview, all levels of type of respondent

$i = 2$,   different interviewer, personal interview, all levels of type of respondent.

From these data we may compute estimates of the unemployment rate. The parent sample estimator is

$$\hat{R}_{ti} = \hat{Y}_{ti} / \hat{X}_{ti},$$

whereas the estimator obtained by deleting the α-th rotation group is

$$\hat{R}_{ti(\alpha)} = \left(\frac{1}{7} \sum_{\alpha' \neq \alpha} \hat{Y}_{ti\alpha'} \right) \Big/ \left(\frac{1}{7} \sum_{\alpha' \neq \alpha} \hat{X}_{ti\alpha'} \right).$$

The α-th pseudovalue is

$$\hat{R}_{ti\alpha} = 8 \hat{R}_{ti} - 7 \hat{R}_{ti(\alpha)},$$

giving Quenouille's estimator

$$\hat{\bar{R}}_{ti} = \frac{1}{8} \sum_\alpha \hat{R}_{ti\alpha}$$

and the jackknife estimator of variance

$$v_1(\hat{\bar{R}}_{ti}) = \frac{1}{8(7)} \sum_\alpha (\hat{R}_{ti\alpha} - \hat{\bar{R}}_{ti})^2.$$

Table 4.7.3. Methods Test Panel Data for June 1978

Rotation Group	Same Interviewer-Personal Interview		Different Interviewer-Personal Interview	
	Unemployed	Civilian Labor Force	Unemployed	Civilian Labor Force
1	653,309.0	10,557,167.9	337,922.0	13,759,507.8
2	1,335,573.6	11,803,414.0	642,213.3	14,405,360.8
3	2,098,283.4	14,102,458.4	1,950,354.4	13,137,112.4
4	482,274.9	15,693,673.4	583,764.2	11,854,271.8
5	3,807,077.2	21,945,593.2	1,143,009.6	15,409,860.5
6	1,082,333.3	14,845,887.2	1,895,684.9	1,631,197.8
7	3,232,428.9	16,629,849.8	1,775,146.9	16,691,216.8
8	2,161,324.4	15,731,836.1	2,076,453.3	21,450,202.0
Parent	1,856,575.6	15,163,735.0	1,300,568.6	15,377,441.3

Table 4.7.4. Methods Test Panel Data for July 1978

Rotation Group	Same Interviewer-Personal Interview		Different Interviewer-Personal Interview	
	Unemployed	Civilian Labor Force	Unemployed	Civilian Labor Force
1	1,364,587.5	11,422,869.2	664,597.9	14,984,653.4
2	893,859.8	12,116,054.1	658,780.0	12,351,011.5
3	2,585,787.0	15,883,591.8	502,733.3	12,501,153.4
4	1,853,348.0	17,469,637.2	774,157.0	12,463,610.8
5	1,510,153.5	18,084,554.8	22,713.2	14,398,900.0
6	737,868.7	16,555,481.4	1,234,190.4	13,979,647.0
7	3,868,439.3	17,896,783.2	1,206,327.3	12,642,314.9
8	1,387,958.9	14,352,667.5	1,732,732.1	20,436,900.0
Parent	1,771,250.3	15,472,704.9	849,528.9	14,219,773.9

Table 4.7.5 presents the pseudovalues. Table 4.7.6 illustrates the computation of a pseudovalue and the computation of $\hat{\bar{R}}_{ti}$ and $v_1(\hat{\bar{R}}_{ti})$.

In the MTP it was also desired to estimate the mean unemployment rate across months and the difference in unemployment rate between treatment combinations. The June–July mean unemployment rate for treatment combination i may be estimated by

$$\hat{R}_{\cdot i} = (\hat{R}_{1i} + \hat{R}_{2i})/2.$$

Table 4.7.5. Pseudovalues for U.S. Unemployment Rate

Rotation Group	June 1978		July 1978	
	Same Interviewer– Personal Visit	Different Interviewer– Personal Visit	Same Interviewer– Personal Visit	Different Interviewer– Personal Visit
1	0.082	0.032	0.118	0.043
2	0.115	0.047	0.084	0.054
3	0.147	0.138	0.164	0.043
4	0.027	0.058	0.105	0.062
5	0.201	0.074	0.077	0.001
6	0.074	0.118	0.039	0.088
7	0.202	0.109	0.233	0.091
8	0.138	0.103	0.098	0.098

The estimator obtained by deleting the α-th rotation group is

$$\hat{R}_{\cdot i(\alpha)} = (\hat{R}_{1i(\alpha)} + \hat{R}_{2i(\alpha)})/2,$$

and the pseudovalue is

$$\hat{R}_{\cdot i\alpha} = 8\hat{R}_{\cdot i} - 7\hat{R}_{\cdot i(\alpha)}.$$

The reader will note that the pseudovalue $\hat{R}_{\cdot i\alpha}$ may also be obtained as the mean of the pseudovalues $\hat{R}_{1i\alpha}$ and $\hat{R}_{2i\alpha}$. Quenouille's estimator and the jackknife estimator of variance are

$$\hat{\bar{R}}_{\cdot i} = \frac{1}{8} \sum_{\alpha} \hat{R}_{\cdot i\alpha}$$

and

$$v_1(\hat{\bar{R}}_{\cdot i}) = \frac{1}{8(7)} \sum_{\alpha} (\hat{R}_{\cdot i\alpha} - \hat{\bar{R}}_{\cdot i})^2,$$

respectively. Table 4.7.7 presents the pseudovalues for the two treatment combinations discussed before, and Table 4.7.8 presents some illustrative computations.

An estimator of the difference in mean unemployment rate between treatment combinations is

$$\Delta\hat{R}_{\cdot} = \hat{R}_{\cdot 1} - \hat{R}_{\cdot 2}.$$

After deleting the α-th rotation group, the estimator is

$$\Delta\hat{R}_{\cdot(\alpha)} = \hat{R}_{\cdot 1(\alpha)} - \hat{R}_{\cdot 2(\alpha)}.$$

Table 4.7.6. Computation of Unemployment Rate and Jackknife
Estimate of Variance

To illustrate the computation of a pseudovalue, we consider June 1978 data
($t = 1$), rotation group 1 ($\alpha = 1$), and the same interviewer–personal interview
treatment combination ($i = 1$). From data in Table 4.7.3 we have

$$\hat{R}_{11} = \hat{Y}_{11}/\hat{X}_{11}$$

$$= 1{,}856{,}575.6/15{,}163{,}735.0$$

$$= 0.1224$$

and

$$\hat{R}_{11(1)} = \left(\frac{1}{7}\sum_{\alpha \neq 1}\hat{Y}_{11\alpha}\right)\Big/\left(\frac{1}{7}\sum_{\alpha \neq 1}\hat{X}_{11\alpha}\right)$$

$$= 2{,}028{,}470.7/15{,}821{,}816.0$$

$$= 0.1282.$$

The corresponding pseudovalue is

$$\hat{R}_{111} = 8(0.1224) - 7(0.1282)$$

$$= 0.082.$$

The pseudovalues, including this example, appear in Table 4.7.5.
   Using these data, we compute

$$\hat{R}_{11} = \frac{1}{8}\sum_{\alpha}\hat{R}_{11\alpha} = 0.123$$

and

$$v_1(\hat{\bar{R}}_{11}) = \frac{1}{8(7)}\sum_{\alpha}(\hat{R}_{11\alpha} - \hat{\bar{R}}_{11})^2$$

$$= 0.000474.$$

We also compute

$$\hat{\bar{R}}_{12} = 0.085$$

$$v_1(\hat{\bar{R}}_{12}) = 0.000177$$

$$\hat{\bar{R}}_{21} = 0.115$$

$$v_1(\hat{\bar{R}}_{21}) = 0.000444$$

$$\hat{\bar{R}}_{22} = 0.060$$

$$v_1(\hat{\bar{R}}_{22}) = 0.000130.$$

Table 4.7.7. Pseudovalues for June–July 1978 Mean U.S. Unemployment Rate

Rotation Group	Same Interviewer–Personal Interview	Different Interviewer–Personal Interview	Column 2 Minus Column 3
1	0.100	0.038	0.062
2	0.099	0.051	0.049
3	0.155	0.090	0.065
4	0.066	0.060	0.006
5	0.139	0.037	0.102
6	0.057	0.103	−0.047
7	0.218	0.100	0.118
8	0.118	0.100	0.018

Table 4.7.8. Computation of Mean Unemployment Rate, Difference in Mean Unemployment Rates, and Jackknife Estimates of Variance

To illustrate the computation of a pseudovalue for the June–July mean unemployment rate, we consider rotation group 1 ($\alpha = 1$) and the same interviewer-personal interview treatment combination ($i = 1$). The parent sample estimate is

$$\hat{R}_{.1} = (\hat{R}_{11} + \hat{R}_{21})/2$$

$$= (0.1224 + 0.1145)/2$$

$$= 0.1185,$$

the estimate obtained by deleting the first replicate is

$$\hat{R}_{.1(1)} = (\hat{R}_{11(1)} + \hat{R}_{21(1)})/2$$

$$= (0.1282 + 0.1140)/2$$

$$= 0.1211,$$

and the corresponding pseudovalue is

$$\hat{R}_{.11} = 8(0.1185) - 7(0.1211)$$

$$= 0.100.$$

The pseudovalues, including this example, appear in Table 4.7.7.
Using these data, we compute

$$\hat{\bar{R}}_{.1} = \frac{1}{8}\sum_{\alpha} \hat{R}_{.1\alpha} = 0.119$$

and

$$v_1(\hat{\bar{R}}_{.1}) = \frac{1}{8(7)}\sum_{\alpha} (\hat{R}_{.1\alpha} - \hat{\bar{R}}_{.1})^2$$

$$= 0.000338.$$

Table 4.7.8. (*Cont.*)

___

We now illustrate the computation of the difference in mean unemployment rate between treatment combinations. The parent sample estimate is

$$\Delta \hat{R}_{.} = \hat{R}_{.1} - \hat{R}_{.2}$$

$$= 0.1185 - 0.0722$$

$$= 0.0463,$$

and the estimate obtained by deleting the first replicate is

$$\Delta \hat{R}_{.(1)} = \hat{R}_{.1(1)} - \hat{R}_{.2(1)}$$

$$= 0.1211 - 0.0771$$

$$= 0.0440.$$

The corresponding pseudovalue is

$$\Delta \hat{R}_{.1} = 8(0.0463) - 7(0.0440)$$

$$= 0.0624.$$

The pseudovalues, including this example, appear in the last column of Table 4.7.7. Using these data, we compute

$$\overline{\Delta \hat{R}}_{.} = \frac{1}{8} \sum_{\alpha} \Delta \hat{R}_{.\alpha} = 0.047$$

and

$$v_1(\overline{\Delta \hat{R}}_{.}) = \frac{1}{8(7)} \sum_{\alpha} (\Delta \hat{R}_{.\alpha} - \overline{\Delta \hat{R}}_{.})^2$$

$$= 0.000355.$$

The normal-theory 95% confidence interval is

$$(\overline{\Delta \hat{R}}_{.} \pm 1.96\{v_1(\overline{\Delta \hat{R}}_{.})\}^{1/2}) = (0.010, 0.084),$$

and we may declare the treatment difference to be just significantly different from zero. Student's t theory would not result in a significant difference at the 95% level.

___

Then we have the pseudovalue

$$\Delta \hat{R}_{.\alpha} = 8\Delta \hat{R}_{.} - 7\Delta \hat{R}_{.(\alpha)},$$

Quenouille's estimator

$$\Delta \hat{R}_{.} = \frac{1}{8} \sum_{\alpha} \Delta \hat{R}_{.\alpha},$$

and the jackknife estimator of variance

$$v_1(\Delta \hat{R}_{.}) = \frac{1}{8(7)} \sum_{\alpha} (\Delta \hat{R}_{.\alpha} - \overline{\Delta \hat{R}}_{.})^2 .$$

Pseudovalues and illustrative computations are presented in Tables 4.7.7 and 4.7.8, respectively.

4.8. Example: Estimating the Size of the U.S. Population

This example is concerned with estimating the size of the U.S. population. The method of estimation is known variously as "dual-system" or "capture–recapture" estimation. Although our main purpose is to illustrate the use of the jackknife method for variance estimation, we first describe, in general terms, the two-sample capture–recapture problem.

Let N denote the total number of individuals in a certain population under study. We assume that N is *unobservable* and to be estimated. This situation differs from the model encountered in classical survey sampling, where the size of the population is considered *known* and the problem is to estimate other parameters of the population. We assume that there are two lists or frames; that each covers a portion of the total population; but that the union of the two lists fails to include some portion of the population N. We further assume the lists are independent in the sense that whether or not a given individual is present on the first list is independent of the individual's presence or absence on the second list.

The population may be viewed as follows:

	Second List		
First List	Present	Absent	
Present	N_{11}	N_{12}	$N_{1\cdot}$
Absent	N_{21}	—	
	$N_{\cdot 1}$		

The size of the (2,2) cell is unknown, and thus the total size N is also unknown. Assuming that these data are generated by a multinomial probability law and that the two lists are independent, the maximum likelihood of estimator of N is

$$\hat{N} = \frac{N_{1\cdot}N_{\cdot 1}}{N_{11}}.$$

See Bishop, Fienberg, and Holland (1975) or Marks, Seltzer, and Krotki (1974) for the derivation of this estimator.

In practice, the two lists must be matched to one another in order to determine the cell counts N_{11}, N_{12}, N_{21}. This task will be difficult, if not impossible, when the lists are large. Difficulties also arise when the two lists are not compatible with computer matching. In certain circumstances these problems can be dealt with (but not eliminated) by drawing samples from either or both lists, and subsequently matching only the sample cases. Dealing with samples instead of entire lists cuts down on work, and presumably on matching difficulties. Survey estimators may then be constructed for N_{11}, N_{12}, and N_{21}. This is the situation considered in the present example.

In this example, the February 1978 Current Population Survey (CPS) is the sample from the first list, and the Internal Revenue Service (IRS) tax returns filed in 1978 comprise the second list. The population N to be estimated is the U.S. adult population age 14–64.

The CPS is a household survey that is conducted monthly by the U.S. Bureau of the Census for the U.S. Bureau of Labor Statistics. The main purpose of the CPS is to gather information about characteristics of the U.S. labor force. The CPS sampling design is quite complex, using geographic stratification, multiple stages of selection, and unequal selection probabilities. The CPS estimators are equally complex, employing adjustments for nonresponse, two stages of ratio estimation, and one stage of composite estimation. For exact details of the sampling design and estimation scheme, the reader should see Hanson (1978). The CPS is discussed further in Section 5.5.

Each monthly CPS sample is actually comprised of eight distinct subsamples (known as rotation groups). Each rotation group is itself a national probability sample, comprised of a sample of households from each of the CPS primary sampling units (PSUs). The rotation groups might be considered random groups (nonindependent), as defined in Section 2.4, except that the ultimate cluster rule (see rule (iii), Section 2.4) was not followed in their construction.

In this example, the design variance of a dual-system or capture–recapture estimator will be estimated by the jackknife method operating on the CPS rotation groups. Because the ultimate cluster rule was not followed, this method of variance estimation will tend to omit the between component of variance. The omission is probably negligible, however, because the between component is considered to be a minor portion (about 5%) of the total CPS variance.

The entire second list (IRS) is used in this application. After matching the CPS sample to the second list, we produce the following data:

IRS

CPS	Present	Absent	
Present	$\hat{N}_{11}$	$\hat{N}_{12}$	$\hat{N}_{1\cdot}$
Absent	$\hat{N}_{21}$	—	
	$N_{\cdot 1}$		

and

IRS

CPS	Present	Absent	
Present	$\hat{N}_{11\alpha}$	$\hat{N}_{12\alpha}$	$\hat{N}_{1\cdot\alpha}$
Absent	$\hat{N}_{21\alpha}$	—	
	$N_{\cdot 1}$		

for $\alpha = 1, \ldots, 8$. The symbol "ᴧ" is being used here to indicate a survey estimator prepared in accordance with the CPS sampling design. The subscript "α" is used to denote an estimator prepared from the α-th rotation group, whereas the absence of "α" denotes an estimator prepared from the parent CPS sample. The total population on the IRS list, $N_{\cdot 1}$, is based on a complete count of that list: $N_{\cdot 1}$ is *not* an estimate prepared from a sample. The data are presented in Table 4.8.1.

Because each random group is a one-eighth sample of the parent CPS sample, we have

$$\hat{N}_{11} = \frac{1}{8} \sum_{\alpha=1}^{8} \hat{N}_{11\alpha},$$

$$\hat{N}_{12} = \frac{1}{8} \sum_{\alpha=1}^{8} \hat{N}_{12\alpha},$$

$$\hat{N}_{21} = \frac{1}{8} \sum_{\alpha=1}^{8} \hat{N}_{21\alpha},$$

$$\hat{N}_{1\cdot} = \frac{1}{8} \sum_{\alpha=1}^{8} \hat{N}_{1\cdot\alpha}.^2$$

The dual-system or capture–recapture estimator of N for this example is

$$\hat{N} = \frac{\hat{N}_{1\cdot} N_{\cdot 1}}{\hat{N}_{11}},$$

and the estimator obtained by deleting the α-th rotation group is

$$\hat{N}_{(\alpha)} = \frac{\hat{N}_{1\cdot(\alpha)} N_{\cdot 1}}{\hat{N}_{11(\alpha)}},$$

[2] Actually, these are only approximate equalities. The reader will recall that estimators $\hat{\theta}$ and $\hat{\bar{\theta}}$ are equal if the estimators are linear. In the present example, all of the estimators are nonlinear, involving ratio and nonresponse adjustments. Thus, the parent sample estimators, such as $\hat{N}_{11}$, are not in general equal to the mean of the random groups estimators $\sum \hat{N}_{11\alpha}/8$. Because the sample sizes are large, however, there should be little difference in this example.

Table 4.8.1. Data from the 1978 CPS–IRS Match Study

Rotation Group α	Matched Population $\hat{N}_{11\alpha}$	CPS Total Population $\hat{N}_{1\cdot\alpha}$	$\hat{N}_{(\alpha)}$	$\hat{N}_\alpha$
1	107,285,040	133,399,520	144,726,785	143,087,553
2	105,178,160	132,553,952	144,447,797	145,040,467
3	110,718,448	139,055,744	144,518,186	144,547,744
4	103,991,496	132,390,240	144,243,095	146,473,380
5	106,818,488	131,627,520	144,910,512	141,801,461
6	106,636,928	133,095,536	144,647,594	143,641,892
7	105,338,552	133,324,528	144,359,733	145,656,914
8	103,349,328	131,061,688	144,323,897	145,907,770
Parent	106,164,555	133,313,591		

Note: The size of the IRS list is $N._1 = 115,090,300$.

where

$$\hat{N}_{11(\alpha)} = \frac{1}{7} \sum_{\alpha' \neq \alpha} \hat{N}_{11\alpha'}$$

$$\hat{N}_{1\cdot(\alpha)} = \frac{1}{7} \sum_{\alpha' \neq \alpha} \hat{N}_{1\cdot\alpha'}.$$

The pseudovalues are defined by

$$\hat{N}_\alpha = 8\hat{N} - 7\hat{N}_{(\alpha)},$$

Quenouille's estimator by

$$\hat{\hat{N}} = \frac{1}{8} \sum_{\alpha=1}^{8} \hat{N}_\alpha,$$

and the jackknife estimator of variance by

$$v_1(\hat{\hat{N}}) = \frac{1}{8(7)} \sum_{\alpha=1}^{8} (\hat{N}_\alpha - \hat{\hat{N}})^2.$$

The conservative estimator of variance is

$$v_2(\hat{\hat{N}}) = \frac{1}{8(7)} \sum_{\alpha=1}^{8} (\hat{N}_\alpha - \hat{N})^2.$$

In this problem we are estimating the design variance of N, given $\hat{N}_{11}$, N_{21}, N_{12}, $N_1.$, and $N._1$. Some illustrative computations are given in Table 4.8.2.

To conclude the example, we comment on the nature of the adjustment for nonresponse and on the ratio adjustments. The CPS uses a combination of "hot deck" methods and "weighting-class" methods to adjust for missing data. The adjustments are applied within the eight rotation groups within

Table 4.8.2. Computation of Pseudovalues, Quenouille's Estimator, and the Jackknife Estimator of Variance

The estimates obtained by deleting the first rotation group are

$$\hat{N}_{11(1)} = \frac{1}{7} \sum_{\alpha' \neq 1} \hat{N}_{11\alpha'}$$

$$= 106,004,489$$

$$\hat{N}_{1\cdot(1)} = \frac{1}{7} \sum_{\alpha' \neq 1} \hat{N}_{1\cdot\alpha'}$$

$$= 133,301,311$$

$$\hat{N}_{(1)} = \frac{\hat{N}_{1\cdot(1)} N_{\cdot 1}}{\hat{N}_{11(1)}}$$

$$= 144,726,785.$$

The estimate obtained from the parent CPS sample is

$$\hat{N} = \frac{\hat{N}_1 . N_{\cdot 1}}{\hat{N}_{11}}$$

$$= \frac{(133,313,591)(115,090,300)}{106,164,555}$$

$$= 144,521,881.$$

Thus, the first pseudovalue is

$$\hat{N}_1 = 8\hat{N} - 7\hat{N}_{(1)}$$

$$= 143,087,553.$$

The remaining $\hat{N}_{(\alpha)}$ and $\hat{N}_\alpha$ are presented in the last two columns of Table 4.8.1. Quenouille's estimator and the jackknife estimator of variance are

$$\hat{\bar{N}} = \frac{1}{8} \sum_{\alpha=1}^{8} \hat{N}_\alpha$$

$$= 144,519,648$$

and

$$v_1(\hat{\bar{N}}) = \frac{1}{8(7)} \sum_{\alpha=1}^{8} (\hat{N}_\alpha - \hat{\bar{N}})^2$$

$$= 3.1284 \times 10^{11},$$

respectively.

The conservative estimator of variance is

$$v_2(\hat{\bar{N}}) = \frac{1}{8(7)} \sum_{\alpha=1}^{8} (\hat{N}_\alpha - \hat{N})^2$$

$$= 3.1284 \times 10^{11}.$$

cells defined by clusters of primary sampling units (PSUs) by race by type
of residence. Because the adjustments are made within the eight rotation
groups, the original principles of jackknife estimation (and of random group
and balanced half-sample estimation) are being observed and the jackknife
variances should properly include the components of variability due to the
nonresponse adjustment.

On the other hand, the principles are violated as regards the CPS ratio
estimators. To illustrate the violation we consider the first of two stages of
CPS ratio estimation. Ratio factors (using 1970 census data as the auxiliary
variable) are computed within region,[3] by kind of PSU, by race, by type of
residence. In this example, the ratio factors computed for the parent sample
estimators were also used for each of the random group estimators. This
procedure violates the original principles of jackknife estimation, which
call for a separate ratio adjustment for each random group. See Section 2.8
for additional discussion of this type of violation. The method described
here greatly simplifies the task of computing the estimates because only one
set of ratio adjustment factors is required instead of nine sets (one for the
parent sample and one each for the eight rotation groups). Some components
of variability, however, may be missed by this method.

[3] The four census regions are Northeast, North Central, South, and West.

CHAPTER 5

Generalized Variance Functions

5.1. Introduction

In this chapter we discuss the possibility of a simple mathematical relationship connecting the variance or relative variance of a survey estimator to the expectation of the estimator. If the parameters of the model can be estimated from past data or from a small subset of the survey items, then variance estimates can be produced for all survey items simply by evaluating the model at the survey estimates, rather than by direct computations. We shall call this method of variance estimation the method of *Generalized Variance Functions* (GVF).

In general, GVFs are applicable to surveys in which the publication schedule is extraordinarily large, giving, for example, estimates for scores of characteristics, for each of several demographic subgroups of the total population, and possibly for a number of geographic areas. For surveys in which the number of published estimates is manageable, we usually prefer a direct computation of variance for each survey statistic, as discussed in other chapters of this book. The primary reasons for considering GVFs include the following:

1. Even with modern computers it is usually more costly and time consuming to estimate variances than to prepare the survey tabulations. If many, perhaps thousands, of basic estimates are involved, then the cost of a direct computation of variance for each one may be excessive.
2. Even if the cost of direct variance estimation can be afforded, the problems of publishing all of the survey statistics and their corresponding standard errors may be unmanageable. The presentation of individual standard errors would essentially double the size of tabular publications.

3. In surveys where statistics are published for many characteristics and a great many subpopulations, it may be impossible to anticipate the various combinations of results (e.g. ratios, differences, etc.) which may be of interest to users. GVFs may provide a mechanism for the data user to estimate standard errors for these custom-made combinations of the basic tabulations.
4. Variance estimates are themselves subject to error. In effect, GVFs simultaneously estimate variances for groups of statistics, rather than individually estimating the variance statistic-by-statistic. It may be that some additional stability is imparted to the variance estimates when they are so estimated (as a group rather than individually). At present, however, there is no theoretical basis for this claim of additional stability.

An example where GVFs have considerable utility is the Current Population Survey (CPS), a national survey conducted monthly for the purpose of providing information about the U.S. labor force. A recent publication from this survey (cf. U.S. Department of Labor (1976)) contained about 30 pages of tables, giving estimated totals and proportions for numerous labor force characteristics for various demographic subgroups of the population. Literally thousands of individual statistics appear in these tables, and the number of subgroup comparisons that one may wish to consider number in the tens of thousands. Clearly, a direct computation of variance for each CPS statistic is not feasible.

The GVF methods discussed in this chapter are mainly applicable to the problem of variance estimation for an estimated proportion or for an estimate of the total number of individuals in a certain domain or subpopulation. There have been a few attempts, not entirely successful, to develop GVF techniques for quantitative characteristics. Section 5.6 gives an illustration of this work.

5.2. Choice of Model

As noted in the introduction, a GVF is a mathematical model describing the relationship between the variance or relative variance of a survey estimator and its expectation. In this section we present a number of possible models and discuss their rationale.

Let $\hat{X}$ denote an estimator of the total number of individuals possessing a certain attribute and let $X = E\{\hat{X}\}$ denote its expectation. The form of the estimator is left unspecified: it may be the simple Horvitz–Thompson estimator, it may involve poststratification, it may be a ratio or regression estimator, and so on. To a certain extent, the sampling design is also left unspecified. However, many of the applications of GVFs involve household surveys, where the design features multiple stage sampling within strata.

We let

$$\sigma^2 = \text{Var}\{\hat{X}\}$$

denote the variance of $\hat{X}$ and

$$V^2 = \text{Var}\{\hat{X}\}/X^2$$

the relative variance (or relvariance). Most of the GVFs to be considered are based on the premise that the relative variance V^2 is a decreasing function of the magnitude of the expectation X.

A simple model which exhibits this property is

$$V^2 = \alpha + \beta/X, \tag{5.2.1}$$

with $\beta > 0$. The parameters α and β are unknown and to be estimated. They depend upon the population, the sampling design, the estimator, and the x-characteristic itself. Experience has shown that Model (5.2.1) often provides an adequate description of the relationship between V^2 and X. In fact, the Census Bureau has used this model for its Current Population Survey since 1947 (cf. Hansen, Hurwitz, and Madow (1953) and Hanson (1978)).

In an attempt to achieve an even better fit to the data than is possible with (5.2.1), we may consider the models

$$V^2 = \alpha + \beta/X + \gamma/X^2, \tag{5.2.2}$$

$$V^2 = (\alpha + \beta X)^{-1}, \tag{5.2.3}$$

$$V^2 = (\alpha + \beta X + \gamma X^2)^{-1}, \tag{5.2.4}$$

and

$$\log(V^2) = \alpha - \beta \log(X). \tag{5.2.5}$$

Edelman (1967) presents a long list of models which he has investigated empirically.

Unfortunately, there is very little theoretical justification for any of the models discussed above. There is some limited justification for Model (5.2.1), and this is summarized in the following paragraphs:

1. Suppose that the population is composed of N clusters, each of size M. A simple random sample of n clusters is selected, and each elementary unit in the selected clusters is enumerated. Then, the variance of the Horvitz–Thompson estimator $\hat{X}$ of the population total X is

$$\sigma^2 = (NM)^2 \frac{N-n}{N-1} \frac{PQ}{nM}\{1 + (M-1)\rho\},$$

where $P = X/NM$ is the population mean per element, $Q = 1 - P$ and ρ denotes the intraclass correlation between pairs of elements in the same cluster. See, e.g., Cochran (1977, pp. 240–243). The relative variance of $\hat{X}$ is

$$V^2 = \frac{N-n}{N-1}\frac{Q}{P(nM)}\{1+(M-1)\rho\},$$

and assuming that the first stage sampling fraction is negligible, we may write

$$V^2 = \frac{1}{X}\frac{NM\{1+(M-1)\rho\}}{nM} - \frac{\{1+(M-1)\rho\}}{nM}.$$

Thus, for this simple sampling scheme and estimator, (5.2.1) provides an acceptable model for relating V^2 to X. If the value of the intraclass correlation is constant (or approximately so) for a certain class of survey statistics, then (5.2.1) may be useful for estimating the variances in the class.

2. Kish (1965) and others have popularized the notion of *design effects*. If we assume an arbitrary sampling design leading to a sample of n units from a population of size N, then the design effect for $\hat{X}$ is defined by

$$\text{Deff} = \sigma^2/\{N^2 PQ/n\},$$

where $P = X/N$ and $Q = 1 - P$. This is the variance of $\hat{X}$ given the true sampling design divided by the variance given simple random sampling. Thus, the relative variance may be expressed by

$$V^2 = Q(Pn)^{-1}\text{Deff}$$

$$= -\text{Deff}/n + (N/n)\,\text{Deff}/X. \tag{5.2.6}$$

Assuming that Deff may be considered independent of the magnitude of X within a given class of survey statistics, (5.2.6) is of the form of Model (5.2.1) and may be useful for estimating variances in the class.

3. Suppose that it is desired to estimate the proportion

$$R = X/Y,$$

where Y is the total number of individuals in a certain subpopulation and X is the number of those individuals with a certain attribute. If $\hat{X}$ and $\hat{Y}$ denote estimators of X and Y, respectively, then the natural estimator of R is $\hat{R} = \hat{X}/\hat{Y}$. Utilizing a Taylor series approximation (cf. Chapter 6) and assuming $\hat{Y}$ and $\hat{R}$ are uncorrelated, we may write

$$V_R^2 \doteq V_X^2 - V_Y^2, \tag{5.2.7}$$

where V_R^2, V_X^2, and V_Y^2 denote the relative variances of $\hat{R}$, $\hat{X}$, and $\hat{Y}$, respectively. If Model (5.1.1) holds for both V_X^2 and V_Y^2, then (5.2.7) gives

$$V_R^2 \doteq \beta/X - \beta/Y$$

$$= \frac{\beta}{Y}\frac{(1-R)}{R},$$

and hence

$$\text{Var}\{\hat{R}\} \doteq (\beta/Y)R(1 - R). \tag{5.2.8}$$

Equation (5.2.8) has the important property that the variance of an estimator

$$\hat{R}' = \hat{X}'/\hat{Y}$$

of a proportion

$$R' = X'/Y$$

which satisfies

$$R' = 1 - R$$

is identical to the variance of the estimator $\hat{R}$ of R. Thus, for example, $\text{Var}\{\hat{R}\} = \text{Var}\{1 - \hat{R}\}$. Tomlin (1974) justifies Model (5.2.1) on the basis that it is the only known model that possesses this important property.

In spite of a lack of rigorous theory to justify (5.2.1) or any other model, GVF models have been successfully applied to numerous real surveys in the past 30 years. In the following sections we shall demonstrate the exact manner in which such models are used to simplify variance calculations.

5.3. Grouping Items Prior to Model Estimation

The basic GVF procedure for variance estimation is summarized in the following steps:

1. Group together all survey statistics that follow a common model, e.g., $V^2 = \alpha + \beta/X$. This may involve grouping similar items from the same survey; the same item for different demographic or geographic sub-groups; or the same survey statistic from several prior surveys of the same population. The third method of grouping, of course, is only possible with repetitive or recurring surveys.
2. Compute a direct estimate $\hat{V}^2$ or V^2 for several members of the group of statistics formed in Step 1. The variance estimating techniques discussed in the other chapters of this book may be used for this purpose.
3. Using the data $(\hat{X}, \hat{V}^2)$ from Step 2, compute estimates, say $\hat{\alpha}, \hat{\beta}, \hat{\gamma}$, etc., of the model parameters α, β, γ, etc. Several alternative fitting methodologies might be used here, and this topic is discussed in Section 5.4.
4. An estimator of the relative variance of a survey statistic $\hat{X}$ for which a direct estimate $\hat{V}^2$ was not computed is now obtained by evaluating the model at the point $(\hat{X}; \hat{\alpha}, \hat{\beta}, \hat{\gamma}, \ldots)$. For example, if Model (5.2.1) is used, then the GVF estimate of V^2 is

$$\tilde{V}^2 = \hat{\alpha} + \hat{\beta}/\hat{X}.$$

5. To estimate the relative variance of an estimated proportion $\hat{R} = \hat{X}/\hat{Y}$, where $\hat{Y}$ is an estimator of the total number of individuals in a certain subpopulation and $\hat{X}$ is an estimator of the number of those individuals with a certain attribute, use

$$\tilde{V}_R^2 = \tilde{V}_X^2 - \tilde{V}_Y^2.$$

Often, $\hat{X}$ and $\hat{Y}$ will be members of the same group formed in Step 1. If this is the case and Model (5.2.1) is used, then the estimated relative variance becomes

$$\tilde{V}_R^2 = \hat{\beta}(\hat{X}^{-1} - \hat{Y}^{-1}).$$

Considerable care is required in performing Step 1. The success of the GVF technique depends critically on the grouping of the survey statistics, i.e., on whether all statistics within a group behave according to the same mathematical model. In terms of the first justification for Model (5.2.1) given in the last section, this implies that all statistics within a group should have a common value of the intraclass correlation ρ. The second justification given in the last section implies that all statistics within a group should have a common design effect, Deff.

From the point of view of data analysis and model confirmation, it may be important to begin with provisional groups of statistics based on past experience and expert opinion. Scatter plots of $\hat{V}^2$ versus $\hat{X}$ should then be helpful in forming the "final" groups. One simply removes from the provisional group those statistics that appear to follow a different model than the majority of statistics in the group, and adds other statistics, originally outside the provisional group, that appear consonant with the group model.

From a substantive point of view, the grouping will often be successful when the statistics (1) refer to the same basic demographic or economic characteristic, (2) refer to the same race-ethnicity group, and (3) refer to the same level of geography.

5.4. Methods for Fitting the Model

As noted in Section 5.2, there is no rigorous theoretical justification for Model (5.2.1) or for any other model that relates V^2 to X. Because we are unable to be quite specific about the model and its attending assumptions, it is not possible to construct, or even to contemplate, optimum estimators of the model parameters α, β, γ, etc. Discussions of optimality would require an exact model and an exact statement of the error structure of the estimators $\hat{V}^2$ and $\hat{X}$. In the absence of a completely specified model, we shall simply seek to achieve a good empirical fit to the data $(\hat{X}, \hat{V}^2)$ as we consider alternative fitting methodologies.

To describe the various methodologies that might be used, we let $g(\cdot)$ denote the functional relationship selected for a specific group of survey statistics, i.e.,

$$V^2 = g(X; \alpha, \beta, \gamma, \ldots).$$

A natural fitting methodology is ordinary least squares (OLS). That is, $\hat{\alpha}$, $\hat{\beta}$, $\hat{\gamma}$, ... are those values of α, β, γ, ... that minimize the sum of squares

$$\sum \{\hat{V}^2 - g(\hat{X}; \alpha, \beta, \gamma, \ldots)\}^2, \tag{5.4.1}$$

where the sum is taken over all statistics $\hat{X}$ for which a direct estimate $\hat{V}^2$ of V^2 is available. When $g(\cdot)$ is linear in the parameters, simple closed form expressions for $\hat{\alpha}$, $\hat{\beta}$, $\hat{\gamma}$, ... are available. For nonlinear $g(\cdot)$ some kind of iterative search is usually required.

The OLS estimators are often criticized because the sum of squares (5.4.1) gives too much weight to the small estimates $\hat{X}$, whose corresponding $\hat{V}^2$ are usually large and unstable. A better procedure might be to give the least reliable terms in the sum of squares a reduced weight. One way of achieving this weighting is to work with the sum of squares

$$\sum \hat{V}^{-4}\{\hat{V}^2 - g(\hat{X}; \alpha, \beta, \gamma, \ldots)\}^2, \tag{5.4.2}$$

which weights inversely to the observed $\hat{V}^4$. Alternatively, we may weight inversely to the square of the fitted relvariances, i.e., minimize the sum of squares

$$\sum \{g(\hat{X}; \alpha, \beta, \gamma, \ldots)\}^{-2}\{\hat{V}^2 - g(\hat{X}; \alpha, \beta, \gamma, \ldots)\}^2. \tag{5.4.3}$$

In the case of (5.4.3), it is usually necessary to consider some kind of iterative search, even when the function $g(\cdot)$ is linear in the parameters. The minimizing values from (5.4.2) may be used as starting values in an iterative search scheme.

As noted earlier, there is little in the way of theory to recommend (5.4.1), (5.4.2), or (5.4.3). The best one can do in an actual problem is to try each of the methods, choosing the one which gives the "best" empirical fit to the data.

One obvious danger to be avoided, regardless of which estimation procedure is used, is the possibility of negative variance estimates. To describe this, suppose that Model (5.2.1) is to be used, i.e.,

$$g(X; \alpha, \beta) = \alpha + \beta/X.$$

In practice, the estimator $\hat{\alpha}$ of the parameter α may be negative, and if $\hat{X}$ is sufficiently large for a particular item, the estimated relative variance

$$\tilde{V}^2 = \hat{\alpha} + \hat{\beta}/\hat{X}$$

may be negative. One way of avoiding this undesirable situation is to introduce some kind of restriction on the parameter α, and then proceed with the estimation via (5.4.1), (5.4.2), or (5.4.3) subject to the restriction.

For example, in the Current Population Survey (CPS) Model (5.2.1) is applied to a poststratified estimator of the form

$$\hat{X} = \sum_a \frac{\hat{X}_a}{\hat{Y}_a} Y_a,$$

where $\hat{X}_a$ denotes an estimator of the number of individuals with a certain attribute in the a-th age–sex–race domain, $\hat{Y}_a$ denotes an estimator of the total number of individuals in the domain, and Y_a denotes the *known* total number of individuals in the domain. If T denotes the sum of the Y_a over all domains in which the x-variable is defined, then we may impose the restriction that the relative variance of T is zero, i.e.,

$$\alpha + \beta / T = 0.^{1}$$

Thus,

$$\alpha = -\beta / T$$

and we fit the one parameter model

$$V^2 = \beta(X^{-1} - T^{-1}). \tag{5.4.4}$$

The estimated β will nearly always be positive, and in this manner the problem of negative estimates of variance is avoided. In the next section, we consider the CPS in some detail.

5.5. Example: The Current Population Survey

The Current Population Survey (CPS) is a large, multistage survey conducted by the U.S. Bureau of the Census for the purpose of providing information about the U.S. labor force. Due to the large amount of data published from the CPS it is not practical to make direct computations of the variance for each and every statistic. GVFs are used widely in this survey in order to provide variance estimates at reasonable cost which can be easily used by data analysts.

Before describing the usage of GVFs, we give a brief description of the CPS sampling design and estimation procedure. Under the CPS design of the early 1970s, the U.S. was divided into 1924 primary sampling units PSU was chosen with probability proportional to size. Additionally, the or one or more contiguous counties. The PSUs were grouped into 376 strata, 156 of which contained only one PSU, which was selected with certainty. The remaining PSUs were grouped into 220 strata with each stratum contain-

---

[1] For example, if $\hat{X}$ is an estimator of total Black unemployed, then T is the sum of the Y_a over all age–sex–Black domains.

ing two or more PSUs. The 156 are referred to as self-representing PSUs while the remaining $1924 - 156 = 1768$ are referred to as nonself-representing PSUs. Within each of the 220 strata one nonself-representing PSU was chosen with probability proportional to size. Additionally, the 220 strata were grouped into 110 pairs; one stratum was selected at random from each pair; and one PSU was selected independently with probability proportional to size from the selected stratum. Thus, in the CPS design three nonself-representing PSUs were selected from each of 110 stratum pairs, although in 25 stratum pairs a selected PSU was duplicated. The sample selection of the nonself-representing PSUs actually utilized a controlled selection design in order to provide a sample in every state. Within each selected PSU, segments (with an average size of four households) were chosen so as to obtain a self-weighting sample of households, i.e., so that the overall probability of selection was equal for every household in the U.S. The final sample consisted of 461 PSUs comprising 923 counties and independent cities. Approximately 47,000 households were eligible for interview every month.

The CPS estimation procedure involves a nonresponse adjustment, two stages of ratio estimation, the formation of a composite estimate that takes into account data from previous months, and an adjustment for seasonal variation. The variances are estimated directly for about 100 CPS statistics using the Taylor series method (see Chapter 6).

The interested reader should consult Hanson (1978) for a comprehensive discussion of the CPS sampling design and estimation procedure.

The CPS statistics for which variances are estimated directly are chosen on the basis of user interest, including certain key unemployment statistics, and on the need to obtain well-fitting GVFs that pertain to all statistics. In most cases a GVF of the form (5.2.1) is utilized, experience having shown this to be a useful model. The statistics are divided into six groups with Model (5.2.1) fitted independently in each group. Thus, different estimated parameters are obtained for each of the six groups. The groups are:

1. Agriculture Employment
2. Total or Nonagriculture Employment
3. Males Only in Total or Nonagriculture Employment
4. Females Only in Total or Nonagriculture Employment
5. Unemployment
6. Unemployment for Black and Other Races.

A separate agricultural employment group is used because the geographic distribution of persons employed in agriculture is somewhat different than that of persons employed in nonagricultural industries. Separate curves are fit for the other groups because statistics in different groups tend to differ in regard to the clustering of persons within segments. In general, as mentioned previously, the grouping aims to collect together statistics with similar intraclass correlations or similar design effects.

To illustrate the GVF methodology, we consider the items used in the total employment group. A list of the items is given in Table 5.5.1. The July 1974 estimates and variance estimates for the items are also given in Table 5.5.1. A plot of the log of the estimated relvariance versus the log of the estimate is given in Figure 5.5.1. The log–log plot is useful since the data will form a concave downward curve if a GVF of the form (5.2.1) is appropriate. Other types of plots can be equally useful.

The parameters of (5.2.1) for the total employment group were estimated using an iterative search procedure to minimize the weighted sum of squares in (5.4.3). This resulted in

$$\hat{\alpha} = -0.0000175 \,(0.0000015)$$

$$\hat{\beta} = 2087 \,(114),$$

where figures in parentheses are the least squares estimated standard errors. The observed R^2 was 0.96.

The normal practice in the CPS is to use data for an entire year in fitting the GVF. This is thought to increase the accuracy of the estimated parameters and to help remove seasonal effects from the data. To illustrate, estimates and estimated variances were obtained for the characteristics listed in Table

Table 5.5.1. Characteristics and July 1974 Estimates and Estimated Variances Used to Estimate the GVF for the Total Employment Group

Characteristic	Estimate	Variance $\times 10^{-10}$
Total civilian labor force	93,272,212	4.4866
Total employed—Nonagriculture	83,987,974	5.0235
Employed—Nonagriculture:		
Wage and salary	77,624,607	5.2885
Worked 35 + hours	57,728,781	5.8084
Blue collar civilian labor force	30,967,968	4.3166
Wage and salary workers—Manufacturing	21,286,788	4.3197
Wage and salary workers—Retail trade	12,512,476	1.9443
Worked 1–34 hours, usually full-time	4,969,964	1.1931
Self-employed	5,873,093	0.9536
Worked 1–14 hours	3,065,875	0.5861
Wage and salary workers—Construction	4,893,046	0.7911
Worked 1–34 hours, economic reasons	3,116,452	0.6384
With a job, not at work	11,136,887	2.5940
Worked 1–34 hours, usually full-time, economic reasons	1,123,992	0.2935
With a job, not at work, salary paid	6,722,185	1.6209
Wage and salary—Private household workers	1,386,319	0.1909

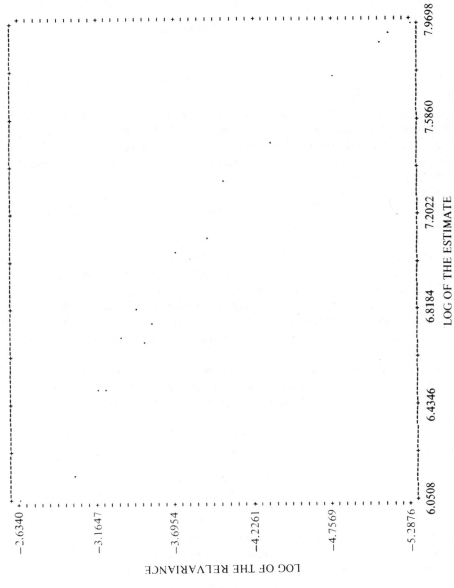

Figure 5.5.1. Log–log plot of estimated relvariance versus estimate for total employment characteristics, July 1974.

5.5.1 for each month from July 1974 through June 1975. The log–log plot of the estimated relvariance versus the estimate is presented in Figure 5.5.2. Once again a concave downward curve is obtained, suggesting a GVF of the form (5.2.1). Minimizing the weighted sum of squares in (5.4.3) yields

$$\hat{\alpha} = -0.000164 \, (0.000004)$$

$$\hat{\beta} = 2020 \, (26).$$

The R^2 on the final iteration was 0.97. A total of 192 observations were used for this regression.

Another example of GVF methodology concerns CPS data on population mobility. Such data, contrary to the monthly collection of CPS labor force data, are only collected in March of each year. For these data one GVF of the form (5.2.1) is fit to all items that refer to movers within a demographic subpopulation. Table 5.5.2 provides a list of the specific items used in estimating the GVF (i.e., those characteristics for which a direct variance estimate is available). The data are presented in Table 5.5.3, where the notation T denotes the known population in the appropriate demographic subgroup, as in (5.4.4). For example, in the first row of the table data are presented for "total movers, 18 to 24 years old," and $T = 25,950,176$ denotes the true total population 18 to 24 years old. Data are presented for both 1975 and 1976, giving a total of 66 observations. The 1975 data represent movers between 1970 and 1975 (a 5-year reference period) whereas the 1976 data represent movers between 1975 and 1976 (a 1-year reference period). The difference in reference periods explains the relatively higher degree of mobility for 1975.

Figure 5.5.3 plots log of the estimated relvariance versus log of the estimate. As in the case of the CPS labor force data, we notice a concave downward pattern in the data, thus tending to confirm the model specification (5.2.1).

Minimizing the weighted sum of squares in (5.4.3) yields the estimated coefficients

$$\hat{\alpha} = -0.000029 \, (0.000013)$$

$$\hat{\beta} = 2196 \, (72).$$

On the final iteration we obtained $R^2 = 93.5\%$.

As a final illustration, we fit the GVF in (5.4.4) to the mobility data in Table 5.5.3. The reader will recall that this model specification attempts to protect against negative estimated variances (particularly for large $\hat{X}$). Minimizing the weighted sum of squares in (5.4.3) yields

$$\hat{\beta} = 2267 \, (64),$$

with $R^2 = 95.0\%$ on the final iteration.

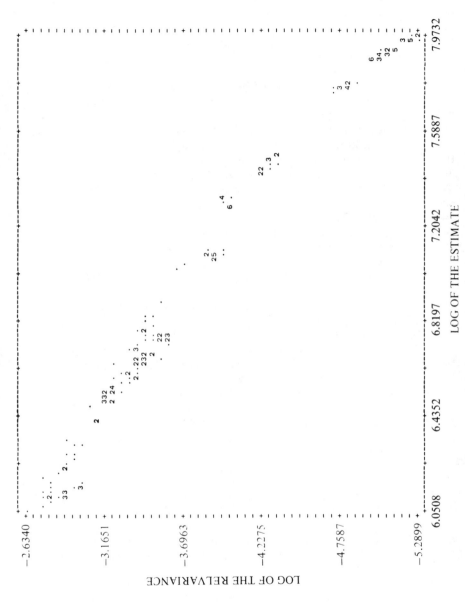

Figure 5.5.2. Log-log plot of estimated relvariance versus estimate for total employment characteristics, July 1974 to June 1975.

Table 5.5.2. Items Used to Estimate the GVF for Movers Within a Demographic Subpopulation

Item Code	Description
23	Total movers, 18 to 24 years old
29	Total movers 16+, never married
27	Total White household heads, movers within same SMSA
38	Total movers, different county, 4 years college
53	Total movers 16+, never married; professional, technical and kindred workers
66	Total movers within and between balance of SMSA, within same SMSA, head 25 to 34 years old with own children under 18
79	Total male movers 16+, within same SMSA, laborers except farm
82	Total movers from central cities to balance of SMSAs, within same SMSA, 18 to 24 years old, 4 years high school
91	Total Black male movers into South
59	Total Black movers, family heads, within same SMSA
70	Total Black employed male movers 16+, within same SMSA
98	Total Black movers, within same SMSA, 4 years high school
44	Total Black movers, family heads
54	Total employed Black male movers
60	Total Black family heads, movers, without public assistance
25	Total female movers, married, spouse present
33	Total White male movers 16+, employed, within same SMSA
36	Total male movers 16+, income $15,000 to $24,999
49	Total female movers, different county, 16 to 24 years old, never married
61	Total male movers within same SMSA, 18+, 4 years college
64	Total male movers 16+, married, spouse present, unemployed
71	Total female movers within and between balance of SMSAs, within same SMSA, 25 to 34 years old, employed
72	Total female movers 16+, clerical and kindred workers, outside SMSAs at both dates
77	Total female movers within same SMSA, 35 to 44 years old, married, spouse present
83	Total employed male movers into South
84	Total male movers 16+, same county, never married, unemployed
85	Total male movers within same SMSA, 16 to 24 years old, married, wife present, with income of $1000 to $9999
87	Total male movers 16+, from central cities to balance of SMSAs, within same SMSA, with income of $7000 to $9999
99	Total Black female movers 16+, not in labor force
100	Total male movers 16+, between SMSAs, with income of $10,000 to $14,999
76	Total female movers into South, 16 to 64 years old
52	Total Black movers 25+, within and between central cities within same SMSA
92	Total Black male movers from Northeast to South
80	Total movers into South, age 25+, 4 years of high school or less

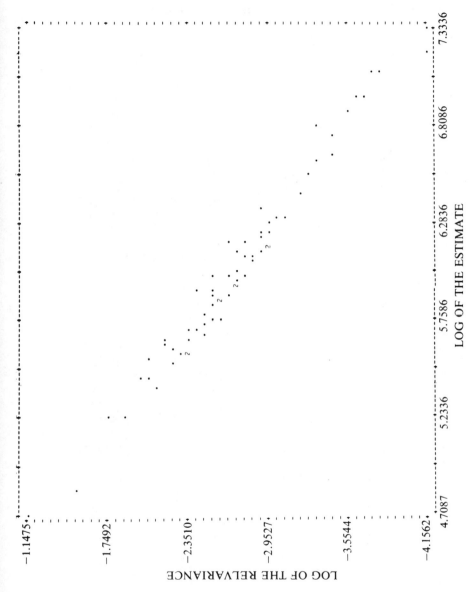

Figure 5.5.3. Log-log plot of estimated relvariance versus estimate for movers in a demographic subpopulation.

Table 5.5.3. Estimates and Variance Estimates Used in Fitting the GVF for Movers Within a Demographic Subpopulation

Item Code	Year	$\hat{X}$	$\hat{\sigma}^2$	T
23	75	15,532,860	0.1685 + 11	25,950,176
23	76	9,076,725	0.1900 + 11	26,624,613
29	75	12,589,844	0.2930 + 11	150,447,325
29	76	6,311,067	0.1978 + 11	153,177,617
27	75	12,447,497	0.2580 + 11	168,200,691
27	76	5,585,767	0.1121 + 11	180,030,064
38	75	3,582,929	0.7451 + 10	192,444,762
38	76	1,310,635	0.3373 + 10	207,149,736
53	75	1,482,186	0.3538 + 10	150,447,325
53	76	784,174	0.1876 + 10	153,177,617
66	75	1,308,198	0.2101 + 10	63,245,759
66	76	540,045	0.9804 + 09	76,352,429
79	75	895,618	0.1579 + 10	70,995,769
79	76	367,288	0.6925 + 09	72,344,487
82	75	449,519	0.8537 + 09	25,950,176
82	76	272,015	0.6405 + 09	26,624,613
91	75	167,514	0.4991 + 09	11,272,923
91	76	69,165	0.1447 + 09	12,704,849
59	75	1,712,954	0.3287 + 10	24,244,071
59	76	721,547	0.1318 + 10	27,119,672
70	75	1,224,067	0.2529 + 10	17,567,954
70	76	503,651	0.1144 + 10	8,215,984
98	75	980,992	0.2115 + 10	12,747,526
98	76	415,049	0.1112 + 10	13,174,373
44	75	2,718,933	0.4504 + 10	24,244,071
44	76	1,002,446	0.1684 + 10	27,119,672
54	75	1,993,665	0.3641 + 10	11,272,923
54	76	786,787	0.1383 + 10	99,856,466
60	75	2,082,244	0.3590 + 10	24,244,071
60	76	732,689	0.1291 + 10	27,119,672
25	75	21,558,732	0.3243 + 11	100,058,547
25	76	7,577,287	0.1608 + 11	107,293,270
33	75	9,116,884	0.1807 + 11	63,071,059
33	76	4,152,975	0.8778 + 10	64,128,503
36	75	4,556,722	0.7567 + 10	70,995,769
36	76	1,637,731	0.3472 + 10	72,344,487
49	75	1,676,173	0.3391 + 10	17,591,865
49	76	839,009	0.2114 + 10	17,884,083
61	75	1,222,591	0.2230 + 10	66,804,777
61	76	673,099	0.1382 + 10	68,142,678
64	75	1,419,308	0.2381 + 10	70,995,769
64	76	563,687	0.9383 + 09	72,344,487
71	75	893,820	0.1571 + 10	15,316,481

Table 5.5.3 (*Cont.*)

Item Code	Year	$\hat{X}$	$\hat{\sigma}^2$	T
71	76	484,469	0.8372 + 09	15,883,656
72	75	1,038,820	0.2079 + 10	79,451,556
72	76	461,248	0.1439 + 10	80,833,130
77	75	1,428,205	0.2368 + 10	11,614,088
77	76	378,148	0.5966 + 09	11,712,165
83	75	1,016,787	0.2855 + 10	92,386,215
83	76	271,068	0.7256 + 09	99,856,466
84	75	497,031	0.9248 + 09	70,995,769
84	76	251,907	0.4909 + 09	72,344,487
85	75	926,436	0.1675 + 10	16,657,453
85	76	582,028	0.9195 + 09	17,049,814
87	75	376,700	0.6399 + 09	70,995,769
87	76	175,184	0.4420 + 09	72,344,487
99	75	1,894,400	0.4023 + 10	9,643,244
99	76	605,490	0.1157 + 10	9,974,863
100	75	1,160,351	0.2045 + 10	70,995,769
100	76	332,044	0.6936 + 09	72,344,487
76	75	1,480,028	0.5013 + 10	67,047,304
76	76	402,103	0.9359 + 09	68,083,789
52	75	2,340,338	0.6889 + 10	12,747,526
52	76	824,359	0.2492 + 10	13,655,438
92	76	51,127	0.1861 + 09	12,704,849
80	76	348,067	0.1120 + 10	118,243,720

5.6. Example: The School Staffing Survey[2]

Much of this chapter has dealt with methods for "generalizing" variances based upon model specification (5.2.1) or related models. Survey statisticians have also attempted to generalize variances by using design effects and other ad hoc methods. In all cases the motivation has been the same: the survey publication schedule is exceedingly large, making it impractical to compute or publish direct variance estimates for all survey statistics, and some simple, user-friendly method is needed to generalize variances from a few statistics to all of the survey statistics. It is not feasible for us to recount all of the ad hoc methods that have been attempted. To illustrate the range of other possible methods, we discuss one additional example.

[2] This example derives from a Westat, Inc. project report prepared by Chapman and Hansen (1972).

The School Staffing Survey (SSS) was a national survey of public elementary and secondary schools carried out in 1969-70. A stratified random sample of about 5600 schools was selected from the SSS universe of about 80,000 schools. Thirty strata were defined based on the following three-way stratification:

(1) Level (elementary and secondary)
(2) Location (large city, suburban, other)
(3) Enrollment (five size classes).

There were a large number of estimates of interest for the SSS. Estimates of proportions, totals, and ratios were made for many pupil, teacher, and staff characteristics. Examples include (1) the proportion of schools with a school counselor, (2) the number of schools that offer Russian classes, (3) the number of teachers in special classes for academically gifted pupils, (4) the number of pupils identified as having reading deficiencies, (5) the ratio of pupils to teachers, and (6) the ratio of academically gifted pupils to all pupils. Survey estimates were made for the entire population and for a large number of population subgroups.

Because the number of estimates of interest for the SSS was very large, it was not feasible to calculate and publish variances for all survey estimates. Consequently, procedures were developed to allow for the calculation of approximate variance estimates as simple functions of the survey estimates. Procedures were developed for three basic types of statistics: (1) proportions, (2) totals, and (3) ratios. Details are given by Chapman and Hansen (1972). Summarized below is the development of the generalized variance procedure for a population or subpopulation total X.

The methodology was developed in terms of the relative variance, or relvariance, V^2, of the estimated total, $\hat{X}$. The relvariance was used instead of the variance because it is a more stable quantity from one statistic to another. For the SSS sampling design, the relvariance of $\hat{X}$ can be written as

$$V^2 = \bar{X}^{-2} \sum_{h=1}^{L} a_h S_h^2, \qquad (5.6.1)$$

where

$\bar{X}$ = the population mean,

L = the number of strata in the population or subpopulation for which the estimate is made,

N_h = the number of schools in stratum h,

N = the number of schools in the population,

n_h = the number of schools in the sample from stratum h,

S_h^2 = the variance, using an $N_h - 1$ divisor, among the schools in stratum h,

$$a_h = \frac{N_h^2}{N^2} \frac{N_h - n_h}{N_h n_h}.$$

A direct sample estimate, $\hat{V}^2$, of V^2, can be calculated as

$$\hat{V}^2 = (\bar{x}^{-2}) \sum_{h=1}^{L} a_h s_h^2, \qquad (5.6.2)$$

where s_h^2 = the ordinary sample variance for stratum h and $\bar{x} = \hat{X}/N$.

The estimated relvariance per unit, v_h^2, is introduced into the right-hand side of (5.6.2) by writing

$$\hat{V}^2 = (\bar{x}^{-2}) \sum_{h=1}^{L} a_h (\bar{x}_h/\bar{x}_h)^2 s_h^2 = \sum_{h=1}^{L} a_h (\bar{x}_h/\bar{x})^2 v_h^2 \qquad (5.6.3)$$

where $v_h^2 = s_h^2/\bar{x}_h^2$.

For many of the characteristics for which estimated totals were of interest, the ratio $\bar{x}_h/\bar{x}$ was approximately equal to the corresponding ratio of mean school enrollments. That is, for many characteristics,

$$\bar{x}_h/\bar{x} \doteq \bar{e}_h/\bar{e}, \qquad (5.6.4)$$

where

$\bar{e}_h$ = the sample mean school enrollment for stratum h,

$$\bar{e} = \sum_{h=1}^{L} (N_h/N) \bar{e}_h.$$

This gives

$$\hat{V}^2 = \sum_{h=1}^{L} b_h v_h^2, \qquad (5.6.5)$$

where $b_h = a_h (\bar{e}_h/\bar{e})^2$.

The most important step in the derivation of the generalized variance estimator was the "factoring out" of an average stratum relvariance from (5.6.5). The estimator $\hat{V}^2$ can be written in the form

$$\hat{V}^2 = \overline{v^2} b (1 + \delta), \qquad (5.6.6)$$

where

$$\overline{v^2} = (L^{-1}) \sum_{h=1}^{L} v_h^2,$$

$$b = \sum_{h=1}^{L} b_h,$$

$$\delta = \rho_{bv^2} V_b V_{v^2}, \qquad (5.6.7)$$

ρ_{bv^2} = the simple correlation between the L pairs of b_h and v_h^2 values,

V_b = the coefficient of variation (i.e., square root of the relvariance) of the L values of b_h,

V_{v^2} = the coefficient of variation of the L values of v_h^2.

It seems reasonable to expect that the correlation between the b_h and v_h^2 values would generally be near zero. Consequently, the following approximate variance estimate is obtained from (5.6.6) by assuming that this correlation is zero:

$$\tilde{V}^2 = \overline{v^2}b. \tag{5.6.8}$$

For the SSS, tables of b values were constructed for use in (5.6.8). An extensive examination of v_h^2 values was conducted for a number of survey characteristics, and guidelines were developed for use in obtaining an approximate value of $\overline{v^2}$. The guidelines consisted of taking $\overline{v^2}$ to be one of two values, 0.2 or 0.7, depending upon the characteristic of interest.

Although this generalized variance estimation procedure was not tested extensively, some comparisons were made between the generalized estimates in (5.6.8) and the standard estimates in (5.6.2). The generalized estimates were reasonably good for the test cases. When differences existed between the two estimates, the generalized estimate was usually slightly larger (i.e., somewhat conservative).

CHAPTER 6

Taylor Series Methods

6.1. Introduction

In sample surveys of both simple and complex design, it is often desirable or necessary to employ estimators that are nonlinear in the observations. Ratios, differences of ratios, correlation coefficients, regression coefficients, and poststratified means are common examples of such estimators. Exact expressions for the sampling variances of nonlinear estimators are not usually available, and moreover, neither are simple, unbiased estimators of the variance.

One useful method of estimating the variance of a nonlinear estimator is to approximate the estimator by a linear function of the observations. Then, variance formulae appropriate to the specific sampling design can be applied to the linear approximation. This leads to a biased, but typically consistent, estimator of the variance of the nonlinear estimator.

This chapter discusses in detail these linearization methods, which rely on the validity of Taylor series or binomial series expansions. The methods to be discussed are old and well-known: no attempt is made to assign priority to specific authors. In Section 6.2, the linearization method is presented for the infinite population model, where a considerable body of supporting theory is available. The remainder of the chapter applies the methods of Section 6.2 to the problems of estimation in finite populations.

It should be emphasized at the outset, that the Taylor series methods cannot act alone in estimating variances. That is, Taylor series per se does not produce a variance estimator. It merely produces a linear approximation to the survey statistic of interest. Then other methods, such as those described elsewhere in this book, are needed to estimate the variance of the linear approximation.

6.2. Linear Approximations in the Infinite Population

In this section we introduce some fairly rigorous theory regarding Taylor series approximations in the context of the infinite population model. The reason for doing so is that rigorous theory about these matters is lacking, to some extent, in the context of the classical finite population model. Our plan is to provide a brief but rigorous review of the methods in this section, and then in the next section (6.3) to show how the methods are adapted and applied to finite population problems.

The concept of *order in probability*, introduced by Mann and Wald (1943), is useful when discussing Taylor series approximations. For convenience, we follow the development given in Fuller (1976). The ideas to be presented apply to random variables, and are analogous to the concepts of order (e.g., 0 and o) discussed in mathematical analysis.

Let $\{\mathbf{Y}_n\}$ be a sequence of p-dimensional random variables and $\{r_n\}$ a sequence of positive real numbers.

Definition 6.2.1. We say $\mathbf{Y}_n$ is at most of order in probability (or is bounded in probability by) r_n and write

$$\mathbf{Y}_n = 0_p(r_n)$$

if, for every $\varepsilon > 0$, there exists a positive real number M_ε such that

$$P\{|Y_{jn}| \geq M_\varepsilon r_n\} \leq \varepsilon, \qquad j = 1, \dots, p$$

for all n.   □

Using Chebyshev's inequality, it can be shown that any random variable with finite variance is bounded in probability by the square root of its second moment about the origin. This result is stated without proof in the following theorem.

Theorem 6.2.1. *Let* Y_{1n} *denote the first element of* $\mathbf{Y}_n$ *and suppose that*

$$E\{Y_{1n}^2\} = 0(r_n^2),$$

i.e., $E\{Y_{1n}^2\}/r_n^2$ *is bounded. Then*

$$Y_{1n} = 0_p(r_n).$$

PROOF. See e.g., Fuller (1976).   □

For example, suppose that Y_{1n} is the sample mean of n independent $(0, \sigma^2)$ random variables. Then, $E\{Y_{1n}^2\} = \sigma^2/n$ and Theorem 6.2.1 shows that

$$Y_{1n} = 0_p(n^{-1/2}).$$

The variance expressions to be considered rest on the validity of Taylor's theorem for random variables, and the approximations employed may be quantified in terms of the order in probability concept. Let $g(\mathbf{y})$ be a real-valued function defined on p-dimensional Euclidian space with continuous partial derivatives of order 2 in an open sphere containing $\mathbf{Y}_n$ and $\mathbf{a}$. Then, by Taylor's theorem

$$g(\mathbf{Y}_n) = g(\mathbf{a}) + \sum_{j=1}^{p} \frac{\partial g(\mathbf{a})}{\partial y_j} (Y_{jn} - a_j) + R_n(\mathbf{Y}_n, \mathbf{a}) \qquad (6.2.1)$$

where

$$R_n(\mathbf{Y}_n, \mathbf{a}) = \sum_{j=1}^{p} \sum_{i=1}^{p} \frac{1}{2!} \frac{\partial^2 g(\ddot{\mathbf{a}})}{\partial y_j \, \partial y_i} (Y_{jn} - a_j)(Y_{in} - a_i),$$

$\partial g(\mathbf{a})/\partial y_j$ is the partial derivative of $g(\mathbf{y})$ with respect to the j-th element of $\mathbf{y}$ evaluated at $\mathbf{y} = \mathbf{a}$, $\partial^2 g(\ddot{\mathbf{a}})/\partial y_j \, \partial y_i$ is the second partial derivative of $g(\mathbf{y})$ with respect to y_j and y_i evaluated at $\mathbf{y} = \ddot{\mathbf{a}}$, and $\ddot{\mathbf{a}}$ is on the line segment joining $\mathbf{Y}_n$ and $\mathbf{a}$. The following theorem establishes the size of the *remainder* $R_n(\mathbf{Y}_n, \mathbf{a})$.

Theorem 6.2.2. *Let*

$$\mathbf{Y}_n = \mathbf{a} + 0_p(r_n),$$

where $r_n \to 0$ as $n \to \infty$. Then $g(\mathbf{Y}_n)$ may be expressed by (6.2.1) where $R_n(\mathbf{Y}_n, \mathbf{a}) = 0_p(r_n^2)$.

PROOF. See, e.g., Fuller (1976).   □

A univariate version of (6.2.1) follows by letting $p = 1$.

In stating these results, the reader should note that we have retained only the linear terms in the Taylor series expansion. This was done to simplify the presentation, and because only the linear terms are used in developing the variance and variance estimating formulas. The expansion, however, may be extended to a polynomial of order $s - 1$, whenever $g(\cdot)$ has s continuous derivatives. See, e.g., Fuller (1976).

We now state the principal result of this section.

Theorem 6.2.3. *Let*

$$\mathbf{Y}_n = \mathbf{a} + 0_p(r_n),$$

where $r_n \to 0$ as $n \to \infty$, and let

$$E\{\mathbf{Y}_n\} = \mathbf{a}$$

$$E\{(\mathbf{Y}_n - \mathbf{a})(\mathbf{Y}_n - \mathbf{a})'\} = \mathbf{\Sigma}_n < \infty.$$

Then the asymptotic variance of $g(\mathbf{Y}_n)$ to order r_n^3 is

$$\bar{\mathrm{E}}\{(g(\mathbf{Y}_n) - g(\mathbf{a}))^2\} = \mathbf{d}\mathbf{\Sigma}_n\mathbf{d}' + 0_p(r_n^3) \tag{6.2.2}$$

where $\mathbf{d}$ is a $1 \times p$ vector with typical element

$$d_j = \frac{\partial g(\mathbf{a})}{\partial y_j}.$$

(The notation, $\bar{\mathrm{E}}$, in (6.2.2) should be interpreted to mean that $(g(\mathbf{Y}_n) - g(\mathbf{a}))^2$ can be written as the sum of two random variables, say X_n and Z_n, where $\mathrm{E}\{X_n\} = \mathbf{d}\mathbf{\Sigma}_n\mathbf{d}'$ and $Z_n = 0_p(r_n^3)$. This does not necessarily mean that $\mathrm{E}\{(g(\mathbf{X}_n) - g(\mathbf{a}))^2\}$ exists for any finite n.)

PROOF. Follows directly from Theorem 6.2.2 since

$$g(\mathbf{Y}_n) - g(\mathbf{a}) = \mathbf{d}(\mathbf{Y}_n - \mathbf{a}) + 0_p(r_n^2). \qquad \square$$

If $\mathbf{Y}_n$ is the mean of n independent random variables, then a somewhat stronger result is available.

Theorem 6.2.4. *Let $\{\mathbf{Y}_n\}$ be a sequence of means of n independent, p-dimensional random variables, each with mean $\mathbf{a}$, covariance matrix $\mathbf{\Sigma}$, and finite fourth moments. If $g(\mathbf{y})$ possesses continuous derivatives of order 3 in a neighborhood of $\mathbf{y} = \mathbf{a}$, then the asymptotic variance of $g(\mathbf{Y}_n)$ to order n^{-2} is*

$$\bar{\mathrm{E}}\{(g(\mathbf{Y}_n) - g(\mathbf{a}))^2\} = (1/n)\mathbf{d}\mathbf{\Sigma}\mathbf{d}' + 0_p(n^{-2}).$$

PROOF. See, e.g., Fuller (1976). $\qquad \square$

The above theorems generalize immediately to multivariate problems. Suppose that $g_1(\mathbf{y}), g_2(\mathbf{y}), \ldots,$ and $g_q(\mathbf{y})$ are real-valued functions defined on p-dimensional Euclidian space with continuous partial derivatives of order 2 in a neighborhood of $\mathbf{a}$, where $2 \le q < \infty$.

Theorem 6.2.5. *Given the conditions of Theorem 6.2.3, the asymptotic covariance matrix of $\mathbf{G}(\mathbf{Y}_n) = [g_1(\mathbf{Y}_n), \ldots, g_q(\mathbf{Y}_n)]'$ to order r_n^3 is*

$$\bar{\mathrm{E}}\{(\mathbf{G}(\mathbf{Y}_n) - \mathbf{G}(\mathbf{a}))(\mathbf{G}(\mathbf{Y}_n) - \mathbf{G}(\mathbf{a}))'\} = \mathbf{D}\mathbf{\Sigma}_n\mathbf{D}' + 0_p(r_n^3), \tag{6.2.3}$$

where $\mathbf{D}$ is a $q \times p$ matrix with typical element

$$d_{ij} = \frac{\partial g_i(\mathbf{a})}{y_j}. \qquad \square$$

Theorem 6.2.6. *Given the conditions of Theorem 6.2.4, the asymptotic covariance matrix of $\mathbf{G}(\mathbf{Y}_n)$ or order n^{-2} is*

$$\bar{\mathrm{E}}\{(\mathbf{G}(\mathbf{Y}_n) - \mathbf{G}(\mathbf{a}))(\mathbf{G}(\mathbf{Y}_n) - \mathbf{G}(\mathbf{a}))'\} = (1/n)\mathbf{D}\mathbf{\Sigma}\mathbf{D}' + 0_p(n^{-2}). \qquad \square$$

A proof of Theorems 6.2.5 and 6.2.6 may be obtained by expanding each function $g_i(\mathbf{Y}_n)$, $i = 1, \ldots, q$, in the Taylor series form (6.2.1).

Theorems 6.2.3 and 6.2.4 provide approximate expressions for the variance of a single nonlinear statistic $g(\cdot)$, while Theorems 6.2.5 and 6.2.6 provide approximate expressions for the covariance matrix of a vector nonlinear statistic $\mathbf{G}(\cdot)$. In the next section, we show how these results may be adapted to the classical finite population model, and then show how to provide estimators of variance.

6.3. Linear Approximations in the Finite Population

We consider a given finite population N, let $\mathbf{Y} = (Y_1, \ldots, Y_p)'$ denote a p-dimensional vector of population parameters, and let $\hat{\mathbf{Y}} = (\hat{Y}_1, \ldots, \hat{Y}_p)'$ denote a corresponding vector of estimators based on a sample s of size $n(s)$. The form of the estimators $\hat{Y}_i$, $i = 1, \ldots, p$, depends on the sampling design generating the sample s. In most applications of Taylor series methods, the Y_i denote population totals or means for p different survey characteristics and the $\hat{Y}_i$ denote standard estimators of the Y_i. Usually the $\hat{Y}_i$ are unbiased for the Y_i, though in some applications they may be biased but consistent estimators. To emphasize the functional dependence on the sample size, we might have subscripted the estimators by $n(s)$, i.e.,

$$\hat{\mathbf{Y}}_{n(s)} = (\hat{Y}_{1,n(s)}, \ldots, \hat{Y}_{p, n(s)})'.$$

For notational convenience, however, we delete the explicit subscript $n(s)$ from all variables, whenever no confusion will result.

We suppose that the population parameter of interest is $\theta = g(\mathbf{Y})$ and adopt the natural estimator $\hat{\theta} = g(\hat{\mathbf{Y}})$. The main problems to be addressed in this section are 1) finding an approximate expression for the design variance of $\hat{\theta}$ and 2) constructing a suitable estimator of the variance of $\hat{\theta}$.

If $g(\mathbf{y})$ possesses continuous derivatives of order 2 in an open sphere containing $\hat{\mathbf{Y}}$ and $\mathbf{Y}$, then by (6.2.1) we may write

$$\hat{\theta} - \theta = \sum_{j=1}^{p} \frac{\partial g(\mathbf{Y})}{\partial y_j} (\hat{Y}_j - Y_j) + R_{n(s)}(\hat{\mathbf{Y}}, \mathbf{Y}),$$

where

$$R_{n(s)}(\hat{\mathbf{Y}}, \mathbf{Y}) = \sum_{j=1}^{p} \sum_{i=1}^{p} (1/2!) \frac{\partial^2 g(\ddot{\mathbf{Y}})}{\partial y_j \, \partial y_i} (\hat{Y}_j - Y_j)(\hat{Y}_i - Y_i)$$

and $\ddot{\mathbf{Y}}$ is between $\hat{\mathbf{Y}}$ and $\mathbf{Y}$. As we shall see, this form of Taylor's theorem is useful for approximating variances in finite population sampling problems.

In the finite population, it is customary to regard the remainder $R_{n(s)}(\hat{\mathbf{Y}}, \mathbf{Y})$ as an "unimportant" component of the difference $g(\hat{\mathbf{Y}}) - g(\mathbf{Y})$, relative to the linear terms in the Taylor series expansion. Thus, the mean square error (MSE) of $\hat{\theta}$ is given approximately by

$$
\begin{aligned}
\text{MSE}\{\hat{\theta}\} &= \text{E}\{(g(\hat{\mathbf{Y}}) - g(\mathbf{Y}))^2\} \\
&\doteq \text{Var}\left\{\sum_{j=1}^{p} \frac{\partial g(\mathbf{Y})}{\partial y_j}(\hat{Y}_j - Y_j)\right\} \\
&= \sum_{j=1}^{p}\sum_{i=1}^{p} \frac{\partial g(\mathbf{Y})}{\partial y_j}\frac{\partial g(\mathbf{Y})}{\partial y_i}\text{Cov}\{\hat{Y}_j, \hat{Y}_i\} \\
&= \mathbf{d}\boldsymbol{\Sigma}_{n(s)}\mathbf{d}',
\end{aligned}
\tag{6.3.1}
$$

where $\boldsymbol{\Sigma}_{n(s)}$ is the covariance matrix of $\hat{\mathbf{Y}}$ and $\mathbf{d}$ is a $1 \times p$ vector with typical element

$$
d_j = \frac{\partial g(\mathbf{Y})}{\partial y_j}.
$$

This expression is analogous to (6.2.2) in Theorem 6.2.3. We refer to (6.3.1) as the *first-order approximation* to $\text{MSE}\{\hat{\theta}\}$. Second- and higher-order approximations are possible by extending the Taylor series expansion and retaining the additional terms in the approximation. Experience with large, complex sample surveys has shown, however, that the first-order approximation often yields satisfactory results. Users should be warned that the approximation may not be satisfactory for surveys of highly skewed populations.

A multivariate generalization of (6.3.1) is constructed by analogy with (6.2.3). Let

$$
\mathbf{G}(\mathbf{Y}) = [g_1(\mathbf{Y}), \ldots, g_q(\mathbf{Y})]'
$$

denote a q-dimensional parameter of interest, and suppose that it is estimated by

$$
\mathbf{G}(\hat{\mathbf{Y}}) = [g_1(\hat{\mathbf{Y}}), \ldots, g_q(\hat{\mathbf{Y}})]'.
$$

Then the matrix of mean square errors and cross products is given approximately by

$$
\text{E}\{[\mathbf{G}(\hat{\mathbf{Y}}) - \mathbf{G}(\mathbf{Y})][\mathbf{G}(\hat{\mathbf{Y}}) - \mathbf{G}(\mathbf{Y})]'\} \doteq \mathbf{D}\boldsymbol{\Sigma}_{n(s)}\mathbf{D}'.
\tag{6.3.2}
$$

The matrix $\mathbf{D}$ is $q \times p$ with typical element

$$
d_{ij} = \frac{\partial g_i(\mathbf{Y})}{\partial y_j}.
$$

For purposes of variance estimation, we shall substitute sample-based estimates of $\mathbf{d}$ (or $\mathbf{D}$) and $\boldsymbol{\Sigma}_{n(s)}$. Suppose that an estimator, say $\hat{\boldsymbol{\Sigma}}_{n(s)}$, of $\boldsymbol{\Sigma}_{n(s)}$ is available. The estimator, $\hat{\boldsymbol{\Sigma}}_{n(s)}$ should be specified in accordance with the sampling design. Then an estimator of $\text{MSE}\{\hat{\boldsymbol{\theta}}\}$ is given by

$$v(\hat{\theta}) = \hat{\mathbf{d}}\hat{\boldsymbol{\Sigma}}_{n(s)}\hat{\mathbf{d}}', \tag{6.3.3}$$

where $\hat{\mathbf{d}}$ is the $1 \times p$ vector with typical element

$$\hat{d}_j = \frac{\partial g(\hat{\mathbf{Y}})}{\partial y_j}.$$

Similarly, an estimator of (6.3.2) is given by

$$\mathbf{v}(\mathbf{G}(\hat{\mathbf{Y}})) = \hat{\mathbf{D}}\hat{\boldsymbol{\Sigma}}_{n(s)}\hat{\mathbf{D}}' \tag{6.3.4}$$

where $\hat{\mathbf{D}}$ is the $q \times p$ matrix with typical element

$$\hat{d}_{ij} = \frac{\partial g_i(\hat{\mathbf{Y}})}{\partial y_j}.$$

In general, $v(\hat{\theta})$ will not be an unbiased estimator of either the true $\text{MSE}\{\hat{\theta}\}$ or the approximation $\mathbf{d}\boldsymbol{\Sigma}_{n(s)}\mathbf{d}'$. It is, however, a consistent estimator provided that $\hat{\mathbf{Y}}$ and $\hat{\boldsymbol{\Sigma}}_{n(s)}$ are consistent estimators of $\mathbf{Y}$ and $\boldsymbol{\Sigma}_{n(s)}$, respectively. The same remarks hold true for $\mathbf{v}(\mathbf{G}(\hat{\mathbf{Y}}))$. The asymptotic properties of these estimators are discussed in Appendix B.

The reader may have observed that our development has been in terms of the mean square error $\text{MSE}\{\hat{\theta}\}$, while our stated purpose was a representation of the variance $\text{Var}\{\hat{\theta}\}$ and construction of a variance estimator. This apparent dichotomy may seem puzzling at first, but is easily explained. The explanation is that to the order of approximation entertained in (6.3.1), the $\text{MSE}\{\hat{\theta}\}$ and the $\text{Var}\{\hat{\theta}\}$ are identical. Of course the true mean square error satisfies

$$\text{MSE}\{\hat{\theta}\} = \text{Var}\{\hat{\theta}\} + \text{Bias}^2\{\hat{\theta}\}.$$

But to a first approximation, $\text{Var}\{\hat{\theta}\}$ and $\text{Bias}\{\hat{\theta}\}$ are the same order and $\text{Bias}^2\{\hat{\theta}\}$ is of lower order. Therefore, $\text{MSE}\{\hat{\theta}\}$ and $\text{Var}\{\hat{\theta}\}$ are the same to a first approximation. In the sequel, we may write either $\text{Var}\{\hat{\theta}\}$ or $\text{MSE}\{\hat{\theta}\}$ in reference to the approximation, and the reader should not become confused.

For the finite population model, the validity of the above methods is often criticized. At issue is whether the Taylor series used to develop (6.3.1) converges, and if so, at what rate does it converge? For the infinite population model it was possible to establish the order of the remainder in the Taylor series expansion, and it was seen that the remainder was of lower order than the linear terms in the expansion. On this basis, the remainder was ignored in making approximations. For the finite population model, no such results are possible without also assuming a superpopulation model or a sequence of finite populations increasing in size.

To illustrate the potential problems, suppose that $\bar{y}$ and $\bar{x}$ denote sample means based on a simple random sample without replacement of size n. The ratio $R = \bar{Y}/\bar{X}$ of population means is to be estimated by $\hat{R} = \bar{y}/\bar{x}$.

Letting

$$\delta_y = (\bar{y} - \bar{Y})/\bar{Y}$$

$$\delta_x = (\bar{x} - \bar{X})/\bar{X}$$

we can write

$$\hat{R} = R(1 + \delta_y)(1 + \delta_x)^{-1},$$

and expanding $\hat{R}$ in a Taylor series about the point $\delta_x = 0$ gives

$$\hat{R} = R(1 + \delta_y)(1 - \delta_x + \delta_x^2 - \delta_x^3 + \delta_x^4 - + \ldots)$$

$$= R(1 + \delta_y - \delta_x - \delta_y\delta_x + \delta_x^2 \ldots).$$

By the binomial theorem, convergence of this series is guaranteed if and only if $|\delta_x| < 1$. Consequently, the approximate formula for MSE$\{\hat{R}\}$ will be valid if and only if $|\delta_x| < 1$ for all $\binom{N}{n}$ possible samples.

Koop (1972) gives a simple example where the convergence condition is violated. In this example $N = 20$; the unit values are

5, 1, 3, 6, 7, 8, 1, 3, 10, 11, 16, 4, 2, 11, 6, 6, 7, 1, 5, 13;

and $\bar{X} = 6.3$. For one sample of size 4 we find $\bar{x} = (11 + 16 + 11 + 13)/4 = 12.75$, and thus $|\delta_x| > 1$. Samples of size $n = 2$ and $n = 3$ also exist where $|\delta_x| > 1$. However, for samples of size $n \geq 5$ convergence is guaranteed. Koop calls $n = 5$ the *critical sample size*.

Even when convergence of the Taylor series is guaranteed for all possible samples, the series may converge slowly for a substantial number of samples, and the first order approximations discussed here may not be adequate. It may be necessary to include additional terms in the Taylor series when approximating the mean square error. Koop (1968) illustrates with numerical examples, Sukhatme and Sukhatme (1970) give a second order approximation to MSE$\{\hat{R}\}$, and Dippo (1981) derives second order approximations in general.

In spite of the convergence considerations, the first order approximation is used widely in sample surveys from finite populations. Experience has shown that where the sample size is sufficiently large and where the concepts of efficient survey design are successfully applied, the first order Taylor series expansion often provides reliable approximations. Again, we caution the user that the approximations may be unreliable in the context of highly skewed populations.

6.4. A Special Case

An important special case of (6.3.1) is discussed by Hansen, Hurwitz, and Madow (1953). The parameter of interest is of the form

$$\theta = g(\mathbf{Y}) = (Y_1 Y_2 \ldots Y_m)/(Y_{m+1} Y_{m+2} \ldots Y_p),$$

where $1 \leq m \leq p$. A simple example is the ratio

$$\theta = Y_1 / Y_2$$

of $p = 2$ population totals Y_1 and Y_2. To a first-order approximation,

$$
\begin{aligned}
\text{MSE}\{\hat{\theta}\} = \theta^2 \{ & [\sigma_{11}/Y_1^2 + \ldots + \sigma_{mm}/Y_m^2] \\
& + [\sigma_{m+1,m+1}/Y_{m+1}^2 + \ldots + \sigma_{pp}/Y_p^2] \\
& + 2[\sigma_{12}/(Y_1 Y_2) + \sigma_{13}/(Y_1 Y_3) \\
& \quad + \ldots + \sigma_{m-1,m}/(Y_{m-1} Y_m)] \\
& + 2[\sigma_{m+1,m+2}/(Y_{m+1} Y_{m+2}) + \sigma_{m+1,m+3}/(Y_{m+1} Y_{m+3}) \\
& \quad + \ldots + \sigma_{p-1,p}/(Y_{p-1} Y_p)] \\
& - 2[\sigma_{1,m+1}/(Y_1 Y_{m+1}) + \sigma_{1,m+2}/(Y_1 Y_{m+2}) + \ldots \\
& \quad + \sigma_{m,p}/(Y_m Y_p)]\},
\end{aligned}
\tag{6.4.1}
$$

where

$$\sigma_{ij} = \text{Cov}\{\hat{Y}_i, \hat{Y}_j\}$$

is a typical element of $\mathbf{\Sigma}_{n(s)}$. If an estimator $\hat{\sigma}_{ij}$ of σ_{ij} is available for $i, j = 1, \ldots, p$, then we estimate $\text{MSE}\{\hat{\theta}\}$ by

$$
\begin{aligned}
v(\hat{\theta}) = \hat{\theta}^2 \{ & [\hat{\sigma}_{11}/\hat{Y}_1^2 + \ldots + \hat{\sigma}_{mm}/\hat{Y}_m^2] + [\hat{\sigma}_{m+1,m+1}/\hat{Y}_{m+1}^2 + \ldots + \hat{\sigma}_{pp}/\hat{Y}_p^2] \\
& + 2[\hat{\sigma}_{12}/(\hat{Y}_1 \hat{Y}_2) + \ldots + \hat{\sigma}_{m-1,m}/(\hat{Y}_{m-1} \hat{Y}_m)] \\
& + 2[\hat{\sigma}_{m+1,m+2}/(\hat{Y}_{m+1} \hat{Y}_{m+2}) + \ldots + \hat{\sigma}_{p-1,p}/(\hat{Y}_{p-1} \hat{Y}_p)] \\
& - 2[\hat{\sigma}_{1,m+1}/(\hat{Y}_1 \hat{Y}_{m+1}) + \ldots + \hat{\sigma}_{m,p}/(\hat{Y}_m \hat{Y}_p)]\}.
\end{aligned}
\tag{6.4.2}
$$

This expression is easy to remember. All terms (i, j) where $i = j$ pertain to a relative variance and have a coefficient of $+1$, whereas terms (i, j) where $i \neq j$ pertain to a relative covariance and have a coefficient of $+2$ or -2. For the relative covariances, $+2$ is used when both i and j are in the numerator or denominator of θ, and -2 is used otherwise.

In the simple ratio example,

$$\hat{\theta} = \hat{Y}_1 / \hat{Y}_2$$

and

$$v(\hat{\theta}) = \hat{\theta}^2 (\hat{\sigma}_{11}/\hat{Y}_1^2 + \hat{\sigma}_{22}/\hat{Y}_2^2 - 2\hat{\sigma}_{12}/\hat{Y}_1 \hat{Y}_2).$$

6.5. A Computational Algorithm

In certain circumstances, an alternative form of (6.3.1) and (6.3.3) is available. Depending on available software, this form may have some computational advantages since it avoids the computation of the $p \times p$ covariance matrix $\hat{\mathbf{\Sigma}}_{n(s)}$.

We shall assume that $\hat{Y}_j$ is of the form

$$\hat{Y}_j = \sum_i^{n(s)} w_i y_{ij}, \qquad j = 1, \ldots, p,$$

where w_i denotes a *weight* attached to the i-th unit in the sample. By (6.3.1) we have

$$
\begin{aligned}
\mathrm{MSE}\{\hat{\theta}\} &\doteq \mathrm{Var}\left\{ \sum_j^p \frac{\partial g(\mathbf{Y})}{\partial y_j} \hat{Y}_j \right\} \\
&= \mathrm{Var}\left\{ \sum_j^p \frac{\partial g(\mathbf{Y})}{\partial y_j} \sum_i^{n(s)} w_i y_{ij} \right\} \\
&= \mathrm{Var}\left\{ \sum_i^{n(s)} w_i \sum_j^p \frac{\partial g(\mathbf{Y})}{\partial y_j} y_{ij} \right\} \\
&= \mathrm{Var}\left\{ \sum_i^{n(s)} w_i v_i \right\},
\end{aligned}
\tag{6.5.1}
$$

where

$$v_i = \sum_j^p \frac{\partial g(\mathbf{Y})}{\partial y_j} y_{ij}.$$

Thus, by a simple interchange of summations we have converted a p-variate estimation problem into a univariate problem. The new variable v_i is a linear combination of the original variables ($y_{i1}, y_{i2}, \ldots, y_{ip}$).

Variance estimation is now simplified computationally because we only estimate the variance of the single statistic $\sum_i^{n(s)} w_i v_i$ instead of estimating the $p \times p$ covariance matrix $\mathbf{\Sigma}_{n(s)}$. The variance estimator that would have been used in $\hat{\mathbf{\Sigma}}_{n(s)}$ for estimating the diagonal terms of $\mathbf{\Sigma}_{n(s)}$ may be used for estimating the variance of this single statistic. Of course, the $\mathbf{Y}$ in $\partial g(\mathbf{Y})/\partial y_j$ is unknown and must be replaced by a sample-based estimate. Thus, we apply the variance estimating formula to the single variate

$$\hat{v}_i = \sum_j^p \frac{\partial g(\hat{\mathbf{Y}})}{\partial y_j} y_{ij}.$$

The expression in (6.5.1) is due to Woodruff (1971). It is a generalization of the identity Keyfitz (1957) gave for estimating the variance of various estimators from a stratified design with two primaries per stratum.

6.6. Usage with Other Methods

Irrespective of whether (6.3.1) or (6.5.1) is used in estimating $\mathrm{MSE}\{\hat{\theta}\}$, it is necessary to be able to estimate the variance of a single statistic, e.g.

$$\sum_i^{n(s)} w_i y_{ij} \quad \text{or} \quad \sum_i^{n(s)} w_i v_i.$$

For this purpose, we may use the textbook estimator appropriate to the specific sampling design and estimator. It is also possible to employ the methods of estimation discussed in other chapters of this book. For example, we may use the random group technique, balanced half-sample replication, or jackknife. If a stratified design is used and only one primary has been selected from each stratum, then we may estimate the variance using the collapsed stratum technique. When units are selected systematically, we may use one of the biased estimators of variance discussed in Chapter 7.

6.7. Example: Composite Estimators

In Section 2.10 we discussed the Census Bureau's retail trade survey. We observed that estimates of total monthly sales are computed for several, selected kinds of business (KB). Further, it was demonstrated how the random group technique is used to estimate the variance of the Horvitz–Thompson estimator of total sales. The Census Bureau, however, does not publish the Horvitz–Thompson estimates. Rather, composite type estimates are published which utilize the correlation structure between the various simple estimators to reduce sampling variability.

To illustrate this method of estimation, we consider a given four-digit standard industrial classification (SIC) code for which an estimate of total sales is to be published.[1] We consider estimation only for the list sample portion of this survey (the distinction between the list and area samples is discussed in Section 2.10).

During the monthly enumeration, all noncertainty units which are engaged in the specific KB report their total sales for both the current and previous month. As a result of this reporting pattern, *two* Horvitz–Thompson estimators of total sales are available for each month. We shall let

$Y'_{t,\alpha}$ = Horvitz–Thompson estimator of total sales for month t obtained from the α-th random group of the sample reporting in month t

and

$Y''_{t,\alpha}$ = Horvitz–Thompson estimator of total sales for month t obtained from the α-th random group of the sample reporting in month $t + 1$.

---

[1] The "Standard Industrial Classification" (SIC) code is a numeric code of two, three, or four digits which denotes a specific economic activity. Kind-of-business (KB) codes of five or six digits are assigned by the Census Bureau to produce more detailed classifications within certain four digit SIC industries.

Thus

$$Y'_t = \sum_{\alpha=1}^{16} Y'_{t,\alpha}/16$$

and

$$Y''_t = \sum_{\alpha=1}^{16} Y''_{t,\alpha}/16$$

are the two simple estimators of total sales for month t for the noncertainty portion of the list sample. The corresponding simple estimators of total sales, including both certainty and noncertainty units, are

$$Y_{t,0} + Y'_t$$

and

$$Y_{t,0} + Y''_t,$$

where $Y_{t,0}$ denotes the fixed total for certainty establishments for month t. These expressions are analogous to the expression presented in (2.10.2).

One of the composite estimators of total sales which is published, known as the *preliminary composite estimator*, is recursively defined by

$$Y'''_t = Y'_t + \beta_t \frac{X'_t}{X''_{t+1}}(Y'''_{t-1} - Y''_{t-1}),$$

where

X'_t = Horvitz–Thompson estimator of total sales in the three digit KB of which the given four digit KB is a part, for month t, obtained from the sample reporting in month t,

X''_{t-1} = Horvitz–Thompson estimator of total sales in the three digit KB of which the given four digit KB is a part, for month $t - 1$, obtained from the sample reporting in month t,

Y'''_{t-1} = preliminary composite estimator for month $t - 1$,

the β_t denote fixed constants, and $Y'_1, Y''_0, X'_0, X'_1, X''_0$ are the initial values. The values of the β_t employed in this survey are as follows:

t	β_t
1	0.00
2	0.48
3	0.62
4	0.75
5	0.75
$\vdots$	$\vdots$

The reader will recognize that Y_t''' is a function of Y_{t-j}', Y_{t-j-1}'', X_{t-j}', and X_{t-j-1}' for $j = 0, 1, \ldots, t - 1$. In particular, we can write

$$Y_t''' = g(Y_t', Y_{t-1}'', Y_{t-1}', Y_{t-2}'', \ldots, Y_1', Y_0'',$$

$$X_t', X_{t-1}'', X_{t-1}', X_{t-2}'', \ldots, X_1', X_0'') \qquad (6.7.1)$$

$$= Y_t' + \sum_{j=1}^{t-1} \left[\prod_{i=1}^{j} \beta_{t-i+1} \frac{X_{t-i+1}'}{X_{t-i}''} \right] (Y_{t-j}' - Y_{t-j}'').$$

Expression (6.7.1) will be useful in deriving a Taylor series estimator of variance. Towards this end, let

$$Y_{t-j} = E\{Y_{t-j}'\},$$

$$Y_{t-j-1} = E\{Y_{t-j-1}''\},$$

$$X_{t-j} = E\{X_{t-j}'\},$$

and

$$X_{t-j-1} = E\{X_{t-j-1}''\}$$

for $j = 0, 1, \ldots, t - 1$. Then, expanding Y_t''' about the point

$$(Y_t, Y_{t-1}, Y_{t-1}, Y_{t-2}, \ldots, Y_1, Y_0, X_t, X_{t-1}, X_{t-1}, X_{t-2}, \ldots, X_1, X_0)$$

gives the following expression, which is analogous to (6.2.1):

$$Y_t''' \doteq Y_t + (Y_t' - Y_t) + \sum_{j=1}^{t-1} \prod_{i=1}^{j} \beta_{t-i+1} \frac{X_{t-i+1}}{X_{t-1}} (Y_{t-j}' - Y_{t-j}''). \quad (6.7.2)$$

Corresponding to (6.3.1), an approximate expression for the mean square error is

$$\text{MSE}\{Y_t'''\} \doteq \mathbf{d}\Sigma\mathbf{d}', \qquad (6.7.3)$$

where Σ denotes the covariance matrix of $(Y_t', Y_{t-1}'', Y_{t-1}', Y_{t-2}'', \ldots, Y_1', Y_0'', X_t', X_{t-1}'', \ldots, X_1', X_0'')$,

$$\mathbf{d} = (\mathbf{d}_y, \mathbf{d}_x),$$

$$\mathbf{d}_y = \left(1, -\beta_t \frac{X_t}{X_{t-1}}, \beta_t \frac{X_t}{X_{t-1}}, -\prod_{i=1}^{2} \beta_{t-i-1} \frac{X_{t-i+1}}{X_{t-1}}, \prod_{i=1}^{2} \beta_{t-i+1} \frac{X_{t-i+1}}{X_{t-1}}, \ldots, \right.$$

$$\left. -\prod_{i=1}^{t-1} \beta_{t-i+1} \frac{X_{t-i+1}}{X_{t-i}}, \prod_{i=1}^{t-1} \beta_{t-i+1} \frac{X_{t-i+1}}{X_{t-i}}, 0 \right),$$

and $\mathbf{d}_x$ denotes a $(1 \times 2t)$ vector of zeros. Alternatively, corresponding to (6.5.1) the mean square error may be approximated by the variance of the single variate

$$\tilde{Y}_t = Y_t' + \sum_{j=1}^{t-1} \prod_{i=1}^{j} \beta_{t-i+1} \frac{X_{t-i+1}}{X_{t-i}} (Y_{t-j}' - Y_{t-j}'')$$

$$= Y_t' + \beta_t \frac{X_t}{X_{t-1}} (\tilde{Y}_{t-1} - Y_{t-1}''). \qquad (6.7.4)$$

To estimate the variance of Y_t''' we shall employ (6.7.4) and the computational approach given in Section 6.5. In this problem, variance estimation according to (6.7.3) would require estimation of the $(4t \times 4t)$ covariance matrix $\boldsymbol{\Sigma}$. For even moderate values of t, this would seem to involve more computations than the univariate method of Section 6.5.

Define the random group totals

$$\tilde{Y}_{t,\alpha} = Y_{t,\alpha}' + \beta_t \frac{X_t}{X_{t-1}} (\tilde{Y}_{t-1,\alpha} - Y_{t-1,\alpha}'')$$

for $\alpha = 1, \ldots, 16$. If the X_{t-j} were known for $j = 0, \ldots, t-1$, then we would estimate the variance of $\tilde{Y}_t$, and thus of Y_t''', by the usual random group estimator

$$\frac{1}{16(15)} \sum_{\alpha=1}^{16} (\tilde{Y}_{t,\alpha} - \tilde{Y}_t)^2.$$

Unfortunately, the X_{t-j} are not known and we must substitute the sample estimates X_{t-j}' and X_{t-j-1}''. This gives

$$v(Y_t''') = \frac{1}{16(15)} \sum_{\alpha=1}^{16} (Y_{t,\alpha}''' - Y_t''')^2 \qquad (6.7.5)$$

Table 6.7.1. Random Group Estimates $Y_{t,\alpha}'''$ and $Y_{t-1,\alpha}''''$ for August and July 1977 Grocery Store Sales from the List Sample Portion of the Retail Trade Survey

Random Group α	$Y_{t,\alpha}'''$ ($1000)	$Y_{t-1,\alpha}''''$ ($1000)
1	4,219,456	4,329,856
2	4,691,728	4,771,344
3	4,402,960	4,542,464
4	4,122,576	4,127,136
5	4,094,112	4,223,040
6	4,368,000	4,577,456
7	4,426,576	4,427,376
8	4,869,232	4,996,480
9	4,060,576	4,189,472
10	4,728,976	4,888,912
11	5,054,880	5,182,576
12	3,983,136	4,144,368
13	4,712,880	4,887,360
14	3,930,624	4,110,896
15	4,358,976	4,574,752
16	4,010,880	4,081,936

as our final estimator of variance, where

$$Y'''_{t,\alpha} = Y'_{t,\alpha} + \beta_t \frac{X'_t}{X''_{t-1}} (Y'''_{t-1,\alpha} - Y''_{t-1,\alpha})$$

for $\alpha = 1, \ldots, 16$.

To illustrate this methodology, Table 6.7.1 gives the quantities $Y'''_{t,\alpha}$ corresponding to August 1977 grocery store sales. The computations associated with $v(Y'''_t)$ are presented in Table 6.7.2.

The Census Bureau also publishes a second composite estimator of total sales, known as the *final composite estimator*. For the noncertainty portion of a given four digit KB, this estimator is defined by

$$Y''''_t = (1 - \gamma_t) Y''_t + \gamma_t Y'''_t,$$

where the γ_t are fixed constants in the unit interval. This *final* estimator is not available until month $t + 1$ (i.e., when Y''_t becomes available), whereas the *preliminary* estimator is available in month t. To estimate the variance

Table 6.7.2. Computation of Y'''_t and $v(Y'''_t)$ for August 1977 Grocery Store Sales

---

By definition, the estimated noncertainty total for August is

$$Y'''_t = \sum_{\alpha=1}^{16} Y'''_{t,\alpha}/16$$

$$= 4,377,223,$$

where the unit is \$1000. The estimator of variance is

$$v(Y'''_t) = \frac{1}{16(15)} \left[\sum_{\alpha=1}^{16} Y'''^2_{t,\alpha} - 16(Y'''_t)^2 \right]$$

$$= \frac{1}{16(15)} [308,360,448 \cdot 10^6 - 306,561,280 \cdot 10^6]$$

$$= 7,496,801,207.$$

To obtain the total estimate of grocery store sales, we add the noncertainty total Y'''_t to the total from the certainty stratum obtained in Section 2.10

$$Y_{t,0} + Y'''_t = 7,154,943 + 4,377,223$$

$$= 11,532,166.$$

The estimated coefficient of variation associated with this estimator is

$$cv(Y_{t,0} + Y'''_t) = \sqrt{v(Y'''_t)}/(Y_{t,0} + Y'''_t)$$

$$= 0.0075.$$

---

of Y_t'''' we use

$$v(Y_t''') = \frac{1}{16(15)} \sum_{\alpha=1}^{16} (Y_{t,\alpha}'''' - Y_t'''')^2, \qquad (6.7.6)$$

where

$$Y_{t,\alpha}'''' = (1 - \gamma_t) Y_{t,\alpha}'' + \gamma_t Y_{t,\alpha}'''$$

for $\alpha = 1, \ldots, 16$. The development of this estimator of variance is similar to that of the estimator $v(Y_t''')$, and utilizes a combination of the random group and the Taylor series methodologies.

6.8. Example: Simple Ratios

Let Y and X denote two unknown population totals. The natural estimator of the ratio

$$R = Y/X$$

is

$$\hat{R} = \hat{Y}/\hat{X},$$

where $\hat{Y}$ and $\hat{X}$ denote estimators of Y and X. By (6.4.2) the Taylor series estimator of the variance of $\hat{R}$ is

$$v(\hat{R}) = \hat{R}^2 \left[\frac{v(\hat{Y})}{\hat{Y}^2} + \frac{v(\hat{X})}{\hat{X}^2} - 2 \frac{c(\hat{Y}, \hat{X})}{\hat{X}\hat{Y}} \right], \qquad (6.8.1)$$

where $v(\hat{Y})$, $v(\hat{X})$, and $c(\hat{Y}, \hat{X})$, denote estimators of $\mathrm{Var}\{\hat{Y}\}$, $\mathrm{Var}\{\hat{X}\}$, and $\mathrm{Cov}\{\hat{Y}, \hat{X}\}$ respectively. Naturally, the estimators $v(\hat{Y})$, $v(\hat{X})$, and $c(\hat{Y}, \hat{X})$ should be specified in accordance with both the sampling design and the form of the estimators $\hat{Y}$ and $\hat{X}$. This formula for $v(\hat{R})$ is well-known, having appeared in almost all of the basic sampling textbooks. In many cases, however, it is discussed in the context of simple random sampling with $\hat{Y} = N\bar{y}$, $\hat{X} = N\bar{x}$. Equation (6.8.1) indicates how the methodology applies to general sample designs and estimators.

We consider two illustrations of the methodology. The first involves the retail trade survey (cf. Sections 2.10 and 6.7). An important parameter is the *month-to-month trend* in retail sales. In the notation of Section 6.7, this trend is defined by $R_t = (Y_{t,0} + Y_t)/(Y_{t-1,0} + Y_{t-1})$ and is estimated by

$$\hat{R}_t = (Y_{t,0} + Y_t''')/(Y_{t-1,0} + Y_{t-1}''').$$

To estimate the variance of $\hat{R}_t$, we require estimates of $\mathrm{Var}\{Y_t'''\}$, $\mathrm{Var}\{Y_{t-1}'''\}$, and $\mathrm{Cov}\{Y_t''', Y_{t-1}'''\}$. As shown in Section 6.7, the variances are estimated by (6.7.5) and (6.7.6), respectively. In similar fashion, we estimate

Table 6.8.1. Computation of $v(\hat{R}_t)$ for the July–August 1977 Trend in Grocery Store Sales

By definition, the final composite estimate is

$$Y'''_{t-1} = \sum_{\alpha=1}^{16} Y''''_{t-1,\alpha}/16$$

$$= 4{,}503{,}464.$$

The certainty total for July is

$$Y_{t-1,0} = 7{,}612{,}644$$

so that the total estimate of July grocery store sales is

$$Y_{t-1,0} + Y''''_{t-1} = 7{,}612{,}644 + 4{,}503{,}464$$

$$= 12{,}116{,}108.$$

Thus, the estimate of the July–August trend is

$$\hat{R}_t = \frac{Y_{t,0} + Y'''_{t}}{Y_{t-1,0} + Y''''_{t-1}}$$

$$= 11{,}532{,}166/12{,}116{,}108$$

$$= 0.952.$$

The August estimates were derived previously in Table 6.7.2. We have seen that

$$v(Y'''_{t}) = 7{,}496{,}801{,}207,$$

and in similar fashion

$$v(Y''''_{t-1}) = 7{,}914{,}864{,}922$$

$$c(Y'''_{t}, Y''''_{t-1}) = 7{,}583{,}431{,}907.$$

An estimate of the variance of $\hat{R}_t$ is then

$$v(\hat{R}_t) = \hat{R}_t^2 \left[\frac{v(Y'''_{t})}{(Y_{t,0} + Y'''_{t})^2} + \frac{v(Y''''_{t-1})}{(Y_{t-1,0} + Y''''_{t-1})^2} \right.$$

$$\left. -2 \frac{c(Y'''_{t} Y''''_{t-1})}{(Y_{t,0} + Y'''_{t})(Y_{t-1,0} + Y''''_{t-1})} \right]$$

$$= 0.952^2[0.564 \cdot 10^{-4} + 0.539 \cdot 10^{-4} - 2(0.543 \cdot 10^{-4})]$$

$$= 0.154 \cdot 10^{-5}.$$

The corresponding estimated coefficient of variation (CV) is

$$cv(\hat{R}_t) = \sqrt{v(\hat{R}_t)}/\hat{R}_t$$

$$= 0.0013.$$

Table 6.8.2. Selected Annual Expenditures by Family Income Before Taxes, 1972

Expenditure Category	Total	$3,000	$3,000 to $3,999	$4,000 to $4,999	$5,000 to $5,999	$6,000 to $6,999	$7,000 to $7,999	$8,000 to $9,999	$10,000 to $11,999	$12,000 to $14,999	$15,000 to $19,999	$20,000 to $24,999	Over $25,000	Incomplete income reporting
							Income							
Number of families (1,000's)	70,788	10,419	4,382	3,825	3,474	3,292	3,345	6,719	6,719	8,282	9,091	4,369	3,553	3,317
Furniture														
Average annual expenditure	$117.23	$31.86	$35.51	$57.73	$69.94	$88.85	$84.38	$96.31	$128.70	$153.57	$185.97	$218.97	$266.26	$119.16
Standard error	4.806	5.601	8.619	10.888	11.755	12.700	11.516	8.477	9.776	9.689	10.565	15.160	18.514	14.789
Percent reporting	41	20	22	26	30	35	40	46	49	54	55	56	56	35
Small Appliances														
Average annual expenditure	$9.60	$3.48	$4.81	$7.09	$6.53	$8.24	$11.59	$9.75	$9.84	$12.17	$14.60	$12.26	$16.98	$8.20
Standard error	0.580	0.987	1.639	1.899	1.770	1.976	2.258	1.408	1.395	1.357	1.383	1.829	2.336	1.964
Percent reporting	32	16	19	24	28	30	32	35	36	39	42	40	42	30
Health Insurance														
Average annual expenditure	$187.33	$93.87	$129.83	$151.02	$158.13	$171.31	$182.22	$190.97	$196.45	$209.56	$233.43	$242.02	$347.00	$199.67
Standard error	6.015	8.661	14.461	16.882	18.108	18.959	18.667	14.459	14.611	13.792	14.013	19.723	26.230	20.574
Percent reporting	89	77	86	87	87	91	99	90	91	91	92	93	94	90
Boats, Aircraft, Wheel Goods														
Average annual expenditure	$74.82	$7.63	$6.73	$66.11	$12.21	$36.14	$54.73	$58.75	$61.98	$99.37	$135.00	$158.80	$197.58	$100.42
Standard error	8.782	16.036	20.111	47.727	21.513	35.853	39.601	24.029	23.948	23.550	25.713	39.299	49.606	49.579
Percent reporting	15	3	4	8	8	9	11	16	17	23	24	25	24	13
Televisions														
Average annual expenditure	$46.56	$18.60	$28.06	$33.68	$32.75	$51.87	$40.71	$52.47	$41.76	$61.06	$63.70	$73.17	$75.50	$37.28
Standard error	3.425	6.991	11.611	13.616	13.576	16.963	15.427	11.233	10.661	10.340	9.833	14.805	16.670	16.010
Percent reporting	16	10	13	13	14	15	14	17	15	19	20	21	21	12

Source: Table 1a., BLS Report 455-3.

the covariance by

$$c(Y_t''', Y_{t-1}'''') = \frac{1}{16(15)} \sum_{\alpha=1}^{16} (Y_{t,\alpha}''' - Y_t''')(Y_{t-1,\alpha}'''' - Y_{t-1}'''').$$

Thus, the estimator of $\mathrm{Var}\{\hat{R}_t\}$ corresponding to (6.8.1) is

$$\mathrm{Var}(\hat{R}_t) = \hat{R}_t^2 \left[\frac{v(Y_t''')}{(Y_{t,0} + Y_t''')^2} + \frac{v(Y_{t-1}'''')}{(Y_{t-1,0} + Y_{t-1}'''')^2} \right.$$
$$\left. - 2 \frac{c(Y_t''', Y_{t-1}'''')}{(Y_{t,0} + Y_t''')(Y_{t-1,0} + Y_{t-1}'''')} \right].$$

The reader will recall that the certainty cases, $Y_{t,0}$ and $Y_{t-1,0}$ are fixed and hence do not contribute to the sampling variance or covariance. They do, however, contribute to the estimated totals and ratio.

The computations associated with $v(\hat{R}_t)$ for the July–August 1977 trend in grocery store sales are presented in Table 6.8.1.

The second illustration of (6.8.1) concerns the Consumer Expenditure Survey, first discussed in Section 2.11. The principal parameters of interest in this survey were the mean average expenditures per consumer unit (CU) for various expenditure categories. To estimate this parameter for a specific expenditure category, the estimator

$$\hat{R} = \hat{Y}/\hat{X}$$

was used, where $\hat{Y}$ denotes an estimator of total expenditures in the category and $\hat{X}$ denotes an estimator of the total number of CUs. To estimate the variance of $\hat{R}$, the Taylor series estimator $v(\hat{R})$ in (6.8.1) was used, where $v(\hat{Y})$, $v(\hat{X})$, and $c(\hat{X}, \hat{Y})$ were given by the random group technique as described in Section 2.11. Table 6.8.2 gives the estimated mean annual expenditures and corresponding estimated standard errors for several important expenditure categories.

6.9. Example: Difference of Ratios

A common problem is to estimate the difference between two ratios, say

$$\Delta = X_1/Y_1 - X_2/Y_2.$$

For example, Δ may represent the difference between the per capita income of men and women in a certain subgroup of the population. The natural estimator of Δ is

$$\hat{\Delta} = \hat{X}_1/\hat{Y}_1 - \hat{X}_2/\hat{Y}_2,$$

where $\hat{X}_1$, $\hat{X}_2$, $\hat{Y}_1$, and $\hat{Y}_2$ denote estimators of the totals X_1, X_2, Y_1, and Y_2, respectively.

Since $\hat{\Delta}$ is a nonlinear statistic, an unbiased estimator of its variance is generally not available. But using the Taylor series approximation (6.3.1) we have

$$\text{MSE}\{\hat{\Delta}\} = \mathbf{d}\boldsymbol{\Sigma}_{n(s)}\mathbf{d}', \tag{6.9.1}$$

where $\boldsymbol{\Sigma}_{n(s)}$ is the covariance matrix of $\hat{\mathbf{Y}} = (\hat{X}_1, \hat{Y}_1, \hat{X}_2, \hat{Y}_2)'$ and

$$\mathbf{d} = (1/Y_1, -X_1/Y_1^2, -1/Y_2, X_2/Y_2^2).$$

Alternatively, (6.5.1) gives

$$\text{MSE}\{\hat{\Delta}\} \doteq \text{Var}\{\tilde{\Delta}\} \tag{6.9.2}$$

where

$$\tilde{\Delta} = \hat{X}_1/Y_1 - X_1\hat{Y}_1/Y_1^2 - \hat{X}_2/Y_2 + X_2\hat{Y}_2/Y_2^2.$$

The Taylor series estimator of the variance of $\hat{\Delta}$ is obtained by substituting sample estimates of $\mathbf{d}$ and $\boldsymbol{\Sigma}_{n(s)}$ into (6.9.1) or, equivalently, by using a variance estimating formula appropriate to the single variate $\tilde{\Delta}$ and substituting sample estimates for the unknown X_1, Y_1, X_2, and Y_2. In the first case, the estimate is

$$v(\hat{\Delta}) = \hat{\mathbf{d}}\hat{\boldsymbol{\Sigma}}_{n(s)}\hat{\mathbf{d}}',$$
$$\hat{\mathbf{d}} = (1/\hat{Y}_1, -\hat{X}_1/\hat{Y}_1^2, -1/\hat{Y}_2, \hat{X}_2/\hat{Y}_2^2), \tag{6.9.3}$$

and $\hat{\boldsymbol{\Sigma}}_{n(s)}$ is an estimator of $\boldsymbol{\Sigma}_{n(s)}$ which is appropriate to the particular sampling design. In the second case, the variance estimator for the single variate $\tilde{\Delta}$ is chosen in accordance with the particular sampling design.

An illustration of these methods is provided by Tepping's (1976) railroad data. In Section 3.9 we estimated the difference Δ between the revenue/cost ratios of the Seaboard Coast Line Railroad Co. (SCL) and the Southern Railway System (SRS), and used a set of partially balanced half-samples to estimate the variance. We now estimate the variance by using the half-sample replicates to estimate $\hat{\boldsymbol{\Sigma}}_{n(s)}$ in (6.9.3).

Table 6.9.1 gives the half-sample estimates of total costs and total revenues for the two railroads. We shall employ the following notation:

$\hat{X}_{1,\alpha} = \alpha$-th half-sample estimate of total revenue for SCL,

$\hat{Y}_{1,\alpha} = \alpha$-th half-sample estimate of total cost for SCL,

$\hat{X}_{2,\alpha} = \alpha$-th half-sample estimate of total revenue for SRS,

Table 6.9.1. Replicate Estimates of Cost and Revenues, 1975

Replicate (α)	Total Cost (SRS)	Total Cost (SCL)	Total Revenue (SRS)	Total Revenue (SCL)
1	11,366,520	11,689,909	12,177,561	17,986,679
2	11,694,053	12,138,136	12,361,504	18,630,825
3	11,589,783	11,787,835	12,384,145	18,248,708
4	11,596,152	11,928,088	12,333,576	18,262,438
5	11,712,123	11,732,072	12,538,185	18,217,923
6	11,533,638	11,512,783	12,264,452	17,912,803
7	11,628,764	11,796,974	12,247,203	18,054,720
8	11,334,279	11,629,103	12,235,234	18,194,872
9	11,675,569	11,730,941	12,489,930	18,112,767
10	11,648,330	11,934,904	12,552,283	18,394,625
11	11,925,708	11,718,309	12,773,700	18,354,174
12	11,758,457	11,768,538	12,560,133	18,210,328
13	11,579,382	11,830,534	12,612,850	18,330,331
14	11,724,209	11,594,309	12,532,763	18,251,823
15	11,522,899	11,784,878	12,399,054	18,146,506
16	11,732,878	11,754,311	12,539,323	18,717,982

Source: Tepping (1976).

and

$$\hat{Y}_{2,\alpha} = \alpha\text{-th half-sample estimate of total cost for SRS.}$$

The overall estimates are then

$$\hat{X}_1 = 16^{-1} \sum_{\alpha=1}^{16} \hat{X}_{1,\alpha} = 18,266,375$$

$$\hat{Y}_1 = 16^{-1} \sum_{\alpha=1}^{16} \hat{Y}_{1,\alpha} = 11,758,070$$

$$\hat{X}_2 = 16^{-1} \sum_{\alpha=1}^{16} \hat{X}_{2,\alpha} = 12,414,633$$

$$\hat{Y}_2 = 16^{-1} \sum_{\alpha=1}^{16} \hat{Y}_{2,\alpha} = 11,628,627,$$

and the difference in the revenue/cost ratios is estimated by

$$\hat{\Delta} = \hat{X}_1/\hat{Y}_1 - \hat{X}_2/\hat{Y}_2 = 0.486.$$

Using the methodology developed in Chapter 3, we estimate the covariance matrix of $\hat{\mathbf{Y}} = (\hat{X}_1, \hat{Y}_1, \hat{X}_2, \hat{Y}_2)'$ by

$$\hat{\boldsymbol{\Sigma}}_{n(s)} = \sum_{\alpha=1}^{16} (\hat{\mathbf{Y}}_\alpha - \hat{\mathbf{Y}})(\hat{\mathbf{Y}}_\alpha - \hat{\mathbf{Y}})'/16$$

$$= \begin{bmatrix} 41{,}170{,}548{,}000 & 17{,}075{,}044{,}000 & 15{,}447{,}883{,}000 & 13{,}584{,}127{,}000 \\ & 20{,}113{,}190{,}000 & 1{,}909{,}269{,}400 & 4{,}056{,}642{,}500 \\ & & 24{,}991{,}249{,}000 & 18{,}039{,}964{,}000 \\ \text{symmetric} & & & 19{,}963{,}370{,}000 \end{bmatrix}$$

where

$$\hat{\mathbf{Y}}_\alpha = (\hat{X}_{1,\alpha}, \hat{Y}_{1,\alpha}, \hat{X}_{2,\alpha}, \hat{Y}_{2,\alpha})'.$$

Also, we have

$$\hat{\mathbf{d}} = (8.5 \cdot 10^{-8}, -1.3 \cdot 10^{-7}, -8.6 \cdot 10^{-8}, 9.2 \cdot 10^{-8}).$$

Thus, the Taylor series estimate of the variance of $\hat{\Delta}$ is

$$v(\hat{\Delta}) = \hat{\mathbf{d}}\hat{\boldsymbol{\Sigma}}_{n(s)}\hat{\mathbf{d}}'$$
$$= 0.00026.$$

This result compares closely with the estimate $v(\hat{\Delta}) = 0.00029$ prepared in Chapter 3.

Alternatively, we may choose to work with expression (6.9.2). We compute the estimates

$$\hat{\hat{\Delta}}_\alpha = \hat{X}_{1,\alpha}/\hat{Y}_1 - \hat{X}_1\hat{Y}_{1,\alpha}/\hat{Y}_1^2 - \hat{X}_{2,\alpha}/\hat{Y}_2 + \hat{X}_2\hat{Y}_{2,\alpha}/\hat{Y}_2^2,$$

$$\hat{\hat{\Delta}} = 16^{-1}\sum_{\alpha=1}^{16}\hat{\hat{\Delta}}_\alpha = \hat{X}_1/Y_1 - \hat{X}_1\hat{Y}_1/\hat{Y}_1^2 - \hat{X}_2/\hat{Y}_2 + \hat{X}_2\hat{Y}_2/\hat{Y}_2^2 = 0,$$

and then the estimator of variance

$$v(\hat{\Delta}) = \sum_{\alpha=1}^{16}(\hat{\hat{\Delta}}_\alpha - \hat{\hat{\Delta}})^2/16$$

$$= 0.00026.$$

6.10. Example: Exponentials with Application to Geometric Means

Let $\bar{Y}$ denote the population mean of a characteristic y, and let $\bar{y}$ denote an estimator of $\bar{Y}$ based on a sample of fixed size n. We assume an arbitrary sampling design, and $\bar{y}$ need not necessarily denote the sample mean.

We suppose that it is desired to estimate $\theta = e^{\bar{Y}}$. The natural estimator is

$$\hat{\theta} = e^{\bar{y}}.$$

From (6.3.1) we have to a first order approximation

$$\text{MSE}\{\hat{\theta}\} \doteq e^{2\bar{Y}} \text{Var}\{\bar{y}\}. \qquad (6.10.1)$$

Let $v(\bar{y})$ denote an estimator of $\text{Var}\{\bar{y}\}$ which is appropriate to the particular sampling design. Then, by (6.3.3) the Taylor series estimator of variance is

$$v(\hat{\theta}) = e^{2\bar{y}} v(\bar{y}). \qquad (6.10.2)$$

These results have immediate application to the problem of estimating the *geometric mean* of a characteristic x, say

$$\theta = (X_1, X_2 \dots, X_N)^{1/N},$$

where we assume $X_i > 0$ for $i = 1, \dots, N$. Let $Y_i = \ln(X_i)$ for $i = 1, \dots, N$. Then

$$\theta = e^{\bar{Y}}$$

and the natural estimator of θ is

$$\hat{\theta} = e^{\bar{y}}.$$

From (6.10.2) we may estimate the variance of $\hat{\theta}$ by

$$v(\hat{\theta}) = \hat{\theta}^2 v(\bar{y}), \qquad (6.10.3)$$

where the estimator $v(\bar{y})$ is appropriate to the particular sampling design and estimator and is based on the variable $y = \ln(x)$.

To illustrate this methodology, we consider the National Survey of Crime Severity (NSCS). The NSCS was conducted in 1977 by the Census Bureau as a supplement to the National Crime Survey. In this illustration we present data from an NSCS pretest. In the pretest, the respondent was told that a score of 10 applies to the crime, "An offender steals a bicycle parked on the street." The respondent was then asked to score approximately 20 additional crimes, each time comparing the severity of the crime to the bicycle theft. There was no *a priori* upper bound to the scores respondents could assign to the various crimes, and a score of zero was assigned in cases where the respondent felt a crime had not been committed. For each of the survey characteristics x, it was desired to estimate the geometric mean of the scores, after deleting the zero observations.

The sampling design for the NSCS pretest was stratified and highly clustered. To facilitate presentation of this example, however, we shall act as if the NSCS were a simple random sample. Thus, the estimator $\bar{y}$ in (6.10.3) is the sample mean of the characteristic $y = \ln(x)$, and $v(\bar{y})$ is the estimator

$$v(\bar{y}) = \sum_{i=1}^{n} (y_i - \bar{y})^2 / n(n-1),$$

where we have ignored the finite population correction (fpc) and n denotes the sample size after deleting the zero scores. The estimates $\bar{y}$, $v(\bar{y})$, $\hat{\theta}$, $v(\hat{\theta})$ are presented in Table 6.10.2 for 12 items from the NSCS pretest. A description of the items is available in Table 6.10.1.

Table 6.10.1. Twelve Items from the NSCS Pretest

Crime No.	Description
1	An offender steals property worth $10 from outside a building
2	An offender steals property worth $50 from outside a building
3	An offender steals property worth $100 from outside a building
4	An offender steals property worth $1000 form outside a building
5	An offender steals property worth $10,000 from outside a building
6	An offender breaks into a building and steals property worth $10
7	An offender does not have a weapon. He threatens to harm a victim unless the victim gives him money. The victim gives him $10 and is not harmed
8	An offender threatens a victim with a weapon unless the victim gives him money. The victim gives him $10 and is not harmed
9	An offender intentionally injures a victim. As a result, the victim dies
10	An offender injures a victim. The victim is treated by a doctor and hospitalized
11	An offender injures a victim. The victim is treated by a doctor but is not hospitalized
12	An offender shoves or pushes a victim. No medical treatment is required

Table 6.10.2. Estimates of Geometric Means and Associated Variances for NSCS Pretest

Item	$\bar{y}$	$v(\bar{y})$	$\hat{\theta}$	$v(\hat{\theta})$
1	3.548	$0.227 \cdot 10^{-4}$	34.74	$0.274 \cdot 10^{-1}$
2	4.126	$0.448 \cdot 10^{-6}$	61.95	$0.172 \cdot 10^{-2}$
3	4.246	$0.122 \cdot 10^{-5}$	69.82	$0.596 \cdot 10^{-2}$
4	4.876	$0.275 \cdot 10^{-4}$	131.06	$0.472 \cdot 10^{0}$
5	5.398	$0.248 \cdot 10^{-4}$	220.87	$0.121 \cdot 10^{+1}$
6	4.106	$0.242 \cdot 10^{-5}$	60.70	$0.891 \cdot 10^{-2}$
7	4.854	$0.325 \cdot 10^{-4}$	128.31	$0.536 \cdot 10^{0}$
8	5.056	$0.559 \cdot 10^{-4}$	156.94	$0.138 \cdot 10^{+1}$
9	6.596	$0.391 \cdot 10^{-3}$	731.80	$0.209 \cdot 10^{+3}$
10	5.437	$0.663 \cdot 10^{-5}$	229.67	$0.350 \cdot 10^{0}$
11	4.981	$0.159 \cdot 10^{-5}$	145.55	$0.336 \cdot 10^{-1}$
12	3.752	$0.238 \cdot 10^{-4}$	42.61	$0.432 \cdot 10^{-1}$

6.11. Example: Regression Coefficients

This example, due to Tepping (1968), is concerned with estimating the variance of an estimated regression coefficient based on a stratified multi-stage sample. We consider the statistic

$$\hat{\beta} = \frac{\left(\sum_h \sum_i x_{hi} y_{hi}/n\right) - \left(\sum_h \sum_i x_{hi}/n\right)\left(\sum_h \sum_i y_{hi}/n\right)}{\left(\sum_h \sum_i x_{hi}^2/n\right) - \left(\sum_h \sum_i x_{hi}/n\right)^2}$$

$$= \frac{n\hat{W} - \hat{X}\hat{Y}}{n\hat{U} - \hat{X}^2},$$

where x_{hi} and y_{hi} denote the values of the x and y variables attached to the i-th elementary unit in the h-th stratum, $n = \sum_h n_h$ denotes the total number of elementary units in the sample,

$$\hat{W} = \sum_h \hat{W}_h = \sum_h \sum_i x_{hi} y_{hi},$$

$$\hat{U} = \sum_h \hat{U}_h = \sum_h \sum_i x_{hi}^2,$$

$$\hat{X} = \sum_h \hat{X}_h = \sum_h \sum_i x_{hi},$$

and

$$\hat{Y} = \sum_h \hat{Y}_h = \sum_h \sum_i y_{hi}.$$

Letting N, W, U, X, and Y denote the expected values of n, $\hat{W}$, $\hat{U}$, $\hat{X}$, and $\hat{Y}$, we have from (6.3.1)

$$\text{MSE}\{\hat{\beta}\} \doteq \text{Var}\left\{ n\frac{\partial\hat{\beta}}{\partial N} + \hat{W}\frac{\partial\hat{\beta}}{\partial W} + \hat{U}\frac{\partial\hat{\beta}}{\partial U} + \hat{X}\frac{\partial\hat{\beta}}{\partial X} + \hat{Y}\frac{\partial\hat{\beta}}{\partial Y}\right\}, \quad (6.11.1)$$

where

$$\frac{\partial\hat{\beta}}{\partial N} = \frac{X(UY - WX)}{(NU - X^2)^2},$$

$$\frac{\partial\hat{\beta}}{\partial W} = \frac{N}{NU - X^2},$$

$$\frac{\partial\hat{\beta}}{\partial U} = \frac{N(NW - XY)}{(NU - X^2)^2},$$

$$\frac{\partial\hat{\beta}}{\partial X} = \frac{-NUY + 2NWX - X^2Y}{(NU - X^2)^2},$$

and

$$\frac{\partial \hat{\beta}}{\partial Y} = -\frac{X}{NU - X^2}.$$

Further, since sampling is independent in the various strata, (6.11.1) may be written as

$$\text{MSE}\{\hat{\beta}\} \doteq \sum_h \text{Var}\left\{\frac{\partial \hat{\beta}}{\partial N} n_h + \frac{\partial \hat{\beta}}{\partial W} \hat{W}_h + \frac{\partial \hat{\beta}}{\partial U} \hat{U}_h + \frac{\partial \hat{\beta}}{\partial X} \hat{X}_h + \frac{\partial \hat{\beta}}{\partial Y} \hat{Y}_h\right\}$$

$$= \sum_h \mathbf{d} \mathbf{\Sigma}_{n_h} \mathbf{d}' \tag{6.11.2}$$

$$= \sum_h \text{Var}\left\{\sum_i v_{hi}\right\}, \tag{6.11.3}$$

where

$$v_{hi} = \frac{\partial \hat{\beta}}{\partial N} + \frac{\partial \hat{\beta}}{\partial W} x_{hi}y_{hi} + \frac{\partial \hat{\beta}}{\partial U} x_{hi}^2 + \frac{\partial \hat{\beta}}{\partial X} x_{hi} + \frac{\partial \hat{\beta}}{\partial Y} y_{hi},$$

$$\mathbf{d} = \left(\frac{\partial \hat{\beta}}{\partial N}, \frac{\partial \hat{\beta}}{\partial W}, \frac{\partial \hat{\beta}}{\partial U}, \frac{\partial \hat{\beta}}{\partial X}, \frac{\partial \hat{\beta}}{\partial Y}\right),$$

and $\mathbf{\Sigma}_{n_h}$ denotes the covariance matrix of the vector $(n_h, \hat{W}_h, \hat{U}_h, \hat{X}_h, \hat{Y}_h)'$. Note that a term in $\partial \hat{\beta}/\partial N$ appears in the variance expression because the sample size n is a random variable. If n were fixed, this term would not be used.

We may use either (6.11.2) or (6.11.3) to estimate the variance of $\hat{\beta}$. In either case, we replace the derivatives by their sample estimates

$$\hat{\mathbf{d}} = (\hat{d}_1, \hat{d}_2, \hat{d}_3, \hat{d}_4, \hat{d}_5),$$

where

$$\hat{d}_1 = \frac{\hat{X}(\hat{U}\hat{Y} - \hat{W}\hat{X})}{(n\hat{U} - \hat{X}^2)^2},$$

$$\hat{d}_2 = \frac{n}{n\hat{U} - \hat{X}^2},$$

$$\hat{d}_3 = \frac{n(n\hat{W} - \hat{X}\hat{Y})}{(n\hat{U} - \hat{X}^2)^2},$$

$$\hat{d}_4 = \frac{-n\hat{U}\hat{Y} + 2n\hat{W}\hat{X} - \hat{X}^2\hat{Y}}{(n\hat{U} - \hat{X}^2)^2},$$

and

$$\hat{d}_5 = -\frac{\hat{X}}{n\hat{U} - \hat{X}^2}.$$

Then, using (6.11.2) we estimate the variance by

$$v(\hat{\beta}) = \sum_h \hat{\mathbf{d}}\hat{\mathbf{\Sigma}}_{n_h}\hat{\mathbf{d}}'$$

$$= \hat{\mathbf{d}}\left(\sum_h \hat{\mathbf{\Sigma}}_{n_h}\right)\hat{\mathbf{d}}',$$

where $\hat{\mathbf{\Sigma}}_{n_h}$ is an estimator of $\mathbf{\Sigma}_{n_h}$ appropriate to the sampling design in the h-th stratum. Alternatively, using (6.11.3) we may estimate the variance of $\hat{\beta}$ by applying a variance estimating formula to the single variate

$$\hat{v}_{hi} = \hat{d}_1 + \hat{d}_2 x_{hi}y_{hi} + \hat{d}_3 x_{hi}^2 + \hat{d}_4 x_{hi} + \hat{d}_5 y_{hi}.$$

Generally, the variance estimating formula used here would be identical with that used for the diagonal terms of $\hat{\mathbf{\Sigma}}_{n_h}$.

An illustration of variance estimation for multiple regression coefficients is also given in Tepping (1968). Also see Fuller (1975) and the references contained therein for a careful discussion of the regression coefficient in finite population sampling.

CHAPTER 7

Variance Estimation for Systematic Sampling

7.1. Introduction

The method of systematic sampling, first studied by the Madows (1944), is used widely in surveys of finite populations. When properly applied, the method picks up any obvious or hidden stratification in the population, and thus can be more precise than random sampling. In addition, systematic sampling is implemented easily, thus reducing costs.

Since a systematic sample can be regarded as a random selection of one cluster, it is not possible to give an unbiased, or even consistent, estimator of the design variance. A common practice in applied survey work is to regard the sample as random, and, for lack of knowing what else to do, estimate the variance using random sample formulae. Unfortunately, if followed indiscriminately this practice can lead to badly biased estimators and incorrect inferences concerning the population parameters of interest.

In what follows, we investigate several biased estimators of variance (including the random sample formula) with a goal of providing some guidance about when a given estimator may be more appropriate than other estimators. We shall agree to judge the estimators of variance on the basis of their bias, their mean square error (MSE), and the proportion of confidence intervals formed using the variance estimators which contain the true population parameter of interest.

In Sections 7.2 to 7.5, we discuss equal probability systematic sampling. The objective is to provide the survey practitioner some guidance about the specific problem of estimating the variance of the systematic sampling mean, $\bar{y}$. Several alternative estimators of variance are presented, and some theoretical and numerical comparisons are made between eight of them. For nonlinear statistics of the form $\hat{\theta} = g(\bar{y})$, we suggest that the variance estimators be used in combination with the appropriate Taylor series formula.

In the latter half of the chapter, Sections 7.6 to 7.9, we discuss unequal probability systematic sampling. Once again, several alternative estimators of variance are presented and comparisons made between them. This work is in the context of estimating the variance of the Horvitz–Thompson estimator of the population total.

In reading this chapter, it will be useful to keep in mind the following general procedure:

(a) Gather as much prior information as possible about the nature and ordering of the target population.
(b) If an auxiliary variable, closely related to the estimation variable, is available for all units in the population, then try several variance estimators on this variable. This investigation may provide information about which estimator will have the best properties for estimating the variance of the estimation variable.
(c) Use the prior information in (a) to construct a simple model for the population. The results presented in later sections may be used to select an appropriate estimator for the chosen model.
(d) Keep in mind that most surveys are multipurpose and it may be important to use different variance estimators for different characteristics.

Steps (a)–(d) essentially suggest that one know the target population well before choosing a variance estimator, which is exactly the advice most authors since the Madows have suggested before using systematic sampling.

7.2. Alternative Estimators in the Equal Probability Case

In this section, we define a number of estimators of variance that are useful for systematic sampling problems. Each of the estimators is biased, and thus the statistician's goal is to choose the least biased estimator, the one with minimum MSE, or the one with the best confidence interval properties. It is important to have an arsenal of several estimators, because no one estimator is best for all systematic sampling problems.

To concentrate on essentials, we shall assume that the population size N is an integer multiple of the sample size n, i.e., $N = np$, where p is the sampling interval. The reader will observe that the estimators extend in a straightforward manner to the case of general N.

In most cases, we shall let Y_{ij} denote the value of the j-th unit in the i-th possible systematic sample, where $i = 1, \ldots, p$ and $j = 1, \ldots, n$. But on one occasion, we shall employ the single subscript notation, letting Y_t denote the value of the t-th unit in the population, for $t = 1, \ldots, N$.

The systematic sampling mean $\bar{y}$ and its variance are

$$\bar{y} = \sum_{j}^{n} y_{ij}/n$$

and

$$\mathrm{Var}\{\bar{y}\} = (\sigma^2/n)[1 + (n-1)\rho], \tag{7.2.1}$$

respectively, where

$$\sigma^2 = \sum_{i}^{p}\sum_{j}^{n} (Y_{ij} - \bar{Y}_{..})^2/np$$

denotes the population variance,

$$\rho = \sum_{i}^{p}\sum_{j}^{n}\sum_{j\neq j'}^{n} (Y_{ij} - \bar{Y}_{..})(Y_{ij'} - \bar{Y}_{..})/pn(n-1)\sigma^2$$

denotes the intraclass correlation between pairs of units in the same sample, and $\bar{Y}_{..}$ denotes the population mean.

7.2.1. Eight Estimators of Variance

One of the simplest estimators of $\mathrm{Var}\{\bar{y}\}$ is obtained by regarding the systematic sample as a simple random sample. We denote this estimator by

$$v_1(i) = (1 - f)(1/n)s^2, \tag{7.2.2}$$

where

$$s^2 = \sum_{j=1}^{n} (y_{ij} - \bar{y})^2/(n-1)$$

and

$$f = n/N = p^{-1}.$$

The argument (i) signifies the selected sample. This estimator is known to be upward or downward biased as the intraclass correlation coefficient is less than or greater than $-(N-1)^{-1}$.

Another simple estimator of $\mathrm{Var}\{\bar{y}\}$ is obtained by regarding the systematic sample as a stratified random sample with 2 units selected from each successive stratum of $2p$ units. This yields an estimator based on nonoverlapping differences

$$v_3(i) = (1 - f)(1/n) \sum_{j=1}^{n/2} a_{i,2j}^2/n, \tag{7.2.3}$$

where $a_{ij} = \Delta y_{ij} = y_{ij} - y_{i,j-1}$ and Δ is the first difference operator. A related estimator, which aims at increasing the number of "degrees of freedom," is based on overlapping differences

$$v_2(i) = (1 - f)(1/n) \sum_{j=2}^{n} a_{ij}^2/2(n - 1). \tag{7.2.4}$$

Several authors, e.g., Yates (1949), have suggested estimators based upon higher-order contrasts than are present in v_2 and v_3. Examples of such estimators include

$$v_4(i) = (1 - f)(1/n) \sum_{j=3}^{n} b_{ij}^2/6(n - 2), \tag{7.2.5}$$

$$v_5(i) = (1 - f)(1/n) \sum_{j=5}^{n} c_{ij}^2/3.5(n - 4), \tag{7.2.6}$$

and

$$v_6(i) = (1 - f)(1/n) \sum_{j=9}^{n} d_{ij}^2/7.5(n - 8), \tag{7.2.7}$$

where

$$b_{ij} = \Delta a_{ij} = \Delta^2 y_{ij}$$
$$= y_{ij} - 2y_{i,j-1} + y_{i,j-2}$$

is the second difference of the sample data,

$$c_{ij} = \tfrac{1}{2}\Delta^4 y_{ij} + \Delta^2 y_{i,j-1}$$
$$= y_{ij}/2 - y_{i,j-1} + y_{i,j-2} - y_{i,j-3} + y_{i,j-4}/2$$

is a linear combination of second and fourth differences and

$$d_{ij} = \tfrac{1}{2}\Delta^8 y_{ij} + 3\Delta^6 y_{i,j-1} + 5\Delta^4 y_{i,j-2} + 2\Delta^2 y_{i,j-3}$$
$$= y_{ij}/2 - y_{i,j-1} + - \ldots + y_{i,j-8}/2$$

is a linear combination of second, fourth, sixth, and eighth differences. There are unlimited variations on this basic type of estimator. One may use any number of data points in forming the contrast; one may use overlapping, nonoverlapping, or partially overlapping contrasts; and one has considerable freedom in choosing the coefficients, so long as they sum to zero. Then, in forming the estimator one divides the sum of squares by the product of the coefficients and the number of contrasts in the sum.

For example, the sixth estimator v_6 employs overlapping contrasts d_{ij}, there are $(n - 8)$ contrasts in the sum $\sum_{j=9}^{n}$, and the sum of squares of the coefficients is equal to

$$(\tfrac{1}{2})^2 + (-1)^2 + 1^2 + (-1)^2 + 1^2 + (-1)^2 + 1^2 + (-1)^2 + (\tfrac{1}{2})^2 = 7.5.$$

Therefore, one divides the sum of squares $\sum_{j=9}^{n} d_{ij}^2$ by $7.5 (n - 8)$.

Another general class of variance estimators arises by splitting the parent sample into equal-sized systematic subsamples. This may be thought of as a kind of random group estimator. Let k and n/k be integers, and let $\bar{y}_\alpha$ denote the sample mean of the α-th systematic subsample of size n/k, i.e.,

$$\bar{y}_\alpha = \frac{k}{n} \sum_{j=1}^{n/k} y_{i,k(j-1)+\alpha}.$$

An estimator of Var$\{\bar{y}\}$ is then given by

$$v_7(i) = (1-f) \frac{1}{k(k-1)} \sum_{\alpha=1}^{k} (\bar{y}_\alpha - \bar{y})^2. \tag{7.2.8}$$

Koop (1971) has investigated this estimator for the case $k = 2$, giving expressions for its bias in terms of intraclass correlation coefficients.

Finally, another class of estimators can be devised from various assumptions about the correlation between successive units in the population. One such estimator, studied by Cochran (1946), is

$$v_8(i) = (1-f)(s^2/n)[1 + 2/\ln(\hat{\rho}_p) + 2/(\hat{\rho}_p^{-1} - 1)] \qquad \text{if } \hat{\rho}_p > 0$$
$$= (1-f)s^2/n \qquad \text{if } \hat{\rho}_p \leq 0, \tag{7.2.9}$$

where

$$\hat{\rho}_p = \sum_{j=2}^{n} (y_{ij} - \bar{y})(y_{i,j-1} - \bar{y})/(n-1)s^2.$$

The statistic $\hat{\rho}_p$ is an estimator of the correlation ρ_p between two units in the population that are p units apart. This notion of correlation arises from a superpopulation model, wherein the finite population itself is generated by a stochastic superpopulation mechanism and ρ_p denotes the model correlation between, e.g., Y_{ij} and $Y_{i,j+1}$. The particular estimator v_8 is constructed from the assumption $\rho_p = \exp(-\lambda p)$, where λ is a constant. This assumption has been studied by Osborne (1942) and Matern (1947) for forestry and land-use surveys, but it has not received much attention in the context of household and establishment surveys. Since $\ln(\hat{\rho}_p)$ is undefined for nonpositive values of $\hat{\rho}_p$, we have set $v_8 = v_1$, when $\hat{\rho}_p \leq 0$. Variations on the basic estimator v_8 may be constructed by using a positive cutoff on $\hat{\rho}_p$, an estimator other than v_1 below the cutoff, or a linear combination of v_8 and v_1 where the weights, say $\phi(t)$ and $1 - \phi(t)$, depend upon a test statistic for the hypothesis that $\rho_p = 0$.

The eight estimators presented above are certainly not the only estimators of Var$\{\bar{y}\}$. Indeed, we have mentioned several techniques for constructing additional estimators. But these eight are broadly representative of the various classes of variance estimators that are useful in applied systematic sampling problems. The reader who has these eight estimators in their arsenal will have sufficient armament to deal effectively with most applied systematic sampling problems.

Finally, we note in passing that a finite population correction $(1 - f)$ was included in all of our estimators. This is not a necessary component of the estimators since there is no explicit fpc in the variance $\text{Var}\{\bar{y}\}$. Moreover, the fpc will make little difference when the sampling interval p is large and the sampling fraction $f = p^{-1}$ negligible.

7.2.2. A General Methodology

We now present a general methodology for constructing estimators of $\text{Var}\{\bar{y}\}$. This section is somewhat theoretical in nature and the applied reader may wish to skip to Section 7.2.3. The general methodology presented here will not be broadly useful for all systematic sampling problems, but will be useful in specialized circumstances where the statistician is reasonably confident about the statistical model underlying the finite population. In the more usual circumstance where the model is unknown or only vaguely known, we recommend that a choice be made among the eight estimators presented in Section 7.2.1 rather than the methodology presented here.

Let $\mathbf{Y} = (Y_1, Y_2, \ldots, Y_N)$ denote the N-dimensional population parameter, and suppose that $\mathbf{Y}$ is selected from a known superpopulation model $\xi(\cdot\,; \boldsymbol{\theta})$, where $\boldsymbol{\theta}$ denotes a vector of parameters. Let $\mathbf{Y}_i = (Y_{i1}, Y_{i2}, \ldots, Y_{in})$ denote the values in the i-th possible systematic sample.

In order to construct the general estimator, it is necessary to distinguish two expectation operators. We shall let roman E denote expectation with respect to the systematic sampling design, and script $\mathscr{E}$ shall denote the ξ-expectation.

Our general estimator of the variance is the conditional expectation of $\text{Var}\{\bar{y}\}$ given the data $\mathbf{y}_i$ from the observed sample. We denote the estimator by

$$v_*(i) = \mathscr{E}(\text{Var}\{\bar{y}\}|\mathbf{y}_i), \qquad (7.2.10)$$

where the i-th sample is selected.

The estimator v_* is not a design unbiased estimator of variance because

$$E\{v_*(i)\} \neq \text{Var}\{\bar{y}\}.$$

It has two other desirable properties, however. First, the expected (with respect to the model) bias of v_* is zero because

$$E\{v_*(i)\} = \frac{1}{p} \sum_{i=1}^{p} \mathscr{E}(\text{Var}\{\bar{y}\}|\mathbf{Y}_i)$$

and

$$\mathscr{E}E\{v_*(i)\} = \frac{1}{p} \sum_{i=1}^{p} \mathscr{E}\mathscr{E}(\text{Var}\{\bar{y}\}|\mathbf{Y}_i)$$

$$= \frac{1}{p} \sum_{i=1}^{p} \mathscr{E}(\text{Var}\{\bar{y}\})$$

$$= \mathscr{E}(\text{Var}\{\bar{y}\}).$$

Second, in the class of estimators $v(i)$ that are functions of the observed data y_i, v_* minimizes the expected quadratic loss. That is, for any such v, we have

$$\mathscr{E}(v(i) - \text{Var}\{\bar{y}\})^2 \geq \mathscr{E}(v_*(i) - \text{Var}\{\bar{y}\})^2. \qquad (7.2.11)$$

This is a Rao–Blackwell type result. Further, since (7.2.11) is true for each of the p possible samples, it follows that v_* is the estimator of $\text{Var}\{\bar{y}\}$ with minimum expected MSE, i.e.,

$$\mathscr{E}\text{E}(v - \text{Var}\{\bar{y}\})^2 = \frac{1}{p}\sum_{i=1}^{p} \mathscr{E}(v(i) - \text{Var}\{\bar{y}\})^2$$

$$\geq \frac{1}{p}\sum_{i=1}^{p} \mathscr{E}(v_*(i) - \text{Var}\{\bar{y}\})^2$$

$$= \mathscr{E}\text{E}(v_* - \text{Var}\{\bar{y}\})^2.$$

It is easy to obtain an explicit expression for v_*. Following Heilbron (1978), we write

$$\text{Var}\{\bar{y}\} = \frac{1}{p}\sum_{i}^{p} (\bar{Y}_{i\cdot} - \bar{Y}_{\cdot\cdot})^2$$

$$= N^{-2}\mathbf{W}'\mathbf{C}\mathbf{W},$$

where $\bar{Y}_{i\cdot}$ is the mean of the i-th possible systematic sample, $\mathbf{W} = (W_1, \ldots, W_p)'$, $W_i = \mathbf{Y}_i\mathbf{e}$, $\mathbf{e}$ is an $(n \times 1)$ vector of 1's, and $\mathbf{C}$ is a $(p \times p)$ matrix with elements

$$c_{ii'} = p - 1, \qquad i = i'$$

$$= -1, \qquad i \neq i'.$$

The estimator v_* is then given by

$$v_*(i) = N^{-2}\boldsymbol{\omega}_i'\mathbf{C}\boldsymbol{\omega}_i + N^{-2}\text{tr}(\mathbf{C}\,\boldsymbol{\Sigma}_i),$$

where

$$\boldsymbol{\omega}_i = \mathscr{E}\{\mathbf{W}|\mathbf{y}_i)$$

and

$$\boldsymbol{\Sigma}_i = \mathscr{E}\{(\mathbf{W} - \boldsymbol{\omega}_i)(\mathbf{W} - \boldsymbol{\omega}_i)'|\mathbf{y}_i\}$$

are the conditional expectation and conditional covariance matrix of $\mathbf{W}$, respectively.

Although v is optimal in the sense of minimum expected mean square error, it is unworkable in practice because $\boldsymbol{\omega}_i = \boldsymbol{\omega}_i(\boldsymbol{\theta})$ and $\boldsymbol{\Sigma}_i = \boldsymbol{\Sigma}_i(\boldsymbol{\theta})$ will be functions of the parameter $\boldsymbol{\theta}$, which is generally unknown. A natural approximation is obtained by computing a sample-based estimate $\hat{\boldsymbol{\theta}}$ and replacing the unknown quantities $\boldsymbol{\omega}_i$ and $\boldsymbol{\Sigma}_i$ by $\hat{\boldsymbol{\omega}}_i = \boldsymbol{\omega}_i(\hat{\boldsymbol{\theta}})$ and $\hat{\boldsymbol{\Sigma}}_i = \boldsymbol{\Sigma}_i(\hat{\boldsymbol{\theta}})$, respectively. The resulting estimator of variance is

$$\hat{v}_*(i) = N^{-2}\hat{\boldsymbol{\omega}}_i'\mathbf{C}\hat{\boldsymbol{\omega}}_i + N^{-2}\text{tr}(\mathbf{C}\hat{\boldsymbol{\Sigma}}_i). \qquad (7.2.12)$$

Another practical difficulty with v^* is that it is known to be optimal only for the true model ξ. Since ξ is never known exactly, the practicing statistician must make a professional judgment about the form of the model, and then derive v_* based on the chosen form. The "practical" variance estimator $\hat{v}_*$ is then subject not only to errors of estimation (i.e., errors in estimating $\boldsymbol{\theta}$) but also to errors of model specification. Unless one is quite confident about the model ξ and the value of the parameter $\boldsymbol{\theta}$, it may be better to rely upon one of the estimators defined in Section 7.2.1.

7.2.3. Supplementing the Sample

A final class of variance estimators arises when we supplement the systematic sample with either a simple random sample or another (or possibly several) systematic sample(s). We present estimators for both cases in this section. Of course, if there is a fixed survey budget, then the combined size of the original and supplementary samples necessarily must be no larger than the size of the single sample that would be used in the absence of supplementation.

We continue to let $N = np$, where n denotes the size of the initial sample. From the remaining $N - n$ units, we shall draw a supplementary sample of size n' via srs wor. It is presumed that $n + n'$ will be less than or equal to the sample size n employed in Sections 7.2.1 and 7.2.2, because of budgetary restrictions.

We let $\bar{y}_s$ denote the sample mean of the initial systematic sample, and $\bar{y}_r$ the sample mean of the supplementary simple random sample. For estimating the population mean, $\bar{y}$, we shall consider the combined estimator

$$\bar{y}(\beta) = (1 - \beta)\bar{y}_s + \beta\bar{y}_r, \qquad 0 \le \beta \le 1. \tag{7.2.13}$$

We seek an estimator of the variance, $\mathrm{Var}\{\bar{y}(\beta)\}$. Zinger (1980) gives an explicit expression for this variance.

To construct the variance estimator, we define three sums of squares:

$$Q_s = \sum_t^n (y_t - \bar{y}_s)^2,$$

the sum of squares within the initial sample;

$$Q_r = \sum_t^{n'} (y_t - \bar{y}_r)^2,$$

the sum of squares within the supplementary sample; and

$$Q_b = (\bar{y}_s - \bar{y}_r)^2$$

the between sum of squares.

Then, an unbiased estimator of the variance of $\bar{y}(\beta)$ is given by

$$v(\bar{y}(\beta)) = B(Q_s + \lambda Q_r) + DQ_b, \tag{7.2.14}$$

where

$$B = \frac{d_2 a_1(\beta) - d_1 a_2(\beta)}{d_2(n + \lambda c_1) + d_1(n + \lambda c_2)},$$

$$D = \frac{a_1(\beta)(n + \lambda c_2) + a_2(\beta)(n + \lambda c_1)}{d_2(n + \lambda c_1) + d_1(n + \lambda c_2)},$$

$$a_1(\beta) = \beta^2 (N - n - n')/n'(N - n - 1),$$

$$a_2(\beta) = \left(1 - \frac{p\beta}{p - 1}\right)^2 - \frac{\beta^2(N - n - n')}{n'(p - 1)^2(N - n - 1)}.$$

$$c_1 = (n' - 1)(N - n)/(N - n - 1),$$

$$c_2 = n^2(n' - 1)/(N - n)(N - n - 1),$$

$$d_1 = (N - n - n')/n'(N - n - 1),$$

$$d_2 = (n'N^2 - n'N - n^2 - nn')/n'(N - n)(N - n - 1).$$

This estimator is due to Wu (1981), who shows that the estimator is guaranteed nonnegative if and only if

$$\lambda \geq 0$$

and

$$\beta \geq \frac{p - 1}{2p}.$$

The choice of $(\lambda, \beta) = (1, (p - 1)/2p)$ results in the simple form

$$v(\bar{y}(\beta)) = \left(\frac{p - 1}{2p}\right)^2 Q_b. \tag{7.2.15}$$

This estimator omits the two within sums of squares Q_s and Q_r.

The estimator with $\lambda = 1$ and $\beta = \frac{1}{2}$ or $\beta = n'/(n + n')$ was studied by Zinger (1980). For $\beta = \frac{1}{2}$, the estimator $v(\bar{y}(\beta))$ is unbiased and nonnegative. But for the natural weighting $\beta = n'/(n + n')$, the estimator may assume negative values.

Because n' will be smaller than n in most applications, the optimum β for $\bar{y}(\beta)$ will usually be smaller than $(p - 1)/2p$, and thus the optimum β will not guarantee nonnegative estimation of the variance. Evidently, there is a conflict between the two goals of (1) choosing β to minimize the variance of $\bar{y}(\beta)$ and (2) choosing a β that will guarantee a nonnegative unbiased estimator of the variance.

Wu suggests the following strategy for resolving this conflict:

(i) If the optimal β, say β_0, is greater than $(p - 1)/2p$, then use $\bar{y}(\beta_0)$ and $v(\bar{y}(\beta_0))$.

(ii) If $0.2 \le \beta_0 \le (p - 1)/2p$, then use $\bar{y}(\frac{1}{2})$ or $\bar{y}((p - 1)/2p)$ and the corresponding variance estimator $v(\bar{y}(\frac{1}{2}))$ or $v(\bar{y}((p - 1)/2p))$. This strategy will guarantee a positive variance estimator while preserving high efficiency for the estimator of $\bar{Y}$.

(iii) If $\beta_0 < 0.2$, then use $\bar{y}(\beta_0)$ and the truncated estimator of variance
$$v_+(\bar{y}(\beta_0)) = \max\{v(\bar{y}(\beta_0)), 0\}.$$

Wu's strategy for dealing with this conflict is sensible, although in case (iii) a variance estimate of zero is almost as objectionable as a negative variance estimate.

In the remainder of this section, we discuss the situation where the systematic sample is supplemented by one or more systematic samples of the *same* size as the original sample. This is commonly called multiple-start systematic sampling. Wu (1981) discusses a modification of this approach whereby the original systematic sample is supplemented by another systematic sample of *smaller* size, although his approach does not appear to have any important advantages over multiple-start sampling.

Let $N = np$, where n continues to denote the size of an individual systematic sample. We assume k integers are selected at random between 1 and p, generating k systematic samples of size n. It is presumed that the combined sample size, kn, will be less than or equal to the size of a comparable single-start sample, because of budgetary restrictions.

Let the k systematic sampling means be denoted by $\bar{y}_\alpha$, $\alpha = 1, \ldots, k$. We shall consider variance estimation for the combined estimator

$$\bar{y} = \frac{1}{k} \sum_{\alpha=1}^{k} \bar{y}_\alpha.$$

Because each sample is of the same size, note that $\bar{y}$ is also the sample mean of the combined sample of kn units.

There are two situations of interest: the k random starts are selected (1) with replacement or (2) without replacement. In the first case an unbiased estimator of $\text{Var}\{\bar{y}\}$ is given by

$$v_{\text{wr}}(\bar{y}) = \frac{1}{k(k - 1)} \sum_{\alpha=1}^{k} (\bar{y}_\alpha - \bar{y})^2, \qquad (7.2.16)$$

while in the second case the unbiased estimator is

$$v_{\text{wor}}(\bar{y}) = (1 - f) \frac{1}{k(k - 1)} \sum_{\alpha=1}^{k} (\bar{y}_\alpha - \bar{y})^2, \qquad (7.2.17)$$

$$f = k/p.$$

Both of these results follow simply from standard textbook results for srs wr and srs wor.

Both of these estimators bear a strong similarity to the estimator v_7 presented in (7.2.8). They are similar in mathematical form to v_7, but differ in that v_7 relies upon splitting a single-start sample whereas v_{wr} and v_{wor} rely upon a multiple-start sample.

The estimators v_{wr} and v_{wor} also bear a strong similarity to the random group estimator discussed in Chapter 2. The present estimators may be thought of as the natural extension of the random group estimator to systematic sampling. It would be possible to carry these ideas further, grouping at random the k selected systematic samples, preparing an estimator for each group, and estimating the variance of $\bar{y}$ by the variability between the group means. In most applications, however, the number of samples k will be small and we see no real advantage in using fewer than k groups.

We conclude this section by recalling Gautschi (1957), who has examined the efficiency of multiple-start sampling versus single-start sampling of the same size. Not surprisingly, he shows that for populations in "random order" the two sampling methods are equally efficient. For "linear trend" and "autocorrelated" populations, however, he shows that multi-start sampling is less efficient than single-start sampling. We shall give these various kinds of populations concrete definition in the next section. However, it follows once again that there is a conflict between efficient estimation of $\bar{Y}$ and unbiased, nonnegative estimation of the variance. The practicing statistician will need to resolve this conflict on a survey-by-survey basis.

7.3. Theoretical Properties of the Eight Estimators

In many applications, the survey statistician will wish to emphasize efficient estimation of $\bar{Y}$, and thus will prefer single-start systematic sampling to the supplementary techniques discussed in the previous section. The statistician will also wish to employ a fairly robust variance estimator with good statistical properties, e.g., small bias and MSE and good confidence interval coverage rates. Selecting wisely from the eight estimators presented in Section 7.2.1 will be a good strategy in many applied problems.

In this section and the next we shall review the statistical properties of these eight estimators so as to enable the statistician to make wise choices between them. We shall consider a simple class of superpopulation models; introduce the notion of model bias; and use it as a criterion for comparing the eight estimators. We shall also present the results of a small Monte Carlo study that sheds light on the estimators' MSEs and confidence interval coverage properties.

We assume the finite population is generated according to the superpopulation model

$$Y_{ij} = \mu_{ij} + e_{ij}, \tag{7.3.1}$$

where the μ_{ij} denote fixed constants and the errors e_{ij} are $(0, \sigma^2)$ random variables. The *expected bias* and *expected relative bias* of an estimator v_α, for $\alpha = 1, \ldots, 8$, are defined by

$$\mathcal{B}\{v_\alpha\} = \mathscr{E} \mathrm{E}\{v_\alpha\} - \mathscr{E} \operatorname{Var}\{\bar{Y}\}$$

and

$$\mathcal{R}\{v_\alpha\} = \mathcal{B}\{v_\alpha\}/\mathscr{E} \operatorname{Var}\{\bar{y}\},$$

respectively. In this notation, we follow the convention of using roman letters to denote moments with respect to the sampling design (i.e., systematic sampling) and script letters to symbolize moments with respect to the model (7.3.1).

In Sections 7.3.1 to 7.3.4, we compare the expected biases or expected relative biases of the eight estimators using five simple models.

7.3.1. Random Model

Random populations may be represented by

$$\mu_{ij} = \mu, \tag{7.3.2}$$

for $i = 1, \ldots, p$ and $j = 1, \ldots, n$, where the e_{ij} are independent and identically distributed (iid) random variables. For such populations, it is well known that the expected variance is

$$\mathscr{E} \operatorname{Var}\{\bar{y}\} = (1 - f)\sigma^2/n \tag{7.3.3}$$

(see, e.g., Cochran (1946)). Further, it can be shown that the expected relative bias of the first seven estimators of variance is zero. We have been unable to obtain an expression for $\mathcal{B}\{v_8\}$ without making stronger distributional assumptions. However, it seems likely that this expected bias is near zero and, therefore, that each of the eight estimators is equally preferable in terms of the bias criterion.

7.3.2. Linear Trend Model

Populations with linear trend may be represented by

$$\mu_{ij} = \beta_0 + \beta_1[i + (j - 1)p], \tag{7.3.4}$$

where β_0 and β_1 denote fixed (but unknown) constants and the errors e_{ij} are iid random variables. For this model, the expected variance is

$$\mathscr{E} \operatorname{Var}\{\bar{y}\} = \beta_1^2(p^2 - 1)/12 + (1 - f)\sigma^2/n. \tag{7.3.5}$$

The expectations of the eight estimators of variance are given in column 2 of Table 7.3.1. The expression for $\mathscr{E} \mathrm{E}\{v_8\}$ was derived by approximating the expectation of the function $v_8(s^2, \hat{\rho}_p s^2)$ by the same function of the

Table 7.3.1. Expected Values of Eight Estimators of Variance

Estimator	Linear Trend	Stratification Effects	Autocorrelated
		Population Model	
v_1	$(1-f)[\beta_1^2 p^2(n+1)/12 + \sigma^2/n]$	$(1-f)\left[\sum_j^n (\mu_j - \bar{\mu})^2/n(n-1) + \sigma^2/n\right]^b$	$(1-f)(\sigma^2/n)\left\{1 - \dfrac{2}{n-1}\dfrac{(\rho^p - \rho^N)}{(1-\rho^p)} + \dfrac{2}{n(n-1)} \times \left[\dfrac{(\rho^p - \rho^N)}{(1-\rho^p)^2} - (n-1)\dfrac{\rho^N}{(1-\rho^p)}\right]\right\}$
v_2	$(1-f)[\beta_1^2 p^2/2n + \sigma^2/n]$	$(1-f)\left[\sum_j^{n-1} (\mu_j - \mu_{j+1})^2/2n(n-1) + \sigma^2/n\right]$	$(1-f)(\sigma^2/n)(1-\rho^p)$
v_3	$(1-f)[\beta_1^2 p^2/2n + \sigma^2/n]$	$(1-f)\left[\sum_j^{n/2} (\mu_{2j-1} - \mu_{2j})^2/n^2 + \sigma^2/n\right]$	$(1-f)(\sigma^2/n)(1-\rho^p)$
v_4	$(1-f)\sigma^2/n$	$(1-f)\left[\sum_j^{n-2} (\mu_j - 2\mu_{j+1} + \mu_{j+2})^2/6n(n-2) + \sigma^2/n\right]$	$(1-f)(\sigma^2/n)[1 - 4\rho^p/3 + \rho^{2p}/3]$
v_5	$(1-f)\sigma^2/n$	$(1-f)\left[\sum_j^{n-4} (\mu_j/2 - \mu_{j+1} + \mu_{j+2} - \mu_{j+3} + \mu_{j+4}/2)^2/3.5n(n-4) + \sigma^2/n\right]$	$(1-f)(\sigma^2/n)[1 - 12\rho^p/7 + 8\rho^{2p}/7 - 4\rho^{3p}/7 + \rho^{4p}/7]$
v_6	$(1-f)\sigma^2/n$	$(1-f)\left[\sum_j^{n-8} (\mu_j/2 - \mu_{j+1} + \dots + \mu_{j+8}/2)^2/7.5n(n-8) + \sigma^2/n\right]$	$(1-f)(\sigma^2/n)[1 - 28\rho^p/15 + 24\rho^{2p}/15 - 20\rho^{3p}/15 + 16\rho^{4p}/15 - 12\rho^{5p}/15 + 8\rho^{6p}/15 - 4\rho^{7p}/15 + \rho^{8p}/15]$

v_7 $(1-f)[\beta_1^2 p^2(k+1)/12 + \sigma^2/n]$

$$(1-f)(\sigma^2/n)\{1+[2/(k-1)][k(\rho^{kp}-\rho^N)/(1-\rho^{kp})$$
$$-(\rho^p-\rho^N)/(1-\rho^p)]-[2/(k-1)][\{k^2/n\}$$

$$\times\{(\rho^{kp}-\rho^N)/(1-\rho^{kp})^2-(n/k-1)\rho^N/(1-\rho^{kp})\}$$
$$-n^{-1}\{(\rho^p-\rho^N)/(1-\rho^p)^2-(n-1)\rho^N/(1-\rho^p)\}\}]^c$$

$$(1-f)(\sigma^2/n)[1+2/\ln(\rho^p)+2\rho^p/(1-\rho^p)]+0(n^{-2})^d$$

v_8 $(1-f)[\gamma(0)/n]$

$$\times\left[1+\dfrac{2}{1+\dfrac{\ln\{\gamma(1)/\gamma(0)\}}{2}}+\dfrac{2}{\gamma(0)/\gamma(1)-1}\right]^a$$

$$(1-f)n^{-1}(\kappa(0)+\sigma^2)$$

$$\times\left\{1+\dfrac{2}{\ln\dfrac{\kappa(1)}{\kappa(0)+\sigma^2}}+\dfrac{2}{\dfrac{\kappa(0)+\sigma^2}{\kappa(1)}-1}\right\}^e$$

---

[a] $\gamma(1)=\mathscr{C}\,\mathrm{E}\{\hat\rho_p s^2\}=\beta_1^2 p^2(n-3)(n+1)/12-\sigma^2/n$
  $\gamma(0)=\mathscr{C}\,\mathrm{E}\{s^2\}=\beta_1^2 p^2 n(n+1)/12+\sigma^2$.

[b] $\bar\mu=\sum_j^n \mu_j/n$.

[c] $\bar\mu_\alpha$ = mean of a systematic subsample of size n/k of the μ_j.

[d] The approximation follows from elementary properties of the estimated autocorrelation function for stationary time series and requires bounded sixth moments.

[e] $\kappa(0)=(n-1)^{-1}\sum_j^n(\mu_j-\bar\mu)^2$

  $\kappa(1)=(n-1)^{-1}\sum_j^{n-1}(\mu_j-\bar\mu)(\mu_{j+1}-\bar\mu)$.

expectations $\mathscr{E}\, E\{s^2\}$ and $\mathscr{E}\, E\{\hat{\rho}_p s^2\}$, where we have used an expanded notation for v_8. In deriving this result it was also assumed that $\hat{\rho}_p > 0$ with probability one, thus guaranteeing that terms involving the operator $\ln(\cdot)$ are well defined.

From Table 7.3.1 and (7.3.5) we see that the intercept β_0 has no effect on the relative biases of the variance estimators, while the error variance σ^2 has only a slight effect. Similarly, the slope β_1 has little effect on the relative biases, unless β_1 is very small. For populations where p is large and β_1 is not extremely close to 0, the following useful approximations can be derived:

$$\mathscr{R}\{v_1\} = n$$
$$\mathscr{R}\{v_2\} = -(n-6)/n$$
$$\mathscr{R}\{v_3\} = -(n-6)/n$$
$$\mathscr{R}\{v_4\} = -1$$
$$\mathscr{R}\{v_5\} = -1$$
$$\mathscr{R}\{v_6\} = -1$$
$$\mathscr{R}\{v_7\} = k.$$

Thus, from the point of view of relative bias, the estimators v_2 and v_3 are preferred.

The reader will notice that these results differ from Cochran (1977), who suggests v_4 for populations with linear trend. The contrasts defining v_4, v_5, and v_6 eliminate the linear trend, whereas v_2, v_3, and v_8 do not. Eliminating the linear trend is not a desirable property here because the variance is a function of the trend.

7.3.3. Stratification Effects Model

We now view the systematic sample as a selection of one unit from each of n strata. This situation may be represented by the model

$$\mu_{ij} = \mu_j, \tag{7.3.6}$$

for all i and j, where the errors e_{ij} are iid random variables. That is, the unit means μ_{ij} are constant within a stratum of p units. For this model, the expected variance of $\bar{y}$ is

$$\mathscr{E}\, \text{Var}\{\bar{y}\} = (1-f)\sigma^2/n, \tag{7.3.7}$$

and the expectations of the eight estimators of variance are given in column 3 of Table 7.3.1. Once again, the expression for the expectation of v_8 is an approximation, and will be valid when n is large and $\hat{\rho}_p > 0$ almost surely.

From Table 7.3.1 and (7.3.7) we see that each of the first seven estimators has small and roughly equal relative bias when the stratum means μ_j are approximately equal. When the stratum means are not equal, there can be important differences between the estimators and v_1 and v_8 often have the largest absolute relative biases. This point is demonstrated in Table 7.3.2 which gives the expected biases for the examples $\mu_j = j$, $j^{1/2}$, j^{-1}, $\ln(j) + \sin(j)$ with $n = 20$.

Based on these simple examples, we conclude that v_4, v_5, and v_6 provide the most protection against stratification effects. The contrasts used in these estimators tend to eliminate a linear trend in the stratum means, μ_j, which is desirable because the expected variance is not a function of such a trend. Conversely, v_2, v_3, and v_7 do not eliminate the trend. Estimators v_5 and v_6 will be preferred when there is a nonlinear trend in the stratum means. When the means μ_j are equal in adjacent nonoverlapping pairs of strata, estimator v_3 will have smallest expected bias. Estimator v_7 will have smallest expected bias when the μ_j are equal in adjacent nonoverlapping groups of k strata.

7.3.4. Autocorrelated Model

Autocorrelated populations occur in the case where the e_{ij} are not independent, but rather have some nonzero correlation structure. For example, estimator v_8 arises from consideration of the stationary correlation structure $\mathscr{E}\{e_{ij}e_{i'j'}\} = \rho_d\sigma^2$, where d is the distance between the (i, j)-th and (i', j')-th units in the population and ρ_d is a correlation coefficient.

Table 7.3.2. Expected Relative Bias Times σ^2 for Eight Estimators of Variance for the Stratification Effects Model

Estimator	μ_j			
	j	$j^{1/2}$	j^{-1}	$\ln(j) + \sin(j)$
v_1	35.00	1.046	0.050	0.965
v_2	0.50	0.020	0.008	0.235
v_3	0.50	0.022	0.013	0.243
v_4	0.00	0.000	0.001	0.073
v_5	0.00	0.000	0.001	0.034
v_6	0.00	0.000	0.000	0.013
v_7	5.00	0.177	0.022	0.206
v_8	-0.67	-0.396	-0.239	-0.373

Note: $n = 20$, $k = 2$, $\sigma^2 = 100$.

In general, we shall study autocorrelated populations by assuming the y-variable has the time series specification

$$Y_t - \mu = \sum_{j=-\infty}^{\infty} \alpha_j \varepsilon_{t-j} \qquad (7.3.8)$$

for $t = 1, \ldots, pn$, where the sequence $\{\alpha_j\}$ is absolutely summable, and the ε_t are uncorrelated $(0, \sigma^2)$ random variables. The expected variance for this model is

$$\mathscr{E} \operatorname{Var}\{\bar{y}\} = (1 - f)(1/n)\left\{ \gamma(0) - \frac{2}{pn(p-1)} \sum_{h=1}^{pn-1} (pn - h)\gamma(h) \right.$$

$$\left. + \frac{2p}{n(p-1)} \sum_{h=1}^{n-1} (n - h)\gamma(ph) \right\}, \qquad (7.3.9)$$

where

$$\gamma(h) = \mathscr{E}\{(Y_t - \mu)(Y_{t-h} - \mu)\} = \sum_{-\infty}^{\infty} \alpha_j \alpha_{j-h} \sigma^2.$$

By assuming that (7.3.8) arises from a low order autoregressive, moving average process, we may construct estimators of $\operatorname{Var}\{\bar{y}\}$ and study their properties.

For example, a representation for the model underlying v_8 is the first order autoregressive process

$$Y_t - \mu = \rho(Y_{t-1} - \mu) + \varepsilon_t, \qquad (7.3.10)$$

where ρ is the first order autocorrelation coefficient (to be distinguished from the intraclass correlation) and $0 < \rho < 1$. By (7.3.9) the expected variance for this model is

$$\mathscr{E} \operatorname{Var}\{\bar{y}\} = (1 - f)(\sigma^2/n)\left\{ 1 - \frac{2}{(p-1)} \frac{(\rho - \rho^{pn})}{(1 - \rho)} \right.$$

$$+ \frac{2}{pn(p-1)}\left[\frac{(\rho - \rho^{pn})}{(1 - \rho)^2} - (pn - 1)\frac{\rho^{pn}}{(1 - \rho)} \right]$$

$$+ \frac{2p}{(p-1)} \frac{(\rho^p - \rho^{pn})}{(1 - \rho^p)}$$

$$\left. - \frac{2p}{n(p-1)}\left[\frac{(\rho^p - \rho^{pn})}{(1 - \rho^p)^2} - (n - 1)\frac{\rho^{pn}}{(1 - \rho^p)} \right] \right\}. \qquad (7.3.11)$$

Letting n index a sequence with p fixed we obtain the following approximation to the expected variance:

$$\mathscr{E} \operatorname{Var}\{\bar{y}\} = (1 - f)(\sigma^2/n)\left\{ 1 - \frac{2}{(p-1)} \frac{\rho}{(1 - \rho)} + \frac{2p}{(p-1)} \frac{\rho^p}{(1 - \rho^p)} \right\}$$

$$+ 0(n^{-2}). \qquad (7.3.12)$$

The expectations of the eight estimators of variance are presented in column 4 of Table 7.3.1. The expression for v_8 is a large-n approximation, as in (7.3.12), whereas the other expressions are exact. Large-n approximations to the expectations of v_1 and v_7 are given by

$$\mathscr{E}\, E\{v_1\} = (1 - f)\sigma^2/n + 0(n^{-2}) \tag{7.3.13}$$

$$\mathscr{E}\, E\{v_7\} = (1 - f)(\sigma^2/n)\{1 + [2/(k - 1)][k\rho^{pk}/(1 - \rho^{pk}) - \rho^p/(1 - \rho^p)]\}$$
$$+ 0(n^{-2}). \tag{7.3.14}$$

The expectations of the remaining estimators (v_2 to v_6) do not involve terms of lower order than $0(n^{-1})$.

From Table 7.3.1 and (7.3.12)–(7.3.14), it is apparent that each of the eight estimators has small bias for ρ near zero. If p is reasonably large, then v_1 is only slightly biased regardless of the value of ρ, provided ρ is not very close to 1. This is also true of estimators v_2 through v_8. The expectation of the first estimator tends to be larger than those of the other estimators since, e.g.,

$$\mathscr{E}\, E\{v_1\} - \mathscr{E}\, E\{v_4\} \doteq (1 - f)(\sigma^2/n)\{(4/3)\rho^p - (1/3)\rho^{2p}\} \geq 0,$$

$$\mathscr{E}\, E\{v_1\} - \mathscr{E}\, E\{v_2\} \doteq (1 - f)(\sigma^2/n)\rho^p \geq 0.$$

As Cochran (1946) noticed, a good approximation to $-2\rho/p(1 - \rho)$ is given by $2/\ln(\rho^p)$. On this basis, v_8 should be a very good estimator since the expectation $\mathscr{E}\, E\{v_8\}$ is nearly identical with the expected variance in (7.3.12).

Exact statements about the comparative biases of the various estimators depend on the values of ρ and p. In Table 7.3.3 we see that differences between the estimator biases are negligible for small ρ, and increase as ρ increases. For a given value of ρ, the differences decline with increasing sampling interval p. Estimator v_8 tends to underestimate the variance, while the remaining estimators (most notably v_1) tend towards an overestimate. Further, v_8 tends to have the smallest absolute bias, except when ρ is small. When ρ is small, the $\ln(\rho^p)$ approximation is evidently not very satisfactory.

7.3.5. Periodic Populations Model

A simple periodic population is given by

$$\mu_{ij} = \beta_0 \sin\{\beta_1[i + (j - 1)p]\} \tag{7.3.15}$$

with e_{ij} iid $(0, \sigma^2)$. As is well known, such populations are the nemesis of systematic sampling, and we mention them here only to make note of that fact. When the sampling interval is equal to a multiple of the period, $2\pi/\beta_1$, the variance of $\bar{y}$ tends to be enormous while all of the estimators of variance tend to be very small. Conversely, when the sampling interval is equal to

Table 7.3.3. Expected Relative Biases of Eight Estimators for Autocorrelated Populations

First Order Autocorrelation Coefficient ρ	Sampling Interval p	Estimator							
		v_1	v_2	v_3	v_4	v_5	v_6	v_7	v_8
0.01	4	0.678 − 02	0.678 − 02	0.678 − 02	0.678 − 02	0.678 − 02	0.678 − 02	0.678 − 02	−0.103 − 00
	10	0.225 − 02	0.225 − 02	0.225 − 02	0.225 − 02	0.225 − 02	0.225 − 02	0.225 − 02	−0.413 − 01
	30	0.697 − 03	0.697 − 03	0.697 − 03	0.697 − 03	0.697 − 03	0.697 − 03	0.697 − 03	−0.138 − 01
0.10	4	0.797 − 01	0.796 − 01	0.796 − 01	0.795 − 01	0.795 − 01	0.795 − 01	0.795 − 01	−0.155 − 00
	10	0.253 − 01	0.253 − 01	0.253 − 01	0.253 − 01	0.253 − 01	0.253 − 01	0.253 − 01	−0.637 − 01
	30	0.772 − 02	0.772 − 02	0.772 − 02	0.772 − 02	0.772 − 02	0.772 − 02	0.772 − 02	−0.215 − 01
0.50	4	0.957 + 00	0.834 + 00	0.834 + 00	0.796 + 00	0.755 + 00	0.740 + 00	0.841 + 00	−0.194 + 00
	10	0.282 + 00	0.281 + 00	0.281 + 00	0.280 + 00	0.280 + 00	0.280 + 00	0.281 + 00	−0.853 − 01
	30	0.741 − 01	0.741 − 01	0.741 − 01	0.741 − 01	0.741 − 01	0.741 − 01	0.741 − 01	−0.292 − 01
0.90	4	0.104 + 02	0.293 + 01	0.293 + 01	0.207 + 01	0.165 + 01	0.150 + 01	0.590 + 01	−0.200 + 00
	10	0.427 + 01	0.243 + 01	0.243 + 01	0.204 + 01	0.174 + 01	0.163 + 01	0.291 + 01	−0.907 − 01
	30	0.112 + 01	0.103 + 01	0.103 + 01	0.100 + 01	0.974 + 00	0.961 + 00	0.104 + 01	−0.321 − 01
0.99	4	0.118 + 03	0.370 + 01	0.370 + 01	0.220 + 01	0.174 + 01	0.156 + 01	0.599 + 02	−0.200 + 00
	10	0.533 + 02	0.419 + 01	0.419 + 01	0.263 + 01	0.211 + 01	0.190 + 01	0.275 + 02	−0.909 − 01
	30	0.183 + 02	0.402 + 01	0.402 + 01	0.278 + 01	0.225 + 01	0.205 + 01	0.101 + 02	−0.323 − 01

Note: Results ignore terms of order n^{-2}.

an odd multiple of the half period, Var$\{\bar{y}\}$ tends to be extremely small while the estimators of variance tend to be large.

7.3.6. Monte Carlo Results

In this subsection, we shall present some simulations concerning the confidence interval properties and MSEs of the variance estimators. We shall also present simulation results concerning the estimator biases, which generally tend to confirm the analytical results described in the previous several subsections.

We present results for the seven superpopulation models set forth in Table 7.3.4. For each model, 200 finite populations of size $N = 1000$ were generated, and in each population, the bias and MSE of the eight estimators of variance were computed, as well as the proportion of confidence intervals that contained the true population mean. We averaged these quantities over the 200 populations, giving the expected bias, the expected MSE, and the expected coverage rate for each of the eight estimators. The multiplier used in forming the confidence intervals was the 0.025 point of the standard normal distribution. Estimator v_7 was studied with $k = 2$.

The Monte Carlo results for the random population are presented in the row labeled A1 of Tables 7.3.5, 7.3.6, and 7.3.7. Estimator v_1 is the best choice in terms of both minimum MSE and the ability to produce 95% confidence intervals. Estimator v_8 is the only one of the eight estimators that is seriously biased. The variance of the variance estimators is related to the number of "degrees of freedom", and on this basis v_1 is the preferred

Table 7.3.4. Description of the Artificial Populations

Population	Description	n	p	μ_{ij}	e_{ij}
A1	Random	20	50	0	e_{ij} iid $N(0, 100)$
A2	Linear Trend	20	50	$i + (j-1)p$	e_{ij} iid $N(0, 100)$
A3	Stratification Effects	20	50	j	e_{ij} iid $N(0, 100)$
A4	Stratification Effects	20	50	$j + 10$	$e_{ij} = \varepsilon_{ij}$ if $\varepsilon_{ij} \geq -(j+10)$ $\quad = -(j+10)$ otherwise ε_{ij} iid $N(0, 100)$
A5	Autocorrelated	20	50	0	$e_{ij} = \rho e_{i-1,j} + \varepsilon_{ij}$ $e_{11} \sim N(0, 100/(1-\rho^2))$ ε_{ij} iid $N(0, 100)$ $\rho = 0.8$
A6	Autocorrelated	20	50	0	same as A5 with $\rho = 0.4$
A7	Periodic	20	50	$20 \sin\{(2\pi/50)$ $\times [i + (j-1)p]\}$	e_{ij} iid $N(0, 100)$

Table 7.3.5. Relative Bias of Eight Estimators of Var$\{\bar{y}\}$

Population	Estimator of Variance							
	v_1	v_2	v_3	v_4	v_5	v_6	v_7	v_8
A1	0.047	0.046	0.043	0.046	0.049	0.053	0.060	-0.237
A2	19.209	-0.689	-0.688	-0.977	-0.977	-0.977	1.910	-0.449
A3	0.419	0.051	0.049	0.046	0.050	0.054	0.116	-0.443
A4	0.416	0.051	0.047	0.048	0.057	0.067	0.116	-0.441
A5	0.243	0.236	0.234	0.230	0.234	0.243	0.263	-0.095
A6	0.073	0.071	0.069	0.070	0.073	0.075	0.084	-0.217
A7	-0.976	-0.976	-0.976	-0.976	-0.976	-0.976	-0.976	-0.983
EMPINC	-0.184	-0.195	-0.193	-0.191	-0.188	-0.208	-0.158	-0.402
EMPRSA	0.316	0.241	0.239	0.234	0.235	0.234	0.100	-0.280
EMPNOO	0.121	0.123	0.119	0.134	0.151	0.148	0.707	-0.155
INCINC	0.398	0.279	0.290	0.279	0.268	0.219	0.214	-0.256
INCRSA	0.210	-0.139	-0.148	-0.143	-0.156	-0.171	-0.450	-0.748
INCNOO	0.662	0.659	0.650	0.658	0.658	0.660	0.547	0.272
FUELID	-0.191	-0.220	-0.212	-0.223	-0.234	-0.256	-0.517	-0.437
FUELAP	1.953	-0.251	0.104	-0.544	-0.641	-0.698	0.693	-0.601

Table 7.3.6. Relative Mean Square Error (MSE) of Eight Estimators of $\text{Var}\{\bar{y}\}$

Population	Estimator of Variance							
	v_1	v_2	v_3	v_4	v_5	v_6	v_7	v_8
A1	0.158	0.212	0.262	0.272	0.467	0.954	2.322	0.294
A2	369.081	0.476	0.479	0.957	0.957	0.957	3.923	0.204
A3	0.417	0.213	0.263	0.272	0.467	0.954	2.549	0.442
A4	0.386	0.200	0.249	0.261	0.464	0.973	2.555	0.441
A5	0.377	0.439	0.505	0.509	0.765	1.446	3.363	0.430
A6	0.180	0.236	0.286	0.296	0.491	0.982	2.367	0.307
A7	0.955	0.955	0.955	0.955	0.955	0.955	0.955	0.967
EMPINC	0.060	0.067	0.068	0.068	0.072	0.095	0.897	0.241
EMPRSA	0.142	0.104	0.115	0.109	0.144	0.206	3.706	0.196
EMPNOO	0.051	0.059	0.065	0.066	0.084	0.112	4.846	0.153
INCINC	0.267	0.185	0.192	0.200	0.199	0.200	2.620	0.247
INCRSA	0.120	0.084	0.087	0.091	0.109	0.132	0.601	0.569
INCNOO	0.554	0.574	0.563	0.585	0.613	0.654	4.865	0.383
FUELID	1.173	1.186	1.150	1.199	1.163	1.109	0.746	0.943
FUELAP	16.761	1.969	7.229	0.547	0.513	0.544	14.272	1.455

Table 7.3.7. Proportion of Times That the True Population Mean Fell Within the Confidence Interval Formed Using One of Eight Estimators of Variance

Population	Estimator of Variance							
	v_1	v_2	v_3	v_4	v_5	v_6	v_7	v_8
A1	94	93	93	93	91	86	70	85
A2	100	64	64	17	17	16	100	85
A3	97	93	93	93	91	86	71	77
A4	97	93	93	93	91	86	71	77
A5	96	95	94	94	92	88	73	88
A6	94	94	93	93	91	86	71	86
A7	14	14	14	13	13	13	11	11
EMPINC	90	90	88	90	88	88	64	84
EMPRSA	96	94	94	94	94	96	74	88
EMPNOO	98	98	98	98	98	98	76	92
INCINC	98	94	94	94	94	96	74	88
INCRSA	94	90	90	90	90	88	76	70
INCNOO	98	98	98	98	100	100	76	92
FUELID	88	86	82	84	82	80	60	74
FUELAP	100	90	88	86	84	82	80	76

estimator. The actual confidence levels are lower than the nominal rate in all cases.

For the linear trend population (see row labeled A2), all of the estimators are seriously biased. Estimators v_2, v_3 are more acceptable than the remaining estimators, although each is downward biased and actual confidence levels are lower than the nominal rate of 95%. Because of large bias, v_1 and v_7 are particularly unattractive for populations with linear trend. Although estimator v_8 was designed for autocorrelated populations, we obtained a relatively small bias for this estimator in the context of the linear trend population. As we shall see, however, this estimator is too sensitive to the form of the model to have broad applicability.

The Monte Carlo results for the stratification effects populations are presented in rows labeled A3 and A4. Population A4 is essentially the same as A3, except truncated so as not to permit negative values. Estimators v_2, v_3, and v_4 are preferred here; they have smaller absolute bias and MSE than the remaining estimators. Estimators v_5 and v_6 have equally small bias but larger variance, presumably because of a deficiency in the "degrees of freedom." Primarily because of large bias, estimators v_1, v_7, and v_8 are unattractive for populations with stratification effects.

Results for the autocorrelated populations are in rows A5 and A6. Estimator v_8 performs well in the highly autocorrelated population (A5), but not as well in the moderately autocorrelated population (A6). Even in the presence of high autocorrelation, the actual confidence level associated with v_8 is low. Any one of the first four estimators is recommended for low autocorrelation.

Row A7 gives the results of the Monte Carlo study of the periodic population. As was anticipated (because the sampling interval $p = 50$ is equal to the period) all of the eight estimators are badly biased downward, and the associated confidence intervals are completely unusable.

In the next section, we shall present some further numerical results regarding the eight estimators of variance. Whereas the above results were based upon computer simulations, the following results are obtained using real data sets. In Section 7.5 we summarize all of this work, pointing out the strengths and weaknesses of each of the estimators.

7.4. An Empirical Comparison

In this section, we compare the eight estimators of variance using eight real data sets. As in the last section, the comparison is based upon the three criteria

- bias
- mean square error
- confidence interval properties.

The results provide the reader with insights about how the estimators behave in a variety of practical settings.

The first six populations are actually based upon a sample taken from the March 1981 Current Population Survey (CPS). The CPS is a large survey of households which is conducted monthly in the United States. Its primary purpose is to produce descriptive statistics regarding the size of the U.S. labor force, the composition of the labor force, and changes in the labor force over time. For additional details see Hanson (1978).

The populations consist of all persons enumerated in the March 1981 CPS who are age 14+, live in one of the ten largest U.S. cities, and are considered to be members of the labor force (i.e., either employed or unemployed). Each population is of size $N = 13,000$ and each contains exactly the same individuals.

The six CPS populations differ only in respect to the characteristic of interest and in respect to the order of the individuals in the population prior to sampling. For three of the populations, EMPINC, EMPRSA, and EMPNOO, the y-variable is the unemployment indicator

$$y = 1, \quad \text{if unemployed}$$

$$= 0, \quad \text{if employed}$$

while for the remaining three populations, INCINC, INCRSA, and INCNOO, the y-variable is total income. EMPINC and INCINC are ordered by the median income of the census tract in which the person resides. EMPRSA and INCRSA are ordered by the person's race, by sex, by age (white before black before other, male before female, age in natural ascending order). EMPNOO and INCNOO are essentially in a geographic ordering.

The seventh and eighth populations, FUELID and FUELAP, are comprised of 6500 fuel oil dealers. The y-variable is 1972 annual sales in both cases. FUELID is ordered by state by identification number. The nature of the identification number is such that within a given state, the order is essentially random. FUELAP is ordered by 1972 annual payroll. The source for these data is the 1972 Economic Censuses. See, e.g., U.S. Bureau of the Census (1976).

Table 7.4.1 provides a summary description of the eight real populations. As an aid to remembering the populations, notice that they are named so that the first three letters signify the characteristic of interest and the last three letters signify the population order.

The populations INCINC, INCRSA, and INCNOO are depicted in Figures 7.4.1, 7.4.2, and 7.4.3 (these figures actually depict a 51-term centered moving average of the data). The ordering by median income (INCINC) results in an upward trend, possibly linear at first and then sharply increasing at the upper tail of the income distribution. There are rather distinct stratification effects, for the population INCRSA, where the ordering is by race by sex by age. The geographical ordering displays characteristics of a random population.

Table 7.4.1. Description of the Real Populations

Population	Characteristic	Order
FUELID	Annual Sales	(1) State
		(2) Identification Number
FUELAP	Annual Sales	Annual Payroll
EMPINC	Unemployment Indicator	Median Income of Census Tract
EMPRSA	Unemployment Indicator	(1) Race[a]
		(2) Sex
		(3) Age
EMPNOO	Unemployment Indicator	(1) Rotation Group
		(2) Identification Number
INCINC	Total Income	Median Income of Census Tract
INCRSA	Total Income	(1) Race[a]
		(2) Sex
		(3) Age
INCNOO	Total Income	(1) Rotation Group
		(2) Identification Number

[a] White before Black before Other; male before female; age in natural ascending order.

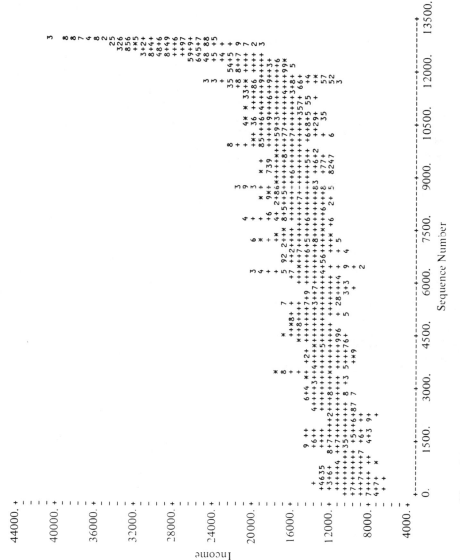

Figure 7.4.1. Plot of total income versus sequence number for population INCINC.

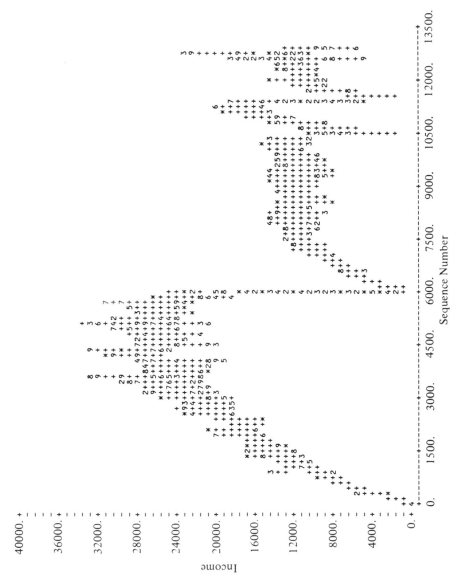

Figure 7.4.2. Plot of total income versus sequence number of population INCRSA.

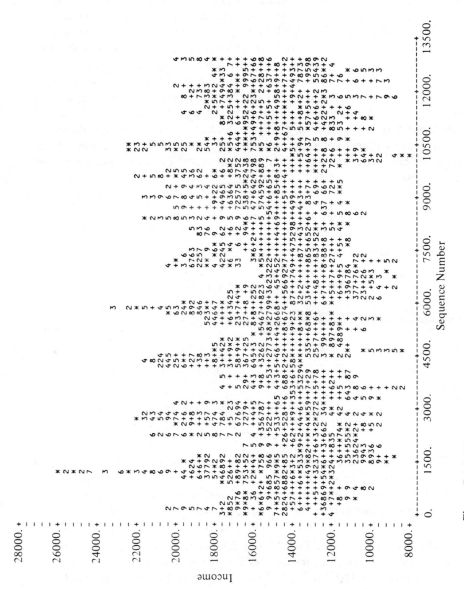

Figure 7.4.3. Plot of total income versus sequence number for population INCNOO.

The unemployment populations EMPINC, EMPRSA, and EMPNOO (see Figures 7.4.4, 7.4.5, and 7.4.6) are similar in appearance to INCINC, INCRSA, and INCNOO, respectively, except that they display negative relationships between the y-variable and the sequence number wherever the income populations display positive relationships, and vice-versa.

The fuel oil population FUELAP (Figure 7.4.8) is similar in appearance to INCINC, except the trend is much stronger in FUELAP than in INCINC. FUELID (Figure 7.4.7) appears to be a random population, or possibly a population with weak stratification effects (due to a state or regional effect).

For each of the eight populations, we have calculated the population mean $\bar{Y}$ and the variance Var$\{\bar{y}\}$. For all possible systematic samples corresponding to $p = 50$ (i.e., $f = n/N = 0.02$), we have also calculated the sample mean $\bar{y}$ and the eight estimators of variance. Utilizing these basic data, we have calculated for each population and each variance estimator, v_α, the *bias*

$$\text{Bias}\{v_\alpha\} = 50^{-1} \sum_{i=1}^{50} v_\alpha(i) - \text{Var}\{\bar{y}\},$$

the *mean square error*

$$\text{MSE}\{v_\alpha\} = 50^{-1} \sum_{i=1}^{50} (v_\alpha(i) - \text{Var}\{\bar{y}\})^2,$$

and the *actual confidence interval probability*

$$50^{-1} \sum_{i=1}^{50} \chi_i,$$

where

$$\chi_i = 1 \qquad \text{if } \bar{Y} \in (\bar{y} \pm 1.96 v_\alpha(i))$$
$$\quad = 0 \qquad \text{otherwise.}$$

The results of these calculations are presented in Tables 7.3.5, 7.3.6, and 7.3.7, respectively. In general, these results mirror those obtained in Section 7.3, where hypothetical superpopulation models were used. In the following paragraphs we summarize the essence of the results presented in the tables.

Populations with a Trend. Populations EMPINC, INCINC, and FUELAP fall generally in this category. Any of the five estimators $v_2, \ldots, v_6$ may be recommended for INCINC. For FUELAP, which has stronger trend than INCINC, v_2 and v_3 are the least biased estimators and also provide confidence levels closest to the nominal rate. The estimator v_1 was shatteringly bad for both of these populations. For EMPINC, which has much weaker trend than INCINC, the first estimator v_1 performed as well as any of the estimators $v_2, \ldots, v_6$.

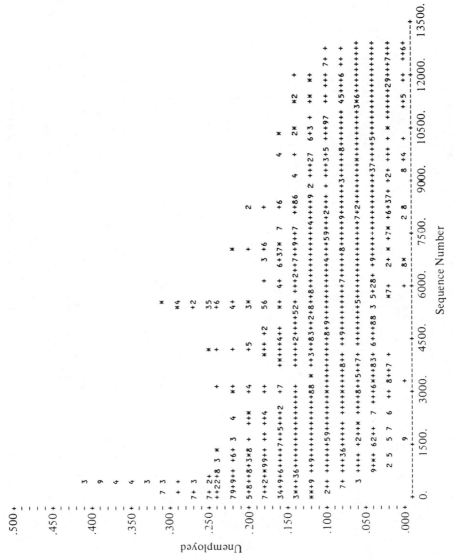

Figure 7.4.4. Plot of proportion unemployed versus sequence number for population EMPINC.

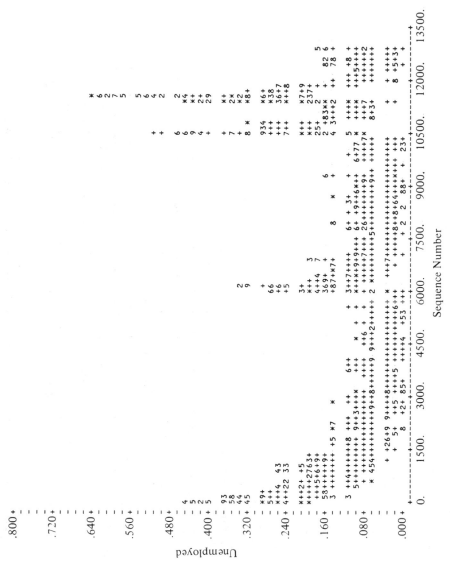

Figure 7.4.5. Plot of proportion unemployed versus sequence number for population EMPRSA.

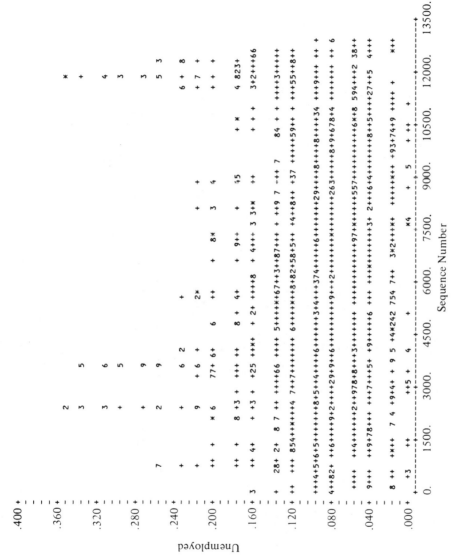

Figure 7.4.6. Plot of proportion unemployed versus sequence number for population EMPNOO.

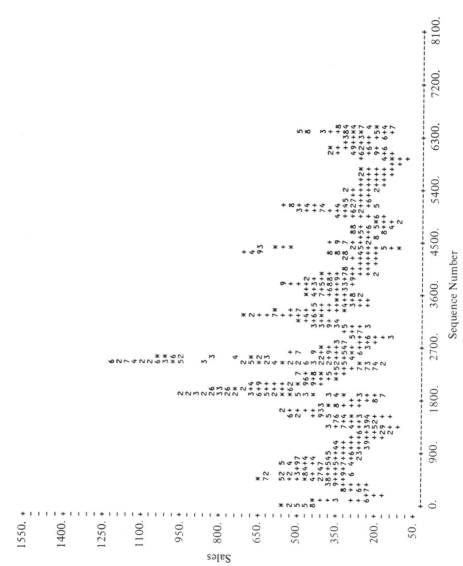

Figure 7.4.7. Plot of 1972 annual sales versus sequence number for population FUELID.

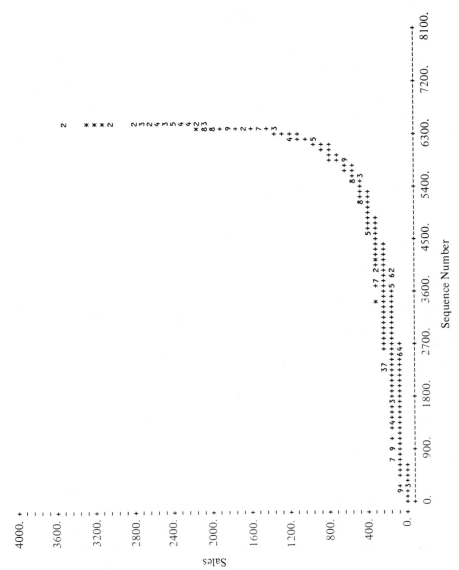

Figure 7.4.8. Plot of 1972 annual sales versus sequence number for population FUELAP.

Populations with Stratification Effects. Any of the three estimators v_2, v_3, v_4 may be recommended for the populations INCRSA and EMPRSA. The absolute bias of v_1 tends to be somewhat larger than the biases of these preferred estimators. All of the preferred estimators are downward biased for INCRSA and, thus, actual confidence levels are too low. Estimator v_6 has larger MSE than the preferred estimators.

Random Populations. Any of the first six estimators may be recommended for INCNOO, EMPNOO, and FUELID. The eighth estimator also performs quite well for these populations, except for FUELID where it has a larger downward bias and corresponding confidence levels are too low.

7.5. Conclusions in the Equal Probability Case

The reader should note that the findings presented in the previous sections apply primarily to surveys of establishments and people. Stronger correlation patterns may exist in surveys of land use, forestry, geology and the like, and the properties of the estimators may be somewhat different in such applications. Additional research is needed to study the properties of the variance estimators in the context of such surveys. With these limitations in mind, we now summarize the numerical and theoretical findings regarding the usefulness of the estimators of variance. The main advantages and disadvantages of the estimators seem to be as follows:

(i) The bias and MSE of the simple random sampling estimator v_1 are reasonably small for all populations which have approximately constant mean μ_{ij}. This excludes populations with a strong trend in the mean or stratification effects. Confidence intervals formed from v_1 are relatively good overall, though are often too wide and lead to true confidence levels exceeding the nominal level.

(ii) In relation to v_1, the estimators v_4, v_5, v_6 based on higher order differences provide protection against a trend, autocorrelation, and stratification effects. They are often good for the approximate random populations as well. v_4 often has the smallest MSE of these three, because the variances of v_5 and v_6 are large when the sample size (and thus the number of differences) is small. In larger samples and in samples with nonlinear trend or complex stratification effects, these estimators should perform relatively well. Confidence intervals are basically good, except when there is a pure linear trend in the mean.

(iii) The bias of v_7 is unpredictable, and its variance is generally too large to be useful. This estimator cannot be recommended on the basis of the work done here. Increasing k, however, may reduce the variance of v_7 enough to make it useful in real applications.

(iv) Estimator v_8 has remarkably good properties for the artificial populations with linear trend or autocorrelation, otherwise it is quite mediocre. Its bias is usually negative, and consequently, confidence intervals formed from v_8 can fail to cover the true population mean at the appropriate nominal rate. This estimator seems too sensitive to the form of the model to be broadly useful in real applications.

(v) The estimators v_2 and v_3 based on simple differences afford the user considerable protection against most model forms studied in this chapter. They are susceptible to bias for populations with strong stratification effects. They are also biased for the linear trend population, but even then the other estimators have larger bias. Stratification effects and trend effects did occur in the real populations, but they were not sufficiently strong effects to defeat the good properties of v_2 and v_3. In the real populations these estimators performed, on average, as well as any of the estimators. Estimators v_2 and v_3 (more degrees of freedom) often have smaller variance than estimators v_4, v_5, and v_6 (fewer degrees of freedom). In very small samples, v_2 might be the preferred estimator.

If an underlying model can be assumed for the finite population of interest and is known approximately or can be determined by professional judgment, then the reader should select an appropriate variance estimator by reference to our theoretical study of the estimator biases, or by reference to our numerical results for similar populations. The summary properties in points (i) to (v) above should be helpful in making an informed choice.

If the model were known exactly, of course, then one may construct an appropriate estimator of variance according to the methodology presented in Section 7.2.2. But true superpopulation models are never known exactly, and moreover are never as simple as the models utilized here. It is thus reasonable to plan to use one of the eight estimators of variance presented in Section 7.2.1. These estimators will not necessarily be optimal for any one specific model, but will achieve good performance for a variety of practical circumstances, and thus will offer a good compromise between optimality given the model and robustness given realistic failures of the model.

On the other hand, if little is known about the finite population of interest, or about the underlying superpopulation model, then, as a good general purpose estimator, we suggest v_2 or v_3. These estimators seem (on the basis of the work presented here) to be broadly useful for a variety of populations found in practice.

7.6. Unequal Probability Systematic Sampling

Unequal probability systematic sampling is one of the most widely used methods of sampling with unequal probabilities without replacement. Its popularity derives from the fact that

- it is an easy sampling scheme to implement either clerically or on a computer;
- if properly applied it can be a πps sampling design, i.e., $\pi_i = np_i$;
- it is applicable to arbitrary sample size n, i.e., its use is not restricted to a certain sample size such as $n = 2$;
- if properly applied it can be quite efficient in the sense of small design variance, picking up any implied or hidden stratification in the population.

As in the case of equal probability systematic sampling, however, the method runs into certain difficulties or dilemmas in regards to the estimation of variances. In the balance of this chapter, we shall discuss the difficulties, define several potentially useful estimators of variance, and examine the range of applicability of the estimators.

Before proceeding, it will be useful to review briefly how to select an unequal probability systematic sample (also called systematic pps sampling). First, the N population units are arranged in a list. They can be arranged at random in the list; they can be placed in a particular sequence; or they can be left in a sequence that they naturally occur. We shall let Y_i denote the value of the estimation variable for the i-th unit in the population and let X_i denote the value of a corresponding auxiliary variable, or "measure of size," thought to be correlated with the estimation variable.

Next, a cumulative measure of size, M_i, is calculated for each population unit. This cumulative size is simply the measure of size of the i-th unit added to the measures of size of all units preceding the i-th unit on the list, i.e.,

$$M_i = \sum_{j=1}^{i} X_j,$$

To select a systematic sample of n units, a selection interval, say I, is calculated as the total of all measures of size divided by n:

$$I = \sum_{i=1}^{N} X_i/n = X/n.$$

The selection interval I is not necessarily an integer, but is typically rounded off to two or three decimal places.

To initiate the sample selection process, a uniform random deviate, say R, is chosen on the half open interval $(0, I]$. The n selection numbers for the sample are then

$$R, R + I, R + 2I, R + 3I, \ldots, R + (n - 1)I.$$

The population unit identified for the sample by each selection number is the *first* unit on the list for which the cumulative size, M_i, is greater than or equal to the selection number. Given this method of sampling, the probability of including the i-th unit in the sample is equal to

$$\pi_i = X_i / I$$

$$= np_i ,$$

where

$$p_i = X_i / X.$$

Thus, systematic pps sampling is a πps sampling scheme.

As an example, suppose a sample of four units is to be selected from the units listed in Table 7.6.1, with probabilities proportional to the sizes indicated in the second column. The cumulative sizes are shown in column 3. The selection interval is $I = 161/4 = 40.25$. Suppose the random start, which would be a random number between 0.01 and 40.25, were $R = 31.68$. The four selection numbers would be 31.68, 71.93, 112.18, 152.43. The corresponding four units selected would be units labelled 4, 8, 11, and 15. The four selection numbers are listed in the last column of the table in the rows representing the selected units.

Prior to the selection of a systematic pps sample, the sizes of the units must be compared to the selection interval. Any unit whose size X_i exceeds the selection interval will be selected with certainty, i.e., with probability 1. Typically, these certainty units are extracted from the list prior to the systematic selection.[1] A new selection interval, based on the remaining sample size and on the sizes of the remaining population units, would be calculated for use in selecting the balance of the sample. Of course, the inclusion probabilities π_i must be redefined as a result of this process.

In identifying certainty selections from the list, units that have a size only slightly less than the selection interval are usually included in the certainty group. In applied survey work, a minimum certainty size cutoff is often established, such as $2I/3$ or $3I/4$, and all units whose size X_i is at least as large as the cutoff are taken into the sample with probability 1. By establishing a certainty cutoff, the survey designer is attempting to control the variance by making certain that large units are selected into the sample.

For the systematic pps sampling design, the Horvitz–Thompson estimator

$$\hat{Y} = \sum_{i=1}^{n} y_i / \pi_i$$

is an unbiased estimator of the population total

$$Y = \sum_{i=1}^{N} Y_i.$$

---

[1] In multi-stage samples for which systematic pps sampling is used at some stage, certainty selections are often not identified prior to sampling. Instead, the systematic sampling proceeds in a routine fashion, even though some units may be "hit" by more than one selection number. Adjustments are made in the subsequent stage of sampling to account for these multiple hits.

Table 7.6.1. Example of Unequal Probability Systematic Sampling

Unit	Size (X_i)	Cum Size (M_i)	Selection Numbers
1	8	8	
2	12	20	
3	11	31	
4	4	35	31.68
5	10	45	
6	15	60	
7	6	66	
8	20	86	71.93
9	11	97	
10	14	111	
11	5	116	112.18
12	9	125	
13	7	132	
14	17	149	
15	12	161	152.43

To estimate the variance of $\hat{Y}$, it is natural to consider, at least provisionally, the variance estimators proposed either by Horvitz and Thompson (1952) or by Yates and Grundy (1953). See Section 1.4 for definitions of these estimators. Unfortunately, both of these estimators run into some difficulty in the context of systematic pps sampling. In fact, neither estimator is unbiased, and in some applications they may be undefined. Most of the difficulties have to do with the joint inclusion probabilities π_{ij}. The π_{ij} will be zero for certain pairs of units, thus defeating the unbiasedness property. Or the π_{ij} may be unknown, thus making it difficult to apply the provisional variance estimators.

In view of these difficulties, we shall broaden our search for variance estimators, including biased estimators that are computationally feasible. We shall define several such estimators of variance in the next section, and in subsequent sections examine whether these estimators have utility for systematic pps designs.

7.7. Alternative Estimators in the Unequal Probability Case

We shall discuss estimators of the variance of $\hat{Y}$. Estimators of the variance of nonlinear statistics of the form $\hat{\theta} = g(\hat{Y})$ may be obtained from the development presented here together with the appropriate Taylor series formula.

An appealing estimator of variance is obtained from the Yates and Grundy formula by substituting an approximation to the $\pi_{ii'}$ developed by Hartley and Rao (1962). The approximation

$$\pi_{ii'} = \frac{n-1}{n}\pi_i\pi_{i'} + \frac{n-1}{n^2}(\pi_i^2\pi_{i'} + \pi_i\pi_{i'}^2) - \frac{n-1}{n^3}\pi_i\pi_{i'}\sum_{j=1}^{N}\pi_j^2$$

is correct to order $0(N^{-3})$ on the conditions that (1) the population listing may be regarded as in random order and (2) π_i is order $0(N^{-1})$. The corresponding estimator of variance

$$v_9 = \frac{1}{n-1}\sum_i^n\sum_{i<i'}^n\left(1 - \pi_i - \pi_{i'} + \sum_{j=1}^{N}\frac{\pi_j^2}{n}\right)\cdot\left(\frac{y_i}{\pi_i} - \frac{y_{i'}}{\pi_{i'}}\right)^2 \qquad (7.7.1)$$

is correct to terms of order $0(N)$. Hartley and Rao also give a better approximation to the $\pi_{ii'}$, correct to order $0(N^{-4})$, and the corresponding variance estimator is correct to order $0(1)$.

If large units are selected into the sample with certainty, then this formula and all symbols contained therein (e.g., N, n, π_i, and $\pi_{i'}$) pertain only to the noncertainty portion of the population. For the certainty cases, the contribution to both the true and estimated variances is identically zero. In fact, these remarks apply generally to all of the variance estimators studied in this section.

In the equal probability situation where N is an integer multiple of n, the probabilities $\pi_i = n/N$ and the estimator v_9 reduces to the simple random sampling estimator v_1 studied in Section 7.2.1.

The estimator v_9 is not an unbiased estimator of the variance $\text{Var}\{\hat{Y}\}$, but it may have useful statistical properties in situations where the population listing can be regarded as random and the approximation involved in $\pi_{ii'}$ is satisfactory.

A second estimator of variance is obtained by treating the sample as if it were a pps with replacement (wr) sample. The estimator is

$$v_{10} = \frac{1}{n(n-1)}\sum_{i=1}^{n}\left(\frac{y_i}{p_i} - \hat{Y}\right)^2. \qquad (7.7.2)$$

This estimator will be biased in the context of systematic pps designs, but the bias may be reasonably small when the population is large, the population listing is in an approximate random order, and none of the population units are disproportionately large. Further, the estimator v_{10} will tend to be conservative (i.e., too large) in situations where systematic pps sampling has smaller true variance than pps with replacement sampling.

A third estimator is obtained by treating the sample as if $n_h = 2$ units were selected from within each of $n/2$ equal-sized strata. The corresponding variance estimator is

$$v_{11} = \frac{1}{n}\sum_{i=1}^{n/2}\left(\frac{y_{2i}}{p_{2i}} - \frac{y_{2i-1}}{p_{2i-1}}\right)^2 \Big/ n. \qquad (7.7.3)$$

Another estimator, which aims to increase the number of "degrees of freedom" is

$$v_{12} = \frac{1}{n} \sum_{i=2}^{n} \left(\frac{y_i}{p_i} - \frac{y_{i-1}}{p_{i-1}} \right)^2 \bigg/ 2(n-1). \qquad (7.7.4)$$

This estimator utilizes overlapping differences, whereas v_{11} utilizes nonoverlapping differences.

A fifth estimator is obtained by application of the random group principle. Let the systematic sample be divided into k systematic subsamples, each of integer size $m = n/k$. Let

$$\hat{Y}_\alpha = \frac{1}{m} \sum_{i=1}^{m} \frac{y_i}{p_i}$$

denote the Horvitz–Thompson estimator of total corresponding to the α-th subsample ($\alpha = 1, \ldots, k$). Then, the variance estimator is defined by

$$v_{13} = \frac{1}{k(k-1)} \sum_{\alpha=1}^{k} (\hat{Y}_\alpha - \hat{Y})^2. \qquad (7.7.5)$$

Alternatively, the systematic sample may be divided into subsamples at random instead of systematically. It can be shown that this form of v_{13} has the same expectation but larger variance than the pps wr estimator v_{10}. See Isaki and Pinciaro (1977).

If desired, each of the estimators $v_{10}, \ldots, v_{13}$ may be multiplied by a finite population correction (fpc) factor, whereas estimator v_9 presumably accounts internally for the without replacement aspect of the sampling design. A computationally simple and potentially useful fpc for systematic pps sampling is

$$\widehat{\text{fpc}} = \left(1 - n^{-1} \sum_{i=1}^{n} \pi_i \right). \qquad (7.7.6)$$

Of course, no exact fpc appears in the true variance for systematic pps sampling. Therefore, use of (7.7.6) should be viewed as a rule of thumb for reducing the estimated variance in applications where systematic pps sampling is thought to be more efficient than pps with replacement sampling.

By now, the reader will have noticed a strong resemblance between the present estimators and those given in Section 7.2.1 for equal probability systematic sampling. In fact, a general method for constructing variance estimators for unequal probability systematic sampling involves using almost any estimator of variance for equal probability sampling, and replacing the values y_i by $z_i = y_i/p_i$ in the definition of the estimator. Aside from the presence or absence of the fpc and from the differences between estimating the population mean or total, the estimator v_{10} corresponds in this way to the estimator v_1 in Section 7.2.1. Likewise, estimators v_{11}, v_{12}, v_{13} correspond in this way to v_3, v_2, and v_7, respectively. It would also be possible to construct unequal probability analogs of v_4, v_5, v_6, and v_8. But we shall leave this work to the reader as an exercise.

Little is known about the exact theoretical properties of these variance estimators, and instead, we offer some general impressions. The behavior of the estimators will depend to a large degree upon the order of the population listing prior to sampling and on any association between that order and the estimation variable. What matters in equal probability systematic sampling is the association between the order and the y-variable itself. What is likely to matter in unequal probability sampling, however, is the association between the order and the z-variable, or in other words, the association between order and the ratio y_i/p_i. By interpreting them in this light, the findings and discussions presented in Sections 7.2 to 7.5 can be used to guide the choice of variance estimator for unequal probability systematic sampling. For example, if a certain estimator possesses good statistical properties in the equal probability case when there is a linear trend in the y-variable, then it may possess good properties in the unequal probability case when there is a linear trend in the ratio y_i/p_i.

Further, the behavior of the estimators will depend upon the fact that the survey design involves without replacement sampling, whereas many of the variance estimators arise from within the context of with replacement sampling. As a result, the estimators will tend to over or underestimate the true variance $\text{Var}\{\hat{Y}\}$ as this variance is less than or greater than the variance under pps wr sampling. Use of the approximate fpc will tend to help matters if the former relationship is known to hold.

Finally, the variance of the variance estimators will tend to be inversely related to the number of "degrees of freedom." This behavior was observed in the preceding sections on equal probability systematic sampling. In small sample sizes the survey statistician should take particular care to choose an estimator of variance with adequate "degrees of freedom" so that the variance of the variance estimator is not so large as to render the estimator unusable.

An altogether different class of variance estimators for $\hat{Y}$ is created by assuming the data are generated by a superpopulation model. One estimator in this class is

$$v_{16} = X^2\{(\hat{\beta}^2 - \hat{\mathcal{V}}ar\{\hat{\beta}\}) \sum_k P(k)(\bar{X}_k - n \sum_{k'} P(k')\bar{X}_{k'})^2 + (N-1)\hat{\sigma}_e^2/Nn\}$$

(7.7.7)

where

$$X = \sum_{i=1}^{N} X_i,$$

$$\hat{\beta} = \frac{\sum_{i=1}^{n} (r_i - \bar{r})(x_i - \bar{x})}{\sum_{i=1}^{n} (x_i - \bar{x})^2},$$

$$\hat{V}a\imath\{\hat{\beta}\} = \hat{\sigma}_e^2 \bigg/ \sum_{i=1}^n (x_i - \bar{x})^2,$$

$$\hat{\sigma}_e^2 = \frac{1}{n-2} \sum_{i=1}^n \{(r_i - \bar{r}) - \hat{\beta}(x_i - \bar{x})\}^2,$$

$$r_i = y_i/x_i,$$

$$\bar{r} = \frac{1}{n} \sum_{i=1}^n r_i,$$

$$\bar{x} = \frac{1}{n} \sum_{i=1}^n x_i,$$

$\bar{X}_k$ = sample mean of the k-th systematic sample, and $P(k)$ = probability of selecting the k-th systematic sample. This estimator is originally due to Hartley (1966), and is obtained by assuming a linear regression model

$$r_i = \alpha + \beta x_i + e_i \tag{7.7.8}$$

between the $r_i = y_i/x_i$ ratio and the measure of size x_i. If we may reasonably assume that the population N is a random sample from a superpopulation wherein (7.7.8) holds with

$$\mathscr{E}\{e_i\} = 0,$$

then v_{16} is an unbiased estimator (with respect to the model) of the design variance Var$\{\hat{Y}\}$.

Extensions of the Hartley method can be created by assuming alternative superpopulation models relating the ratios r_i to x_i. In fact, in numerical work to be described in Section 7.8, we encounter a population wherein a hyperbolic relation between r_i and x_i may be appropriate.

Finally, we note that estimator v_{16} requires calculation of the between sum of squares

$$\sum_k P(k)\bigg(\bar{X}_k - \sum_{k'} P(k')\bar{X}_{k'}\bigg)^2$$

and thus carries a greater computational burden than estimators $v_9, \ldots, v_{13}$. Presumably, a similar burden would accompany any other member of this class of estimators.

7.8. An Empirical Comparison

In this section, we report on a small empirical comparison that was made in order to understand better the properties of the alternative variance estimators. In the absence of firm theoretical results about the estimators, the empirical results should provide the reader with the best available guidance on choosing variance estimators for systematic pps sampling. This material was originally reported by Isaki and Pinciaro (1977).

7.8.1. Description of the Study

We compare the estimators of variance defined in the previous section using four real data sets, each comprised of $N = 5634$ mobile home dealers that were enumerated in the 1972 U.S. Census of Retail Trade. Table 7.8.1 provides a description of the four populations. In populations SALPAY and SALGEO, the estimation variable (y) is 1972 annual sales, whereas in EMPPAY and EMPGEO it is 1972 first quarter employment. The populations also differ by the ordering of the units prior to sampling. For SALPAY and EMPPAY, the units were ordered by decreasing value of 1972 average quarterly payroll (x), and for SALGEO and EMPGEO the ordering was by identification number (this essentially provides a geographic ordering).

As in Section 7.4 the populations are named for the convenience of the reader. The first three letters of the name signify the estimation variable and the last three letters signify the ordering. For example, SALPAY equates to

* estimation variable = SALes
* ordering variable = PAYroll.

The population totals of the sales, employment, and payroll variables are $0.32385 \cdot 10^{10}$ dollars, $0.33213 \cdot 10^5$ employees, and $0.57300 \cdot 10^8$ dollars, respectively.

Figures 7.8.1 to 7.8.4 plot the data in various ways. The figures show

Figures	Plots
7.8.1	sales vs. payroll
7.8.2	employment vs. payroll
7.8.3	sales/payroll ratio vs. payroll
7.8.4	employment/payroll ratio vs. payroll

There is an approximately linear relationship between sales and payroll and between employment and payroll, where in each case the residual variance about the linear relation would appear to increase with payroll.

Table 7.8.1. Description of the Populations Used in the Empirical Comparison

Population	Characteristic	Order
SALPAY	Annual Sales	Average Payroll
SALGEO	Annual Sales	Identification Number
EMPPAY	First Quarter Employment	Average Payroll
EMPGEO	First Quarter Employment	Identification Number

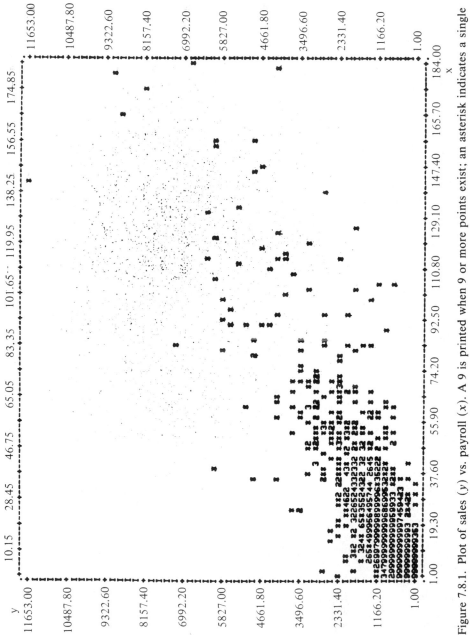

Figure 7.8.1. Plot of sales (y) vs. payroll (x). A 9 is printed when 9 or more points exist; an asterisk indicates a single point.

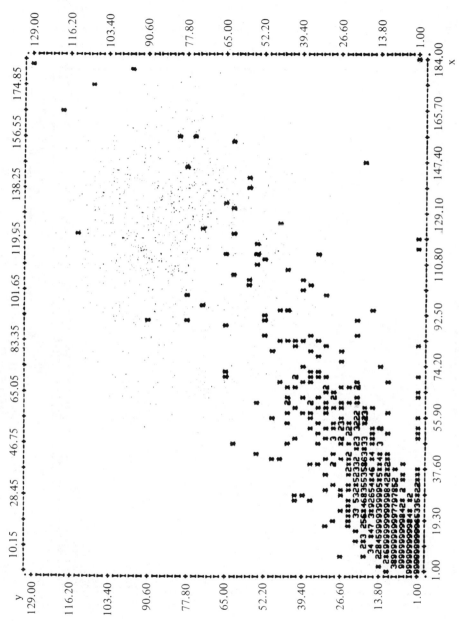

Figure 7.8.2. Plot of employment (y) vs. payroll (x).

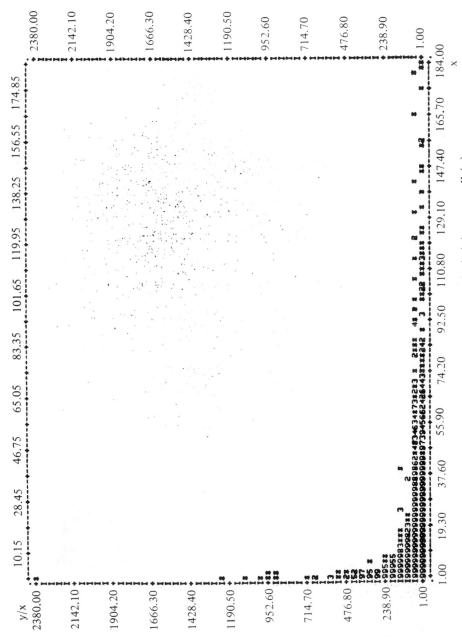

Figure 7.8.3. Plot of sales to payroll ratio (y/x) vs. payroll (x).

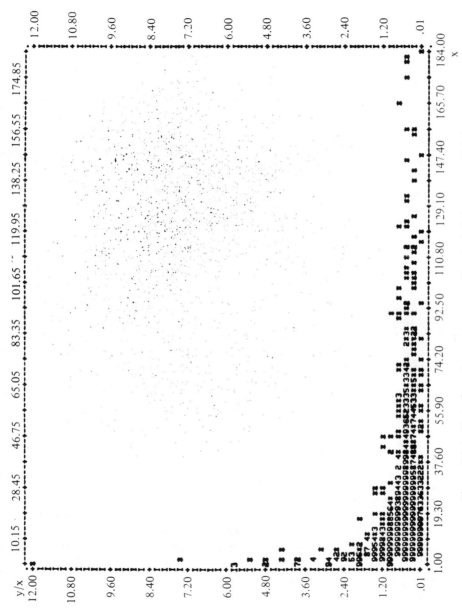

Figure 7.8.4. Plot of employment to payroll ratio (y/x) vs. payroll (x).

The population correlation coefficients are

$$\rho(\text{sales, payroll}) = 0.74$$

$$\rho(\text{employment, payroll}) = 0.75.$$

These data suggest that a systematic pps sampling design, using payroll as the measure of size, would be an efficient scheme for sampling from this population.

In Figures 7.8.3 and 7.8.4 there is an apparent hyperbolic relationship between the sales/payroll ratio and payroll and between the employment/payroll ratio and payroll. These data suggest that the populations should be ordered by payroll in addition to sample selection with probability proportional to payroll. Ordering in this way is a good sampling strategy because it ensures that each potential sample contains a cross-section of units with different values of the y/x ratio. Indeed the numerical work, described in the next subsection, confirms this observation. Had there been a flat relationship between the y/x ratio and payroll, then no additional sampling efficiencies would be gained by ordering by the measure of size; in effect, all of the useful sampling information in the measure of size would be used up by selecting with probability proportional to size.

In the empirical comparisons, we are concerned with the statistical properties of eight estimators of the variance of the Horvitz–Thompson estimator, $\hat{Y}$, of the population total. We study the estimators v_9, v_{10}, v_{11}, v_{12}, v_{13} defined in Section 7.7. The estimator v_{13} is studied both with $k = 5$ groups and $k = 15$ groups. We also study two modified estimators created by appending the approximate finite population correction, $\widehat{\text{fpc}}$. These estimators are defined by

$$v_{14} = \left(1 - \sum_{i=1}^{n} \pi_i/n\right)v_{10}$$

$$v_{15} = \left(1 - \sum_{i=1}^{n} \pi_i/n\right)v_{12},$$

modifying the pps wr estimator and the estimator based upon overlapping differences, respectively.

We do not present results for the Hartley estimator v_{16}. In view of the apparent hyperbolic relationship between the ratios r_i and the measure of size x_i, it would be appropriate in this population to replace x_i by x_i^{-1} throughout the definition of the estimator.

Throughout the study, sample selection is with probability proportional to X_i, where X_i is the 1972 average quarterly payroll of the i-th unit. Results are presented for three sample sizes, including $n = 30, 60, 150$. In advance of the study, 19 large dealers were declared to be certainty units on the basis of large X_i, and were omitted from the study. The population size $N = 5634$ is net of these certainty cases. Also in advance, the actual payroll sizes of the various units were modified slightly so that the total

$$X = \sum_{i=1}^{N} X_i$$

would be perfectly divisible by the sample size n. This modification permits certain computational efficiencies in the conduct of the study (see, e.g., equations 7.8.1, 7.8.2, and 7.8.3), but is not an essential part of the systematic pps method.

We compare the estimators of variance on the basis of their relative .biases, relative mean square errors (MSE), and confidence interval coverage rates. The relative bias of an arbitrary estimator of variance v is given by

$$\text{Rel Bias}\{v\} = \frac{E\{v\} - \text{Var}\{\hat{Y}\}}{\text{Var}\{\hat{Y}\}}.$$

The expectation is obtained computationally as

$$E\{v\} = \sum_{s} v(s)\frac{1}{p}, \tag{7.8.1}$$

where $p = X/n$ denotes the number of potential integer random starts, s denotes the sample associated with a given random start, Σ_s denotes summation over all possible integer random starts, and $v(s)$ denotes the value of the random variable v given a certain random start. In this formulation, note that the samples s are not necessarily distinct. In fact, two or more integer random starts may produce the same sample of units.

The relative MSE of v is given by

$$\text{Rel MSE}\{v\} = \frac{E\{(v - \text{Var}\{\hat{Y}\})^2\}}{(\text{Var}\{\hat{Y}\})^2},$$

where

$$E\{(v - \text{Var}\{\hat{Y}\})^2\} = \sum_{s} (v(s) - \text{Var}\{\hat{Y}\})^2\frac{1}{p}. \tag{7.8.2}$$

And the actual confidence interval coverage percentage is

$$c = \frac{100}{p}\sum_{s} X_s,$$

where

$$X_s = 1, \quad \text{if the true total } Y \text{ satisfies } Y \in (\hat{Y}(s) \pm z\sqrt{v(s)})$$

$$= 0, \quad \text{otherwise.}$$

In this notation, $\hat{Y}(s)$ denotes the value of the Horvitz–Thompson estimator given a specific integer random start and z denotes a tabular value from the standard normal distribution. We present results for $z = 1.96$ and thus investigate the actual coverage properties of nominal 95% confidence intervals.

7.8.2. Results

Table 7.8.2 presents certain summary information with respect to the various populations and sample sizes. The third column gives the design variance of the Horvitz–Thompson estimator given the pps systematic sampling design. Columns four and five compare that variance to the variance that would obtain given a pps wr sampling design. Hartley's (1966) intraclass correlation is equivalent to

$$\text{Intraclass Correlation} = \frac{\text{Var}\{\hat{Y}|\text{pps syst}\} - \text{Var}\{\hat{Y}|\text{pps wr}\}}{(n-1)\,\text{Var}\{\hat{Y}|\text{pps wr}\}}$$

and takes its lower bound of $-(n-1)^{-1}$ when the pps systematic design has zero variance, takes the value zero when the two sampling designs are equally efficient, and increases above zero as pps wr sampling becomes more and more efficient. The column headed Eff represents the efficiency

$$\text{Eff} = \frac{\text{Var}\{\hat{Y}|\text{pps wr}\}}{\text{Var}\{\hat{Y}|\text{pps syst}\}}$$

of pps systematic sampling with respect to pps wr sampling.

These data show clearly that the population ordering by payroll results in the most efficient sampling design. It is more efficient than either the population ordering by geography or the pps wr design. The data also show that the pps wr design tends to be more efficient than the population ordering by geography, although these results are equivocal for the population EMPGEO. These data suggest that the essence of the systematic pps method is to order the population in such fashion as to produce a large negative intraclass correlation, and that this can often be achieved by ordering according to X_i.

Table 7.8.2. Population Parameters

Population	Sample Size	Var$\{\hat{Y}|\text{pps syst}\}$	Intraclass Correlation	Eff
SALPAY	30	$1.427 \cdot 10^{11}$	-0.0071	1.26
	60	$0.718 \cdot 10^{11}$	-0.0034	1.26
	150	$0.276 \cdot 10^{11}$	-0.0016	1.30
SALGEO	30	$1.876 \cdot 10^{11}$	0.0014	0.96
	60	$0.931 \cdot 10^{11}$	0.0006	0.97
	150	$0.411 \cdot 10^{11}$	0.0009	0.88
EMPPAY	30	$1.076 \cdot 10^{7}$	-0.0041	1.13
	60	$0.506 \cdot 10^{7}$	-0.0029	1.21
	150	$0.218 \cdot 10^{7}$	-0.0007	1.12
EMPGEO	30	$1.270 \cdot 10^{7}$	0.0014	0.96
	60	$0.576 \cdot 10^{7}$	-0.0010	1.06
	150	$0.265 \cdot 10^{7}$	0.0006	0.92

The numerical comparison of the variance estimators is presented in Tables 7.8.3 (relative bias), 7.8.4 (relative MSE), and 7.8.5 (confidence interval coverage percentages). For population SALPAY, the estimator v_{11} based upon nonoverlapping differences tends to have the smallest bias. The estimator v_{12} based upon overlapping differences also tends to have small bias. The pps wr estimator v_{10} and the Hartley–Rao estimator v_9 tend to be too big. Fewer groups $k = 5$ seems to produce smaller absolute bias than more groups $k = 15$ in the context of v_{13}. Use of the fpc's in v_{14} and v_{15} is not very helpful; the fpc reduces the bias marginally when the basic estimator is too large and makes matters slightly worse when the basic estimator is too small. In terms of MSE, the estimator based upon overlapping differences v_{12} and its modification v_{15} are the clear winners. There is little to choose between the remaining estimators, except v_{13} with $k = 5$ tends to be worse than with $k = 15$. For the largest sample size, everything tends to even out and all estimators perform about the same. In terms of confidence intervals, all estimators tend to produce actual coverage rates lower than the nominal rate for small sample sizes. The situation is much improved for the larger sample sizes, however. Estimators v_9, v_{10}, and v_{13} with $k = 15$ tend to provide the best confidence intervals (in the sense of smallest departure from nominal levels). v_{13} with $k = 5$ is not as good as with $k = 15$, which is opposite the finding with respect to bias. At large sample sizes, all of the estimators, except v_{13} with $k = 5$, perform similarly. Once again, we see that the fpc's tend not to produce a significant or helpful outcome, but only a marginally different outcome.

The results for population EMPPAY are very nearly identical with those just described for SALPAY. This is to be expected since the structure of the data in these two populations is nearly identical. Compare Figures 7.8.1 to 7.8.4.

Turning next to the geographic ordering, we see in population SALGEO that all of the estimators tend to be too small. Estimators v_9 and v_{10} may be considered slight favorites in the race for smallest bias, but really there is little to choose between the different estimators. Once again, the addition of the fpc's does not seem to offer any significant improvement. There is little difference between the various estimators in terms of their MSE. It is noteworthy, however, that the estimator MSE's are smaller for this population ordering than for the ordering by payroll. Evidently there is a slight conflict between efficient estimation of the Y total (which suggests ordering by measure of size) and efficient estimation of the estimator variance (which suggests against ordering by measure of size). All of the actual confidence levels are lower than the nominal level of 95%. In fact, the actual confidence levels are noticeably lower than they were for the populations ordered by payroll. Estimator v_{13} with $k = 5$ is not as good as the estimator with $k = 15$, a finding which is consistent with the findings for populations SALPAY and EMPPAY. Otherwise, there are few differences between the remaining variance estimators in terms of confidence interval properties.

Table 7.8.3. Relative Bias of Eight Estimators of $\mathrm{Var}\{\hat{Y}\}$

Population	v_9	v_{10}	v_{11}	v_{12}	$v_{13}\,(k=5)$	$v_{13}\,(k=15)$	v_{14}	v_{15}
a. $n = 30$								
SALPAY	0.263	0.271	0.028	-0.143	0.132	0.219	0.249	-0.158
SALGEO	-0.047	-0.041	-0.067	-0.050	-0.088	-0.050	-0.058	-0.067
EMPPAY	0.128	0.139	-0.001	-0.087	0.167	0.122	0.119	-0.104
EMPGEO	-0.050	-0.040	-0.043	-0.042	-0.039	-0.044	-0.057	-0.058
b. $n = 60$								
SALPAY	0.243	0.259	0.007	-0.113	-0.007	0.128	0.215	-0.145
SALGEO	-0.045	-0.033	-0.046	-0.047	-0.066	-0.043	-0.067	-0.049
EMPPAY	0.186	0.209	0.055	-0.002	0.200	0.162	0.166	-0.038
EMPGEO	0.040	0.061	0.072	0.064	0.113	0.043	0.023	0.026
c. $n = 150$								
SALPAY	0.265	0.307	0.037	0.003	0.042	0.129	-0.190	-0.087
SALGEO	-0.152	-0.125	-0.150	-0.152	-0.109	-0.109	-0.202	-0.228
EMPPAY	0.071	0.123	-0.041	-0.062	-0.014	0.059	0.023	-0.146
EMPGEO	-0.122	-0.080	-0.061	-0.060	-0.053	-0.100	-0.162	-0.144

Table 7.8.4. Relative Mean Square Error (MSE) of Eight Estimators of $\text{Var}\{\hat{Y}\}$

Population	v_9	v_{10}	v_{11}	v_{12}	$v_{13}(k=5)$	$v_{13}(k=15)$	v_{14}	v_{15}
a. $n = 30$								
SALPAY	13.91	13.94	11.26	3.75	11.85	13.67	13.46	3.63
SALGEO	8.12	8.11	7.23	8.28	7.57	8.35	7.85	8.02
EMPPAY	2.11	2.12	1.91	0.75	2.54	2.17	2.05	0.72
EMPGEO	1.57	1.57	1.67	1.70	1.74	1.47	1.52	1.65
b. $n = 60$								
SALPAY	6.78	6.82	6.62	2.20	5.76	6.56	6.33	2.00
SALGEO	4.09	4.10	4.08	4.26	4.25	4.10	3.83	3.98
EMPPAY	1.19	1.21	1.01	0.40	1.82	1.29	1.11	0.37
EMPGEO	0.92	0.93	0.92	0.96	1.24	0.85	0.87	0.89
c. $n = 150$								
SALPAY	2.89	2.96	2.08	2.35	2.61	2.99	2.42	1.96
SALGEO	0.90	1.33	1.30	1.33	1.28	1.27	1.14	1.14
EMPPAY	0.42	0.44	0.37	0.32	0.94	0.61	0.35	0.28
EMPGEO	0.23	0.27	0.30	0.29	0.69	0.36	0.25	0.26

Table 7.8.5. Percentage of Times that the True Population Total Fell Within the Confidence Interval Formed Using One of Eight Estimators of Variance

Population	v_9	v_{10}	v_{11}	v_{12}	$v_{13}\ (k=5)$	$v_{13}\ (k=15)$	v_{14}	v_{15}
					a. $n=30$			
SALPAY	93	93	90	90	90	93	93	90
SALGEO	88	88	87	87	83	87	88	87
EMPPAY	95	95	93	93	91	94	95	93
EMPGEO	91	91	90	90	85	90	90	90
					b. $n=60$			
SALPAY	95	95	92	92	88	93	95	91
SALGEO	93	93	92	92	84	91	93	92
EMPPAY	96	96	93	94	91	94	95	93
EMPGEO	93	93	93	93	88	93	93	92
					c. $n=150$			
SALPAY	96	96	95	94	88	94	95	93
SALGEO	91	92	90	90	89	90	89	89
EMPPAY	96	97	96	96	90	95	97	94
EMPGEO	90	91	90	90	86	91	88	89

All of the estimators behave similarly in the context of population
EMPGEO. They all tend towards an underestimate for sample sizes $n = 30$
and 150 and an overestimate for sample size $n = 60$. These results are
consistent with the efficiency comparison in Table 7.8.2: in cases where pps
systematic sampling is more efficient than pps wr sampling then the
estimators of variance tend to be too large, evidently tracking the pps wr
variance, and vice versa. The confidence intervals associated with v_{13} ($k =$
15) tend to be better than those associated with v_{13} ($k = 5$). But all of the
actual confidence levels are too low, particularly for $n = 30$ and 150 where
the variance estimators are too small. There is little to choose between the
remaining estimators.

7.9. Conclusions in the Unequal Probability Case

As in the case of equal probability systematic sampling, we suggest that the
statistician study well the population and its order prior to selecting an
estimator of the variance. By consulting expert opinion and analyzing prior
data sets, the statistician should

- determine the statistical relationship between the estimation variable
(y) and the measure of size (x);
- determine the statistical relationship between the ordering variable and
the ratio ($r = y/x$).

Only after making these determinations can the statistician make an informed selection from among the many alternative variance estimators. And
remember, different estimation variables may bear different relationships
to the measure of size or to the ordering variable, and so each may warrant
a different estimator of variance.

In Section 7.7 we defined a number of alternative estimators of the
variance of the Horvitz–Thompson estimator, $\hat{Y}$, of the population total. A
general method of construction of variance estimators for $\hat{Y}$ is to begin
with a variance estimator for equal probability systematic sampling and
replace y_i in the definition of the estimator by the ratio y_i/p_i. In fact, many
of the estimators defined in Section 7.7 were obtained in this way. For
nonlinear survey statistics, we suggest use of one of the estimators in Section
7.7 together with the appropriate Taylor series formula.

In Section 7.8 we presented numerical evidence relating to the statistical
properties of eight of the variance estimators. The numerical evidence may
be used to guide the selection of a variance estimator. We envision a selection
process consisting of the following features:

- determine the relationships between x and y and between r and the
ordering variable, as mentioned above;

  • find the population in Section 7.7 that most closely resembles the population under study with respect to these relationships;
  • choose an appropriate variance estimator according to the results in Tables 7.8.3, 7.8.4, 7.8.5 and to any special circumstances involved in the particular application.

Now we summarize the main findings regarding the usefulness of the eight estimators studied in Section 7.8. The estimators seem to fall into three basic categories on the basis of the numerical results: (i) v_{13} ($k = 5$) and v_{13} ($k = 15$), (ii) v_9, v_{10}, and v_{14}, and (iii) v_{11}, v_{12}, and v_{15}.

  (i) Although v_{13} tends to display reasonable statistical properties, it is almost never the optimal estimator. There is always another class of estimators that performs somewhat better, and thus we tend not to recommend v_{13} on the basis of the work done to date. This recommendation is consistent with our recommendation regarding v_7 in the context of equal probability systematic sampling. The estimator with fewer groups, $k = 5$, generally has smaller bias, larger MSE, and worse confidence interval properties than the estimator with more groups, $k = 15$. Evidently, the variance of this variance estimator is inversely related to the number of groups. The bias of the variance estimator is smallest when the random group estimators are based upon sample sizes that most closely resemble the full sample size, i.e., when the number of groups is small.

 (ii) v_9, v_{10}, and v_{15} tend to perform best for the geographic ordering. This is because this ordering approximates a random ordering and these three estimators owe their heritage to the assumption of a random order. The estimators tend to be too small in the case of the geographic ordering and too large in the case of the payroll ordering. It seems that the estimators of variance tend to be biased on the same side of Var$\{\hat{Y}|$pps syst$\}$ as the pps wr variance. Confidence intervals formed using v_9, v_{10}, and v_{14} seem to be quite good for all populations studied. In most cases the actual coverage rates are lower than the nominal coverage rate, particularly for the geographic ordering. In general, the use of the fpc in v_{14} results in little improvement in the unmodified estimator v_{10}.

(iii) v_{11}, v_{12}, and v_{15} perform best for the populations ordered by payroll. These estimators are a function of the ordering, whereas estimators v_9, v_{10}, and v_{14} are not. This is a desirable property here, because the true variance of $\hat{Y}$ is greatly reduced by the payroll ordering. v_{12} and v_{15} clearly have the smallest MSE for this ordering. v_{11} also has smaller MSE than the remaining estimators, but not as small as v_{12}. This is because v_{12} is based upon overlapping differences, increasing the "degrees of freedom" relative to v_{11}, and thus reducing the variance of the variance estimator. v_{11}, v_{12}, and v_{15} tend to have the smallest bias for the populations ordered by payroll, and even in the populations ordered by geography the bias is not too bad. These estimators tend to be smaller than v_9, v_{10}, and v_{14} though there are some exceptions. As a consequence, actual confidence interval coverage rates are lower for these estimators than for v_9, v_{10}, and v_{14}. Once again, use of the fpc in v_{15} affords no significant improvements vis-à-vis the unmodified estimator v_{12}.

Taking all of the results together, we recommend

• choose v_9, v_{10}, or v_{14} for any population that is in an approximate random order, and among these choose v_{10} or v_{14} if computational convenience is important;

• choose v_{11}, v_{12}, or v_{15} for any population that is ordered in such a way as to display a trend (in this case hyperbolic) in the ratios y/x, and if sample sizes are small to moderate choose v_{11}; and

• consider using v_9, v_{10}, or v_{14} if a confidence interval for Y is desired, regardless of the population order.

We also recommend additional study of the Hartley (1966) type variance estimators. Although such estimators were not included in the numerical work reported here, they may hold promise for future applications.

CHAPTER 8

Summary of Methods for Complex Surveys

In this book, we have studied several practical methods for variance estimation for complex sample surveys. And at this point, we address briefly the ultimate question: *Which of the various methods can be recommended and under what circumstances can they be recommended?*

We attempt to provide a partial answer to this question by offering some comparative remarks about the random group method (RG), the balanced half-sample method (BHS), the jackknife method (J), the method of generalized variance functions (GVF), and the Taylor series method (TS). Variance estimation issues for systematic sampling designs were treated in Chapter 7 and we shall not repeat that treatment here. In any case, the variance estimators for systematic sampling are not necessarily competitors of the RG, BHS, J, GVF, and TS estimators, but are instead intended for a different class of variance estimation problems.

As was explained in Chapter 1, methods for variance estimation must be compared in terms of statistical factors such as bias, mean square error (MSE), and confidence interval coverage probabilities, and administrative considerations such as timing and cost. Our comments on the RG, BHS, J, GVF, and TS methods shall be in terms of these factors. We shall also comment upon the flexibility of the different variance methods in terms of ability to work with different sampling designs and different estimators.

Before discussing the merits of the individual methods, we note that the accuracy of a variance estimator can be defined in terms of different criteria, including bias, MSE, and confidence interval coverage probabilities. Indeed, it is often the case that different variance estimators turn out to be best given different accuracy criteria. Since the most important purpose of a variance estimator will usually be for constructing confidence intervals for the parameter of interest, θ, or for testing statistical hypotheses about θ, we suggest the confidence interval coverage probability will usually be the most relevant criterion of accuracy.

The bias and MSE criteria of accuracy are important, such as for planning future surveys, but are secondary to the primacy of the confidence interval criterion. Even if this were not the case, it turns out that the bias criterion, and to some extent the MSE criterion, do not lead to any definitive conclusions or recommendations about the different variance estimators. This is because the biases of the RG, BHS, J, and TS estimators of variance are, in almost all circumstances, identical, at least to a first-order approximation. Thus, one has to look to second- and higher-order terms in order to distinguish between the estimators. Because the square of the bias is one component of MSE, this difficulty also carries over to the MSE criterion of accuracy. To a limited extent, the second component of the MSE, i.e., variance, is within the statistician's control because he/she can choose from a range of strategies about the number of random groups, partial versus full balancing, and the like. Thus, the best estimator of variance is not obvious in terms of the bias and MSE criteria.

As a consequence, we prefer to decide which variance estimators to use in different survey applications based upon the confidence interval criterion, administrative considerations of some kind, or the compatibility of the survey design-estimator pair with the variance methods.

As will become clear, the BHS method seems to have some advantages in terms of accuracy and is as good as other methods in terms of flexibility and administrative factors.

8.1. Accuracy

The four methods RG, BHS, J, and TS will normally have identical asymptotic properties (see Appendix B). Thus, we focus here on the finite-sample properties of the methods. The bias and MSE of the RG method will depend on both the number (k) and size (m) of the random groups. Generally speaking, we have found the variance of the variance estimator declines with increasing k while the bias increases. At this writing, it is somewhat unclear as to what is the net effect of these competing forces in the MSE. These remarks also apply to the bias and MSE of the BHS, J, and TS estimators of variance. We observe, however, that the MSE of the RG estimator may be slightly larger than that of BHS, J, and TS in many applications.

In the case of nonindependent random groups, the variance estimators do not properly account for both the between and the within components of variance for a multistage survey design. This problem will be negligible whenever the between component of variance is unimportant, or adjustments can be made to the variance estimators so that the problem will be negligible whenever the within component of variance is unimportant. See Section 2.4.4 for details.

A number of Monte Carlo studies of the variance estimators have been conducted during the past 15 to 20 years. Such studies are important to our understanding of the accuracy of the variance estimators in finite samples. Such studies provide an understanding of the behavior of the estimators in terms of the confidence interval criterion, which we have said is often the most relevant criterion of accuracy.

In Tables 8.1.1 to 8.1.15 we present illustrative results from five Monte Carlo studies. Fortunately, these studies seem to be telling us the same story about the behavior of the variance estimators. We are always hesitant to draw definitive conclusions and formulate general recommendations from one Monte Carlo study to the conceptual set of all present and future survey designs, estimators, characteristics, and populations. But since the five studies are in some agreement and since the range of survey conditions in the various studies is fairly broad, we feel that some generally useful recommendations are warranted.

Tables 8.1.1 to 8.1.3 are abstracted from the extensive study by Frankel (1971b). This study involved data from the U.S. Current Population Survey (CPS) as the finite population of interest. A two-per-stratum, single-stage cluster sample design of households was used. Results were produced for a number of sample sizes; for a number of characteristics such as "number of persons per household under 18," "total income of household," and "age of head of household"; and for a number of parameters such as ratio means, differences of means, regression coefficients, and correlation coefficients. We present results for one sample size (with $L = 12$ strata) for

Table 8.1.1. Relative Bias of Variance Estimators, Averaged Over Characteristics, from Frankel's Study

	Parameter of Interest θ	
Variance Estimator	Simple Regression Coefficients	Ratio Means
TS	−0.023	0.024
BHS(1)	0.102	0.064
BHS(2)	0.116	0.069
BHS(3)	0.109	0.067
BHS(4)	0.061	0.053
J(1)	0.019	0.057
J(2)	0.018	0.019
J(3)	0.0004	0.038
J(4)	0.008	0.033

Table 8.1.2. Relative MSE of Variance Estimators,
Averaged Over Characteristics, from Frankel's Study

	Parameter of Interest θ	
Variance Estimator	Simple Regression Coefficients	Ratio Means
TS	1.15	0.384
BHS(1)	1.43	0.446
BHS(2)	1.51	0.441
BHS(3)	1.45	0.440
BHS(4)	1.35	0.419
J(1)	1.27	0.492
J(2)	1.19	0.414
J(3)	1.20	0.420
J(4)	1.18·	0.407

estimators of simple regression coefficients and of ratio means, and averaged over all of the characteristics presented in the Frankel study. For the TS estimator, Frankel estimated the covariance matrix Σ by the random group method as applied to cluster sampling. The four BHS methods correspond to equations (3.4.1), (3.4.2), (3.4.3), and (3.4.4), respectively. The four J methods do not correspond to any of the J methods presented in Chapter 4, but instead are obtained by a procedure of omitting one observation and duplicating another.

Tables 8.1.4 to 8.1.6 are drawn from the study of poststratified means by Bean (1975). This study used 131,575 people from the U.S. Health Interview Survey as the finite population of interest. Bean's sample design involved two PSUs per stratum selected by pps wr sampling. Five variables/parameters were included in the study, e.g., "average income per person" and "average number of restricted activity days per person per year." We present bias and MSE results for all five items but average over items in presenting Bean's confidence interval results. The variance estimators appearing in this study, i.e., TS, BHS(1), and BHS(3), are defined as they were in the Frankel study.

Data from the Mulry and Wolter (1981) study is presented in Tables 8.1.7 to 8.1.9. This study is described in detail in Appendix C, and here we repeat some of the data for sample size $n = 60$ merely for convenience. This study looked at variance estimators for the sample correlation coefficient using data from the U.S. Consumer Expenditure Survey. For the TS estimator, the covariance matrix Σ was estimated by standard srs wor formulae. The estimators BHS(1) and BHS(4) are defined by equations (3.4.1) and (3.4.4), respectively, where pseudostrata were created by pairing adjacent selections in the srs wor design. The J estimator corresponds to equation (4.2.5) with

Table 8.1.3. Actual Confidence Interval Coverage Probabilities Associated with Variance Estimators, Averaged Over Characteristics, from Frankel's Study

	Nominal Probability Assuming Standard Normal Theory				
Variance Estimator	0.99	0.95	0.90	0.80	0.85
a. Simple Regression Coefficients					
TS	0.966	0.912	0.850	0.744	0.622
BHS(1)	0.975	0.930	0.873	0.770	0.650
BHS(2)	0.970	0.930	0.875	0.778	0.653
BHS(3)	0.973	0.934	0.875	0.773	0.653
BHS(4)	0.970	0.925	0.865	0.765	0.641
J(1)	0.966	0.921	0.856	0.745	0.630
J(2)	0.967	0.910	0.849	0.745	0.624
J(3)	0.968	0.916	0.854	0.750	0.628
J(4)	0.967	0.914	0.851	0.747	0.625
b. Ratio Means					
TS	0.971	0.919	0.865	0.763	0.654
BHS(1)	0.973	0.920	0.869	0.771	0.661
BHS(2)	0.971	0.952	0.872	0.768	0.659
BHS(3)	0.972	0.922	0.870	0.769	0.661
BHS(4)	0.972	0.921	0.869	0.767	0.658
J(1)	0.972	0.921	0.867	0.766	0.659
J(2)	0.970	0.918	0.864	0.759	0.650
J(3)	0.971	0.920	0.866	0.765	0.655
J(4)	0.971	0.920	0.866	0.764	0.655

Table 8.1.4. Bias of Estimators of the Variance of a Poststratified Mean, for Five Characteristics of Interest, from Bean's Study

	Characteristic of Interest				
Variance Estimator	Family Income	Restricted Activity Days	Physician Visits	Hospital Days	Proportion Seeing a Physician
TS	$3.21 \cdot 10^2$	$-1.60 \cdot 10^{-1}$	$1.02 \cdot 10^{-3}$	$-6.85 \cdot 10^{-5}$	$6.95 \cdot 10^{-6}$
BHS(1)	$1.49 \cdot 10^3$	$-4.34 \cdot 10^{-2}$	$2.69 \cdot 10^{-3}$	$1.93 \cdot 10^{-3}$	$1.61 \cdot 10^{-5}$
BHS(2)	$1.64 \cdot 10^3$	$-4.09 \cdot 10^{-2}$	$2.69 \cdot 10^{-3}$	$1.79 \cdot 10^{-3}$	$1.58 \cdot 10^{-5}$

Table 8.1.5. MSE of Estimators of the Variance of a Poststratified Mean, for Five Characteristics of Interest, from Bean's Study

	Characteristic of Interest				
Variance Estimator	Family Income	Activity Days	Restricted Physician Visits	Hospital Days	Proportion Seeing a Physician
TS	$3.26 \cdot 10^8$	0.485	$7.65 \cdot 10^{-4}$	$5.65 \cdot 10^{-5}$	$3.39 \cdot 10^{-9}$
BHS(1)	$3.24 \cdot 10^8$	0.522	$8.38 \cdot 10^{-4}$	$1.24 \cdot 10^{-4}$	$3.73 \cdot 10^{-9}$
BHS(2)	$3.23 \cdot 10^8$	0.507	$8.32 \cdot 10^{-4}$	$1.07 \cdot 10^{-4}$	$3.56 \cdot 10^{-9}$

Table 8.1.6. Actual Confidence Interval Coverage Probabilities Associated with Variance Estimators, Averaged Over Characteristics, from Bean's Study

	Nominal Probability Assuming Standard Normal Theory			
Variance Estimators	0.99	0.95	0.90	0.68
TS	0.974	0.930	0.879	0.652
BHS(1)	0.978	0.937	0.889	0.672
BHS(2)	0.978	0.938	0.890	0.671

group size $m = 1$, and the RG estimator to equation (2.4.3) with group size $m = 5$.

Tables 8.1.10 to 8.1.12 present data from the recent study by Dippo and Wolter (1984). The universe for this study was 14,360 consumer units obtained from the U.S. Consumer Expenditure Survey. Estimators such as ratios and correlation coefficients were studied for a wide range of consumer expenditure items such as flour, candy, and eggs. The sample design involved

Table 8.1.7. Bias of Variance Estimators for the Correlation Coefficient, from Mulry and Wolter's Study

Variance Estimators	Bias
TS	$-0.453 \cdot 10^{-2}$
BHS(1)	$0.072 \cdot 10^{-2}$
BHS(4)	$-0.123 \cdot 10^{-2}$
$J(k = 60)$	$0.320 \cdot 10^{-2}$
$RG(k = 12)$	$-0.068 \cdot 10^{-2}$

Table 8.1.8. MSE of Variance Estimators for the Correlation Coefficient, from Mulry and Wolter's Study

Variance Estimators	MSE
TS	$0.713 \cdot 10^{-4}$
BHS(1)	$1.651 \cdot 10^{-4}$
BHS(4)	$1.085 \cdot 10^{-4}$
$J(k = 60)$	$3.791 \cdot 10^{-4}$
$RG(k = 12)$	$0.508 \cdot 10^{-4}$

Table 8.1.9. Actual Confidence Interval Coverage Probabilities Associated with Variance Estimators, from Mulry and Wolter's Study

	Nominal Probability Assuming Standard Normal Theory	
Variance Estimators	0.95	0.90
TS	0.828	0.746
BHS(1)	0.872	0.816
BHS(4)	0.864	0.796
$J(k = 60)$	0.878	0.817
$RG(k = 12)$	0.881	0.829

Table 8.1.10. Relative Bias of Estimators of Variance, from Dippo and Wolter's Study

	Characteristic of Interest				
Variance Estimators	Flour[a]	Ground Beef[a]	Gasoline[a]	Food at Home[b]	Food Away from Home[b]
$RG(k = 8)$	0.249	0.051	0.079	−0.050	−0.042
$RG(k = 4)$	0.134	0.076	0.080	−0.148	−0.069
$RG(k = 2)$	0.050	0.112	0.106	−0.132	−0.005
BHS(1)	0.055	0.079	−0.082	−0.153	−0.054

[a] Ratio estimator of the average cost per consumer unit for the particular commodity, among consumer units reporting the commodity.

[b] Simple correlation coefficient between the annual consumer unit income before taxes and expenditures on the particular commodity.

Table 8.1.11. Relative MSE of Estimators of Variance from Dippo and Wolter's Study

Variance Estimators	Flour[a]	Ground Beef[a]	Gasoline[a]	Food at Home[b]	Food Away from Home[b]
				Characteristic of Interest	
RG($k=8$)	3.58	4.11	0.552	0.257	0.282
RG($k=4$)	2.32	4.63	0.961	0.458	0.715
RG($k=2$)	3.21	6.29	3.13	1.40	2.60
BHS(1)	2.20	4.28	0.248	0.236	0.640

[a] Ratio estimator of the average cost per consumer unit for the particular commodity, among consumer units reporting the commodity.

[b] Simple correlation coefficient between the annual consumer unit income before taxes and expenditures on the particular commodity.

Table 8.1.12. Actual Confidence Interval Coverage Probabilities Associated with Variance Estimators, from Dippo and Wolter's Study

Variance Estimators	0.99	0.95	0.90	0.68
	Nominal Probability Assuming Standard Normal Theory			
a. Ratio Means[a]				
RG($k=8$)	0.909	0.850	0.799	0.604
RG($k=4$)	0.874	0.803	0.761	0.584
RG($k=2$)	0.744	0.681	0.634	0.493
BHS(1)	0.906	0.840	0.794	0.596
b. Correlation Coefficients[b]				
RG($k=8$)	0.963	0.905	0.853	0.637
RG($k=4$)	0.918	0.843	0.784	0.582
RG($k=2$)	0.752	0.689	0.641	0.492
BHS(1)	0.967	0.902	0.838	0.607

[a] Actual probability averaged over the three ratio means: flour, ground beef, and gasoline.

[b] Actual probability averaged over the two correlation coefficients: ρ (food at home, income) and ρ (food away from home, income).

Table 8.1.13. Bias of Estimators of the Variance of the Regression Estimator, from Deng and Wu's Study

Variance Estimators	Population					
	1	2	3	4	5	6
TS	−1.2	−5.2	−10.0	−5.6	−1.6	−13.8
$J(k = 32)$	0.6	16.8	5.3	7.1	1.8	9.9

$L = 20$ equal-sized strata with srs wor within strata. Three sample sizes $n_h = 6, 12, 24$ were included in the study and our tables present results only for the largest sample size. The random group estimators RG($k = 8$), RG($k = 4$), and RG($k = 2$) correspond to equation (2.4.3) with group sizes $m = 3, 6$, and 12, respectively. The BHS(1) estimator corresponds to equation (3.4.1).

Finally, a selection of results from the recent study by Deng and Wu (1984) is presented in Tables 8.1.13 to 8.1.15. In this study, simple random samples without replacement of size $n = 32$ were used. The estimator of interest was the standard regression estimator of the finite population mean,

$$\hat{\mu} = \bar{y} + \left\{ \sum_{i=1}^{n} (y_i - \bar{y})(x_i - \bar{x}) \middle/ \sum_{i=1}^{n} (x_i - \bar{x})^2 \right\} (\bar{X} - \bar{x}),$$

and this estimator was studied in six small populations, known as

1. Cancer
2. Cities
3. Counties 60
4. Counties 70
5. Hospital
6. Sales.

Table 8.1.14. Root MSE of Estimators of the Variance of the Regression Estimator, from Deng and Wu's Study

Variance Estimators	Population					
	1	2	3	4	5	6
TS	2.91	52.6	13.6	22.0	6.75	24.9
$J(k = 32)$	5.56	84.2	37.1	52.6	11.36	48.1

Table 8.1.15. Actual Confidence Interval Coverage Probabilities Associated with Variance Estimators, from Deng and Wu's Study

Variance Estimators	Nominal Probability Assuming t_{30} Theory		
	0.99	0.95	0.90
a. Population 1			
TS	0.943	0.885	0.805
J($k = 32$)	0.973	0.927	0.876
b. Population 6			
TS	0.915	0.841	0.774
J($k = 32$)	0.984	0.939	0.892

See Royall and Cumberland (1981b) for a complete description of these populations. For the TS estimator, the covariance matrix Σ was obtained by standard srs wor formulae. And for the J estimator, equation (4.2.3) was used with group size $m = 1$ and with a finite population correction factor.

The overall study by Frankel shows generally that

- TS and J may have smaller bias than BHS, but the patterns are not very clear or general;
- TS tends to have the smallest MSE for most simple survey statistics but BHS and J may have smaller MSE for multiple correlation coefficients;
- BHS is clearly best in terms of the confidence interval criterion.

The abstracted data in Tables 8.1.1 to 8.1.3 are generally consistent with these overall conclusions.

Bean's data tell a similar story:

- no one estimator of variance consistently and generally has smallest bias, although in Table 8.1.4 TS tends in this direction;
- TS tends to have the smallest MSE;
- BHS tends to offer the best confidence intervals.

The Mulry and Wolter data and the Dippo and Wolter data conclude similarly that

- TS tends to have good properties in terms of MSE;

- BHS and RG are favored for confidence interval problems;
- actual confidence interval coverage probabilities tend to be too low in all cases.

Tables 8.1.7 to 8.1.12 generally support these conclusions. Appendix C explains that data transformations are sometimes useful in closing the gap between actual and nominal confidence probabilities.

Finally, Deng and Wu's study concludes that

- J and TS tend to be upward and downward biased, respectively;
- TS has smaller MSE than J;
- actual confidence interval probabilities are lower than nominal probabilities;
- J has the better performance regarding confidence interval probabilities.

Note that Deng and Wu use confidence intervals based upon Student's t theory, whereas the four earlier studies used standard normal theory.

Although there are gaps between these studies, we feel that it may be warranted to conclude that the TS method is good, perhaps best in some circumstances, in terms of the MSE and bias criteria, but the BHS method in particular, and secondarily the RG and J methods, are preferable from the point of view of confidence interval coverage probabilities. These results may arise because the TS variance estimator does not generally behave as a multiple of a χ^2-variate independent of the estimated parameter $\hat{\theta}$.

Of course, as is pointed out in Appendix C, each of the replication methods (RG, BHS, and J) may benefit in some circumstances from a transformation of the data.

We turn finally to the accuracy of the GVF method. This method cannot be recommended for any but the very largest sample surveys where administrative considerations may prevail. There is little theory for this method, and the resulting estimators of variance are surely biased. Survey practitioners who have used these methods, however, feel that some additional stability (lower variance) is imparted to variance estimates through use of GVF techniques.

In terms of the confidence interval criterion, the GVF method is clearly inferior to the other methods. Indeed the studentized statistic

$$t_{\text{GVF}} = \frac{\hat{\theta} - \theta}{\sqrt{v_{\text{GVF}}(\hat{\theta})}} = \frac{\hat{\theta} - \theta}{\sqrt{\hat{\theta}^2 a + \hat{\theta} b}}$$

is generally not distributed as a standard normal or as a Student's t random variable, not even asymptotically. To our knowledge, there have been no Monte Carlo studies of the actual confidence interval probabilities associated with the GVF method, and thus we conclude that it is an open question as to whether GVF methods provide "usable" confidence intervals.

8.2. Flexibility

The RG method provides one of the most flexible methods of variance estimation. Almost any estimator $\hat{\theta}$ likely to occur in survey work can be accommodated by the RG method. The RG method is also the most versatile method in terms of dealing with almost any sampling design. This is particularly true for patch-work designs that evolve over time, where the patch-work results from influences such as budget cuts, new objectives, political compromises, and the like.

The BHS method is likely to be as flexible as the random group method in terms of the kinds of estimators that can be accommodated. In terms of sampling designs, BHS is often thought to be restricted to stratified, two-per-stratum designs. Indeed, this is the way in which the BHS method was defined in Chapter 3. By pairing adjacent selections in a random sample design, however, the BHS method can also be applied to nonstratified designs, and thus it can accommodate a wider range of sampling designs than originally thought possible. By more complicated balancing schemes or by collapsing schemes, BHS can also accommodate 3-or-more-per-stratum designs and 1-per-stratum designs.

The J method can accommodate most estimators likely to occur in survey sampling practice. There is an exception to this rule for $(k, m) = (n, 1)$, i.e., delete one observation at a time, where jackknife fails for nondifferentiable statistics like the median. However, if the number deleted is substantially more than one, say $m = 0(\sqrt{n})$, the jackknife still works. Many kinds of sampling designs can be treated by the J method. The J method is likely to be as versatile in this regard as the BHS method, but not quite as versatile as the RG method.

The GVF method is somewhat less flexible than the other methods. It is designed primarily for multi-stage sample surveys of households. Some ad hoc developments have occurred for other applications, but these have not generally been as successful. The method is also applicable primarily to dichotomous variables.

The TS method can generally accommodate any survey estimator of the form $\hat{\theta} = g(\hat{Y})$, which includes most statistics used in survey sampling practice. It is difficult to apply for very complex $g(\cdot)$, but such statistics do not often occur in practice. The TS method can deal with any sample design for which an estimator of the covariance matrix Σ can be given. This makes the TS method as flexible as any in terms of the kinds of sampling designs and estimators that it can accommodate.

8.3. Administrative Considerations

Both the RG method and the BHS method have many advantages in terms of cost, timing, and the like. Software is available for both of these methods (see Appendix E), and the processing costs of both are relatively quite low.

For the BHS method, cost can be reduced in large surveys by resorting to partial balancing. Processing costs for the RG method can be reduced by decreasing the number of random groups, although this has a trade-off against the accuracy of the variance estimator.

The cost characteristics of the J method may rule it out of consideration in most large scale survey applications. Furthermore, none of the existing software packages listed in Attachment E implement the J method, and thus the potential user would need to develop special purpose software. This may not, however, be a difficult task depending on the user's computer environment.

The GVF method is quick, cheap, and easy to use with rotating panel surveys. Publication of the variance estimates is especially convenient because only two parameters, a and b, need to be published, whereas, for the RG, BHS, J, and TS methods, the survey publication will necessarily contain as many variance estimates as there are estimates. The GVF method is implemented easily using existing software packages for regression analysis. The GVF method, of course, is not a stand-alone variance estimation methodology. Some direct variance estimator such as RG, BHS, or J needs to be used to produce the inputs (i.e., the data needed to estimate the coefficients a and b) to the GVF method. In very large scale survey systems, GVF's may have cost, timing and publication advantages over other methods.

The TS method is not a stand-alone method either, but rather must be used in connection with other methods. The RG, BHS, or J methods must be used to estimate the covariance matrix Σ prior to implementing the TS method. Software exists for TS variance estimators (described in Appendix E) but, in general, this software only handles the common survey statistics. For less common and more complicated survey applications, derivatives of the function $\hat{\theta} = g(\hat{Y})$ need to be derived and programmed on the computer, and this issue must be addressed in regards to staffing, cost, and timing of the survey. The cost of the TS method will primarily be a function of the other method that is used in conjunction with the TS method. For example, if the J method is used to produce an estimate of the covariance matrix Σ, then we might expect the TS method to be quite expensive. The TS method will be relatively less expensive when the single variate alternative discussed in Section 6.5 is used as opposed to the full $p \times p$ covariance matrix.

In terms of administrative considerations, we conclude that the J method is more difficult and costly to apply than the other methods. We also conclude that the GVF method is the easiest to apply and publish, although it cannot stand alone. There is probably little basis on which to choose between the TS, BHS, and RG methods, because costs and other administrative considerations are nearly equal in these three methods.

8.4. Summary

The choice of a method for variance estimation involves a complex trade-off or balancing of factors such as accuracy, cost, and flexibility. The statistician will usually need to treat each survey on a case-by-case basis, considering the special circumstances and objectives of the survey. A good deal of judgment is involved in selecting a method for variance estimation, and it will not be surprising if the statistician recommends different methods for different survey applications. Indeed, no one method of variance estimation is best over all.

Hadamard Matrices

The orthogonal matrices used in defining half-sample replicates in Chapter 3 are known in mathematics as Hadamard matrices. A Hadamard matrix $\mathbf{H}$ is a $k \times k$ matrix all of whose elements are $+1$ or -1 which satisfies $\mathbf{H}'\mathbf{H} = k\mathbf{I}$, where $\mathbf{I}$ is the identity matrix of order k. The order k is necessarily 1, 2, or $4t$ with t a positive integer.

Plackett and Burman (1946) presented methods of constructing Hadamard matrices for the following three cases:

1. $k = 4t = p + 1$, where p is an odd prime;
2. $k = 4t = p^r + 1$, where r is an integer and p is an odd prime;
3. $k = 4t = 2(p^r + 1)$, where r is an integer, p an odd prime, and $(p^r + 1)$ is not divisible by 4.

They also presented a simple rule for doubling the size of any Hadamard matrix:

4. If $\mathbf{H}$ is a Hadamard matrix of order k, then

$$\begin{pmatrix} \mathbf{H} & \mathbf{H} \\ \mathbf{H} & -\mathbf{H} \end{pmatrix}$$

is a Hadamard matrix of order $2k$.

Surprisingly, the constructions given by Plackett and Burman include all orders less than 200 (and of course many orders above 200) except 92, 116, 156, 172, 184, 188. Subsequent constructions have been given for these six special cases. See Baumert, Golomb, and Hall (1962) and Hall (1967).

To assist the reader, Hadamard matrices for all orders from $k = 2$ through 100 are provided in Tables A.1 to A.26. These matrices allow one to implement the balanced half-sample method of variance estimation for all sampling designs up to $L = 100$ strata. Fully balanced designs may be constructed for the cases $L = 2, 3, 4, \ldots, 99$ but not for $L = 100$. For sampling designs with $L > 100$ strata, one may

1. construct a partially balanced set of half-sample replicates using the tables or
2. construct a larger Hadamard matrix by the methods of the earlier cited authors.

In the tables, " + " denotes +1 and " − " denotes −1.

A.1   Hadamard Matrix of Order 2

```
1.   + +
2.   + -
```

A.2   Hadamard Matrix of Order 4

```
1.   + + + +
2.   + + - -
3.   + - - +
4.   + - + -
```

A.3   Hadamard Matrix of Order 8

```
1.   + + + + + + + +
2.   + + - + - - + -
3.   + - + - - + - +
4.   + - + + - + - -
5.   + - - + + - + -
6.   + + - - + + - -
7.   + - - + - + + -
8.   + + - - + - - +
```

A.4   Hadamard Matrix of Order 12

```
 1.   + + + + + + + + + + + +
 2.   + + - + + + - - - + - +
 3.   + - + - + + + - - - + +
 4.   + + - + - + + + - - - +
 5.   + - + - + - + + + - - +
 6.   + - - + - + - + + + - +
 7.   + - - - + - + - + + + +
 8.   + + - - - + - + - + + +
 9.   + + + - - - + - + - + +
10.   + + + + - - - + - + - +
11.   + - + + + - - - + - + +
12.   + + - + + + - - - + - +
```

A.5 Hadamard Matrix of Order 16

```
 1.
 2.
 3.
 4.
 5.
 6.
 7.
 8.
 9.
10.
11.
12.
13.
14.
15.
16.
```

A.6 Hadamard Matrix of Order 20

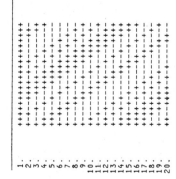

```
 1.
 2.
 3.
 4.
 5.
 6.
 7.
 8.
 9.
10.
11.
12.
13.
14.
15.
16.
17.
18.
19.
20.
```

A.7 Hadamard Matrix of Order 24

```
 1.   ++++++++++++++++++++++++
 2.   +-+-+++---+-+-+-+++---+-
 3.   +--+-+++---++-+-+++---+
 4.   ++--+-+++---++--+-+++---
 5.   +-+--+-+++--+-+---+-++--
 6.   +--+-+-+++-+--+-+-+++-
 7.   +---+--+-+-+++---+--+-+++
 8.   ++---+--+-+-+++---+--+-++
 9.   +++---+--+-+++++---+--+-+
10.   ++++---+--+-+++++---+--+-
11.   +-+++---+--++-+++---+--+
12.   ++-+++---+--+-++-+++---+--
13.   +++++++++++++------------
14.   +-+-+++---+--+-+---+++-+
15.   +--+-+++---+-+-++-+---+++-
16.   ++--+-+++-----++-+---+++
17.   +-+--+-+++---+--+-++-+---++
18.   +--+--+-+++--++-++-+---+
19.   +---+--+-+++-+++-++-+---
20.   ++---+--+-+++-+++-++-+--
21.   +++---+--+-+-+---+++-++-+-
22.   ++++---+--+--+-----+++-++-+
23.   +-+++---+--+-+---+++-++-
24.   ++-++'---+--+----+---+++-++
```

A.8 Hadamard Matrix of Order 28

```
 1.   +-+++++++++++++++++++++++++++
 2.   --+-+-+-+-+-+-+-+-+-+-+-+-+-
 3.   +++-++--++++-------++++--++
 4.   +---+--++-+--+-+-+-++-+--++-
 5.   +++++-++--+++-------++++--
 6.   +-+---+--++-+--+-+-+-++-+--+
 7.   ++--+++-++--++++-------++++
 8.   +--++---+--++-+--+-+-+-++-+-
 9.   ++!+--+++-++--++++--------++
10.   +-+--++---+--++-+--r-+-+-++-
11.   ++++++--+++-++--++++--------
12.   +-+-+--++---+--++-+--+-+-+-+
13.   ++--++++--+++-++--++++------
14.   +--++-+--++---+--++-+--+-+-+
15.   ++----++++--+++-++--++++----
16.   +--+-++-+--++---+--++-+--+-+
17.   ++------++++--+++-++--++++--
18.   +--+-+-++-+--++---+--++-+--+
```

A.8 Hadamard Matrix of Order 28 (continued)

```
19.
20.
21.
22.
23.
24.
25.
26.
27.
28.
```

A.9 Hadamard Matrix of Order 32

```
 1.
 2.
 3.
 4.
 5.
 6.
 7.
 8.
 9.
10.
11.
12.
13.
14.
15.
16.
17.
18.
19.
20.
21.
22.
23.
24.
25.
26.
27.
28.
29.
30.
31.
32.
```

A.10 Hadamard Matrix of Order 36

A.11 Hadamard Matrix of Order 40

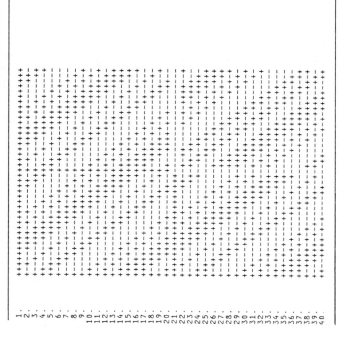

A.12   Hadamard Matrix of Order 44

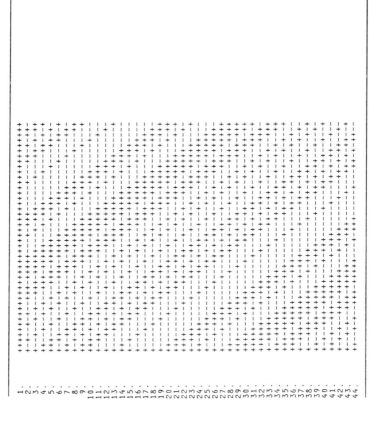

A.13   Hadamard Matrix of Order 48

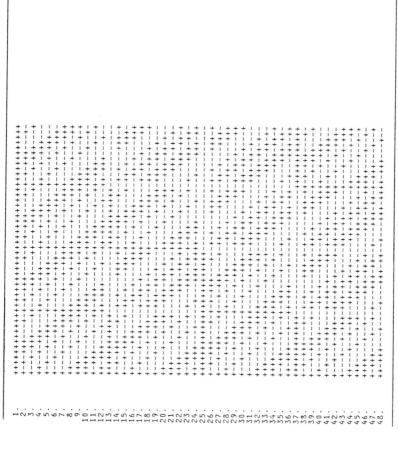

A.14   Hadamard Matrix of Order 52

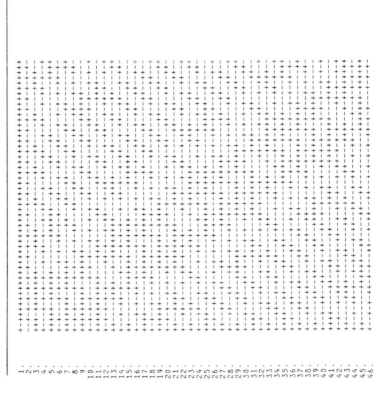

A.14 Hadamard Matrix of Order 52 (continued)

A.15 Hadamard Matrix of Order 56

A.15   Hadamard Matrix of Order 56 (continued)

A.16   Hadamard Matrix of Order 60

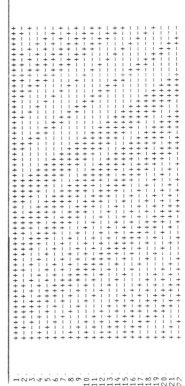

A.16 Hadamard Matrix of Order 60 (continued)

A.17 Hadamard Matrix of Order 64

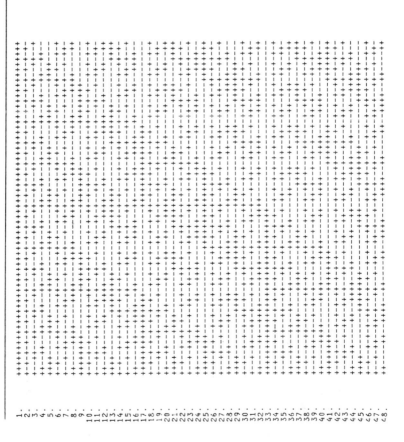

A.17 Hadamard Matrix of Order 64 (continued)

A.18 Hadamard Matrix of Order 68

A.18   Hadamard Matrix of Order 68 (continued)

A.19   Hadamard Matrix of Order 72

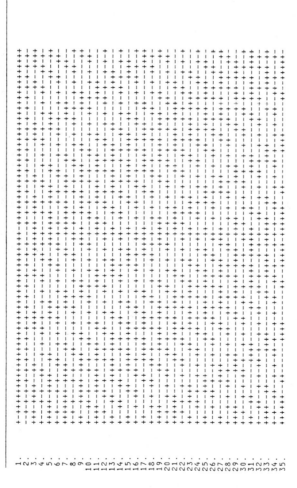

A.19 Hadamard Matrix of Order 72 (continued)

A.20   Hadamard Matrix of Order 76

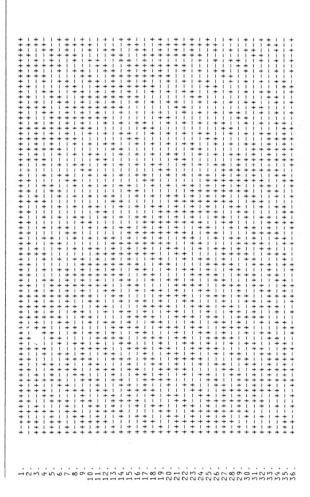

A.20   Hadamard Matrix of Order 76 (continued)

A.21   Hadamard Matrix of Order 80

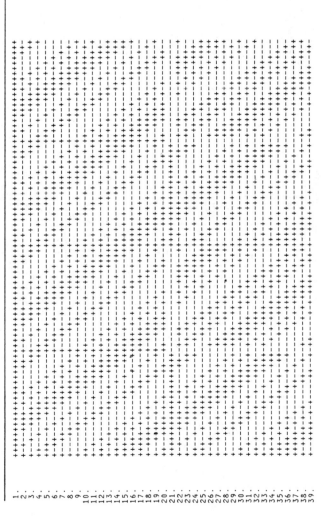

A.21 Hadamard Matrix of Order 80 (continued)

A.22 Hadamard Matrix of Order 84

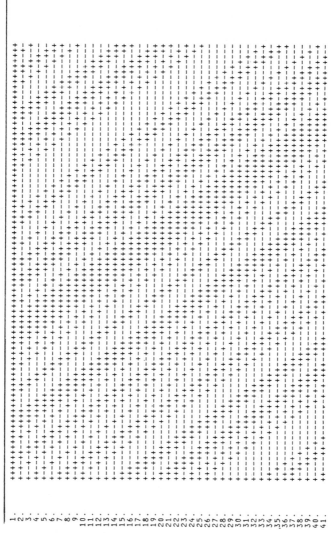

A.22 Hadamard Matrix of Order 84 (continued)

42.
43.
44.
45.
46.
47.
48.
49.
50.
51.
52.
53.
54.
55.
56.
57.
58.
59.
60.
61.
62.
63.
64.
65.
66.
67.
68.
69.
70.
71.
72.
73.
74.
75.
76.
77.
78.
79.
80.
81.
82.
83.
84.

A.23   Hadamard Matrix of Order 88

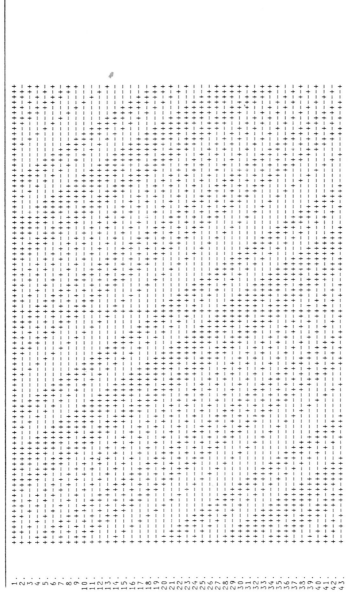

A.23   Hadamard Matrix of Order 88 (continued)

A.24 Hadamard Matrix of Order 92

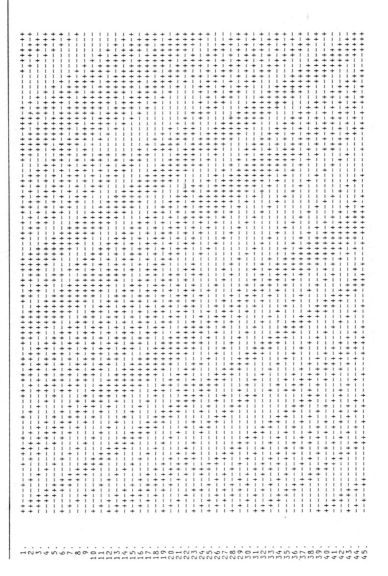

A.24 Hadamard Matrix of Order 92 (continued)

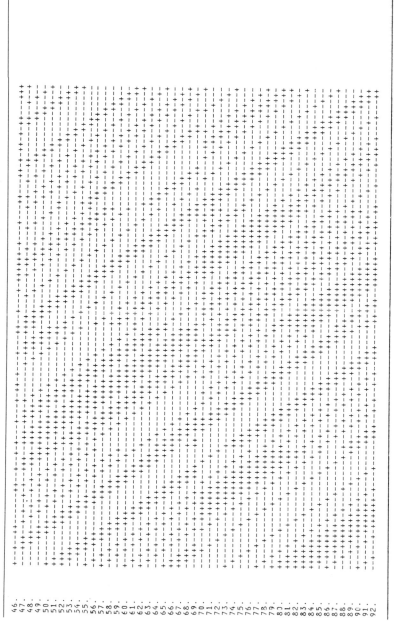

A.25 Hadamard Matrix of Order 96

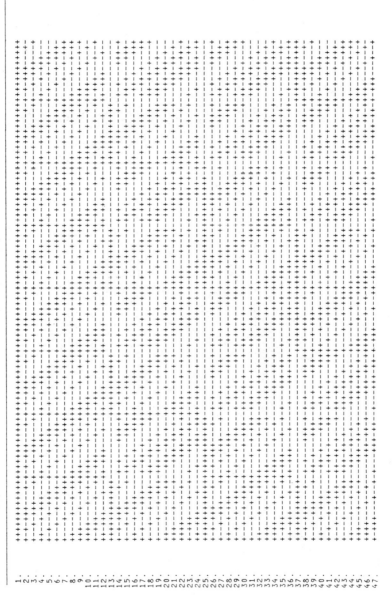

A.25 Hadamard Matrix of Order 96 (continued)

A.26 Hadamard Matrix of Order 100

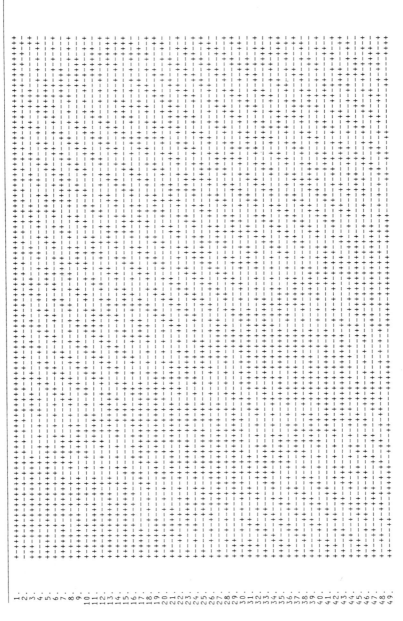

A.26   Hadamard Matrix of Order 100 (continued)

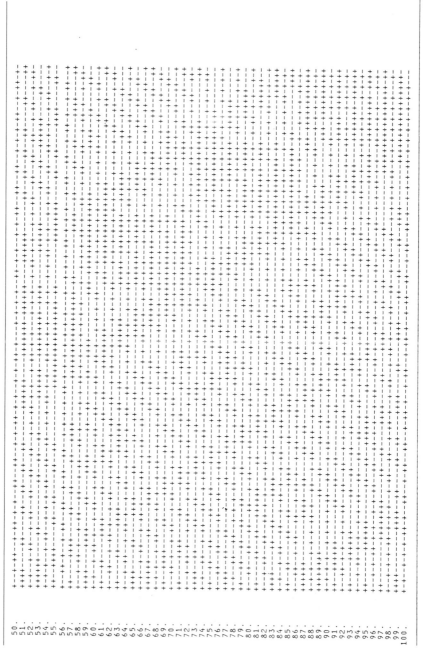

APPENDIX B

Asymptotic Theory of Variance Estimators

B.1. Introduction

Inferences from large, complex sample surveys usually derive from the pivotal quantity

$$t = \frac{\hat{\theta} - \theta}{\sqrt{v(\hat{\theta})}},$$

where θ is a parameter of interest, $\hat{\theta}$ is an estimator of θ, $v(\hat{\theta})$ is an estimator of the variance of $\hat{\theta}$, and it is assumed that t is distributed as (or approximately distributed as) a standard normal random variable $N(0, 1)$. The importance of the variance estimator $v(\hat{\theta})$ and the pivotal quantity t has been stressed throughout this book. For the most part, however, we have concentrated on defining the various tools available for variance estimation and on illustrating their proper use with real data sets. Little has been provided in regards to theory supporting the normality or approximate normality of the pivotal t. This approach was intended to acquaint the reader with the essentials of the methods while not diverting attention to mathematical detail.

In the present appendix we shall provide some of the theoretical justifications. All of the results to be discussed, however, are asymptotic results. There is little small sample theory for the variance estimators and none for the pivotal quantity t. The small sample theory that is available for the variance estimators has been reviewed in the earlier chapters of this book, and it is difficult to envision a small sample distributional theory for t, unless one is willing to postulate a superpopulation model for the target population. See Hartley and Sielken (1975).

Asymptotic results for finite population sampling have been presented by Madow (1948), Erdös and Rényi (1959), Hajek (1960, 1964), Rosen (1972), Holst (1973), Hidiroglou (1974), Fuller (1975), and Fuller and Isaki (1981). Most of these articles demonstrate the asymptotic normality or consistency of estimators of finite population parameters, such as means, totals, regression coefficients, and the like. Because little of this literature addresses the asymptotic properties of variance estimators or of pivotal quantities, we shall not include this work in the present review.

Authors presenting asymptotic theory for variance estimators or pivotal qualities include Nandi and Sen (1963), Krewski (1978b), and Krewski and Rao (1981). Much of our review concentrates on these papers. As we proceed, the reader will observe that all of the results pertain to moderately simple sampling designs, although the estimator $\hat{\theta}$ may be nonlinear and quite complex. It is thus reasonable to ask whether or not these results provide a theoretical foundation for the variance estimating methods in the context of large-scale, complex sample surveys? Our view is that they do provide an implicit foundation in the context of complex sample designs, and that by specifying enough mathematical structure explicit extensions of these results could be given for almost all of the complex designs found in common practice.

We discuss asymptotic theory for two different situations concerning a sequence of samples of increasing size. In the first case, the population is divided into L strata. The stratum sample sizes are regarded as fixed and limiting results are obtained as the number of strata tends to infinity, i.e., as $L \to \infty$. In the second case, the number of strata L is regarded as fixed ($L = 1$ is a special case) and limiting results are obtained as the stratum sample sizes tend to infinity, i.e., as $n_h \to \infty$. We shall begin with the results for case 1, followed by those for case 2.

B.2. Case I: Increasing L

We let $\{\Pi_L\}_{L=1}^{\infty}$ denote a sequence of finite populations, with L strata in Π_L. The value of the i-th unit in the h-th stratum of the L-th population is denoted by

$$\mathbf{Y}_{Lhi} = (Y_{Lhi1}, Y_{Lhi2}, \ldots, Y_{Lhip})',$$

where there are N_{Lh} units in the (L, h)-th stratum. A simple random sample with replacement of size n_{Lh} is selected from the (L, h)-th stratum, and $\mathbf{y}_{Lh1}, \mathbf{y}_{Lh2}, \ldots, \mathbf{y}_{Lhn_h}$ denote the resulting values. The vectors of stratum and sample means are denoted by

$$\bar{\mathbf{Y}}_{Lh} = N_{Lh}^{-1} \sum_{i=1}^{N_{Lh}} \mathbf{Y}_{Lhi}$$

$$= (\bar{Y}_{Lh1}, \ldots, \bar{Y}_{Lhp})'$$

and

$$\bar{\mathbf{y}}_{Lh} = n_{Lh}^{-1} \sum_{i=1}^{n_{Lh}} \mathbf{y}_{Lhi}$$

$$= (\bar{y}_{Lh1}, \ldots, \bar{y}_{Lhp})',$$

respectively. Henceforth, for simplicity of notation, we shall suppress the population index L from all of these variables.

We shall be concerned with parameters of the form

$$\theta = g(\bar{\mathbf{Y}})$$

and corresponding estimators

$$\hat{\theta} = g(\bar{\mathbf{y}}),$$

where $g(\cdot)$ is a real-valued function,

$$\bar{\mathbf{Y}} = \sum_{h=1}^{L} W_h \bar{\mathbf{Y}}_h$$

is the population mean,

$$\bar{\mathbf{y}} = \sum_{h=1}^{L} W_h \bar{\mathbf{y}}_h$$

is the unbiased estimator of $\bar{\mathbf{Y}}$, and

$$W_h = N_h / \sum_{h'} N_{h'} = N_h / N$$

is the proportion of units in the population that belong to the h-th stratum.

The covariance matrix of $\bar{\mathbf{y}}$ is given by

$$\mathbf{\Sigma} = \sum_{h=1}^{L} W_h^2 n_h^{-1} \mathbf{\Sigma}_h,$$

where $\mathbf{\Sigma}_h$ is the $(p \times p)$ covariance matrix with typical element

$$\sigma_{hj,hj'} = N_h^{-1} \sum_{i=1}^{N_h} (Y_{hij} - \bar{Y}_{hj})(Y_{hij'} - \bar{Y}_{hj'}).$$

The textbook (unbiased) estimator of Σ is given by

$$\hat{\Sigma} = \sum_{h=1}^{L} W_h^2 n_h^{-1} \hat{\Sigma}_h, \tag{B.2.1}$$

where $\hat{\Sigma}_h$ is the $(p \times p)$ matrix with typical element

$$\hat{\sigma}_{hj,hj'} = (n_h - 1)^{-1} \sum_{i=1}^{n_h} (y_{hij} - \bar{y}_{hj})(y_{hij'} - \bar{y}_{hj'}).$$

Several alternative estimators of the variance of $\hat{\theta}$ are available, including Taylor series (TS), balanced half-samples (BHS), and jackknife (J). For this appendix, we shall let $v_{TS}(\hat{\theta})$, $v_{BHS}(\hat{\theta})$, and $v_J(\hat{\theta})$ denote these estimators as defined in equations (6.3.3), (3.4.1), and (4.5.3), respectively. In the defining equation for v_{TS} we shall let (B.2.1) be the estimated covariance matrix of $\bar{y}$. As noted in earlier chapters, alternative defining equations are available for both v_{BHS} and v_J and we shall comment on these alternatives later on.

The following four theorems, due to Krewski and Rao (1981), set forth the asymptotic theory for $\bar{y}$, $\hat{\theta}$, v_{TS}, v_{BHS}, v_J, and corresponding pivotals. For simplicity, proofs are omitted from our presentation, but are available in the original reference.

Theorem B.1. *We assume that the sequence of populations is such that the following conditions are satisfied:*

(i) $\sum_{h=1}^{L} W_h \, E\{|y_{hij} - \bar{Y}_{hj}|^{2+\delta}\} = 0(1)$

   for some $\delta > 0$ $(j = 1, \ldots, p)$;

(ii) $\max_{1 \leq h \leq L} n_h = 0(1)$;

(iii) $\max_{1 \leq h \leq L} W_h = 0(L^{-1})$;

(iv) $n \sum_{h=1}^{L} W_h^2 n_h^{-1} \Sigma_h \to \Sigma^*$

   where Σ^* *is a* $(p \times p)$ *positive definite matrix.*
Then, as $L \to \infty$ *we have*

$$n^{1/2}(\bar{y} - \mathbf{Y}) \overset{d}{\to} N(0, \Sigma^*).$$

PROOF. See Krewski and Rao (1981).   □

Theorem B.2. *We assume regularity conditions* (i)–(iv) *and in addition assume*

(v) $\bar{Y} \to \mu$ *(finite)*,
(vi) *the first partial derivatives* $g_j(\cdot)$ *of* $g(\cdot)$ *are continuous in a neighborhood of*
   $\mu = (\mu_1, \ldots, \mu_p)$.
Then, as $L \to \infty$ *we have*

(a) $n^{1/2}(\hat{\theta} - \theta) \overset{d}{\to} N(0, \sigma_\theta^2)$,

(b) $nv_{\mathrm{TS}}(\hat{\theta}) \overset{p}{\to} \sigma_\theta^2$,

   and

(c) $t_{\mathrm{TS}} = \dfrac{\hat{\theta} - \theta}{\{v_{\mathrm{TS}}(\hat{\theta})\}^{1/2}} \overset{d}{\to} N(0, 1)$,

   where

   $$\sigma_\theta^2 = \sum_j \sum_{j'} g_j(\mu) g_{j'}(\mu) \sigma_{jj'}^*$$

   and $\sigma_{jj'}^*$ *is the* (j, j')-*th element of* Σ^*.

PROOF. See Krewski & Rao (1981). □

Theorem B.3. *Given regularity conditions* (i)–(vi),

(a) $nv_{\mathrm{J}}(\hat{\theta}) \overset{p}{\to} \sigma_\theta^2$
   and

(b) $t_{\mathrm{J}} = \dfrac{\hat{\theta} - \theta}{\{v_{\mathrm{J}}(\hat{\theta})\}^{1/2}} \overset{d}{\to} N(0, 1)$.

PROOF. See Krewski and Rao (1981). □

Theorem B.4. *Given regularity conditions* (i)–(vi) *and the restriction* $n_h = 2$ *for all* h, *then*

(a) $nv_{\mathrm{BHS}}(\hat{\theta}) \overset{p}{\to} \sigma_\theta^2$
   and

(b) $t_{\mathrm{BHS}} = \dfrac{\hat{\theta} - \theta}{\{v_{\mathrm{BHS}}(\hat{\theta})\}^{1/2}} \overset{d}{\to} N(0, 1)$.

The results also hold for $n_h = p$ *(for* p *a prime) with the orthogonal arrays of Section 3.7.*

PROOF. See Krewski & Rao (1981). □

In summary, the four theorems show that as $L \to \infty$ both $\bar{y}$ and $\hat{\theta}$ are asymptotically normally distributed; v_{TS}, v_{J}, and v_{BHS} are consistent estimators of the asymptotic variance of $\hat{\theta}$; and the pivotals t_{TS}, t_{J}, and t_{BHS} are asymptotically $N(0, 1)$. The assumptions required in obtaining these results are not particularly restrictive and will be satisfied in most applied problems. Condition (i) is a standard Liapounov-type condition on the $2 + \delta$ absolute moments. Condition (ii) will be satisfied in surveys with large numbers of strata and relatively few units selected per stratum, and condition (iii) when no stratum is disproportionately large. Conditions (iv) and (v) require that both the limit of the covariance matrix multiplied by the sample size n and the limit of the population mean exist. The final condition (vi) will be satisfied by most functions $g(\cdot)$ of interest in finite population sampling.

These results extend in a number of directions.

• The results are stated in the context of simple random sampling with replacement. But the results are also valid for any stratified, multistage design in which the primary sampling units (PSUs) are selected with replacement and in which independent subsamples are taken within those PSUs selected more than once. In this case, the values y_{hij} employed in the theorems become

$$y_{hij} = \hat{y}_{hij}/p_{hi},$$

where $\hat{y}_{hij}$ is an estimator of the total of the j-th variable within the (h, i)-th PSU and p_{hi} is the corresponding per-draw selection probability.

• The results are stated in the context of a single function $\hat{\theta} = g(\bar{y})$. Multivariate extensions of the results can be given that refer to $q \geq 2$ functions

$$\hat{\boldsymbol{\theta}} = \begin{pmatrix} g^1(\bar{y}) \\ g^2(\bar{y}) \\ \vdots \\ g^q(\bar{y}) \end{pmatrix}.$$

For example, $n^{1/2}(\hat{\boldsymbol{\theta}} - \boldsymbol{\theta})$ converges in distribution to a q-variate normal random variable with mean vector $\mathbf{0}$ and covariance matrix $\mathbf{D\Sigma^*D'}$, where

$$\boldsymbol{\theta} = \begin{pmatrix} g^1(\bar{Y}) \\ g^2(\bar{Y}) \\ \vdots \\ g^q(\bar{Y}) \end{pmatrix}$$

and $\mathbf{D}$ is a $(q \times p)$ matrix with typical element

$$D_{ij} = g^i_j(\boldsymbol{\mu}).$$

• The results are stated for one definition of the jackknife estimator of the variance. The same results are valid not only for definition (4.5.3), but also for alternative definitions (4.5.4), (4.5.5), and (4.5.6).

• The results are stated for the case where the variance estimators are based on the individual observations, not on groups of observations. In earlier chapters, descriptions were presented of how the random group method may be applied within strata, of how the jackknife method may be applied to grouped data, of how the balanced half-samples method may be applied if two random groups are formed within each stratum, and of how the Taylor series method can be applied to an estimated covariance matrix $\hat{\Sigma}$ that is based on grouped data. Each of the present theorems may be extended to cover these situations where the variance estimator is based upon grouped data.

B.3. Case II: Increasing n_h

We shall now turn our attention to the second situation concerning the sequence of samples. In this case, we shall require that the number of strata L be fixed, and all limiting results shall be obtained as the stratum sizes and sample sizes tend to infinity, i.e., as $N_h \to \infty$, $n_h \to \infty$. To concentrate on essentials, we shall present the case $L = 1$, and shall drop the subscript h from our notation. All of the results to be discussed, however, extend to the case of general $L \geq 1$.

We let $\{\Pi_N\}_{N=1}^{\infty}$ denote the sequence of finite populations, where N is the size of the N-th population. As before, $\mathbf{Y}_{Ni}$ denotes the p-variate value of the i-th unit $(i = 1, \ldots, N)$ in the N-th population. Also as before, we shall omit the population index in order to simplify the notation.

Arvesen (1969) has established certain asymptotic results concerning the jackknife method applied to U-statistics and functions of U-statistics. After defining what are U-statistics, we shall briefly review Arvesen's results. Extensions of the results to other sampling schemes are discussed next, followed by a discussion of the properties of the other variance estimators.

Let $\mathbf{y}_1, \ldots, \mathbf{y}_n$ denote a simple random sample with replacement from Π. Let $f(\mathbf{y}_1, \ldots, \mathbf{y}_b)$ denote a real-valued statistic, symmetric in its arguments, that is unbiased for some population parameter η, where b is the smallest sample size needed to estimate η.

The *U-statistic* for η is defined by

$$U(\mathbf{y}_1, \ldots, \mathbf{y}_n) = \binom{n}{b}^{-1} \sum_{C_b} f(\mathbf{y}_{i_1}, \ldots, \mathbf{y}_{i_b}),$$

where the summation extends over all possible combinations of $1 \leq i_1 \leq i_2 \leq \ldots \leq i_b \leq n$. The statistic f is the *kernel* of U, and b is the *degree* of f. Krewski (1978b) gives several examples of statistics that are of importance in survey sampling and shows that they are members of the class of U-statistics. For example, the kernel $f(\mathbf{y}_1) = y_{11}$ with $b = 1$ leads to the mean of the first variable while the kernel $f(\mathbf{y}_1, \mathbf{y}_2) = (y_{11} - y_{21})(y_{12} - y_{22})/2$ with $b = 2$ leads to the covariance between the first and second variables.

Nearly all descriptive statistics of interest in survey sampling may be expressed as a U-statistic or as a function of several U-statistics.

We shall let $\mathbf{U} = (U^1, U^2, \ldots, U^q)'$ denote q U-statistics corresponding to kernels $f^1, f^2, \ldots, f^q$ based on $b_1, b_2, \ldots, b_q$ observations, respectively. We shall be concerned with an estimator

$$\hat{\theta} = g(U^1, U^2, \ldots, U^q)',$$

of a population parameter

$$\theta = g(\eta^1, \eta^2, \ldots, \eta^q),$$

where g is a real-valued, smooth function and the $\eta^1, \eta^2, \ldots, \eta^q$ denote the expectations of $U^1, U^2, \ldots, U^q$, respectively.

The jackknife method with $n = mk$ and $k = n$ (i.e., no grouping) utilizes the statistics

$$\hat{\theta}_{(i)} = g(U^1_{(i)}, U^2_{(i)}, \ldots, U^q_{(i)}),$$

where $U^j_{(i)}$ is the j-th U-statistic based upon the sample after omitting the i-th observation ($i = 1, \ldots, n$). The corresponding pseudovalue is

$$\hat{\theta}_i = n\hat{\theta} - (n - 1)\hat{\theta}_{(i)};$$

Quenouille's estimator is

$$\hat{\bar{\theta}} = n^{-1} \sum_{i=1}^{n} \hat{\theta}_i;$$

and a jackknife estimator of variance is

$$v_J(\hat{\bar{\theta}}) = n^{-1}(n - 1)^{-1} \sum_{i=1}^{n} (\hat{\theta}_i - \hat{\bar{\theta}})^2.$$

Then, we have the following theorem, establishing the asymptotic properties of $\hat{\bar{\theta}}$ as $n \to \infty$.

Theorem B.5. *Let the kernels have finite second moments*

$$E\{[f^j(\mathbf{y}_1, \mathbf{y}_2, \ldots, \mathbf{y}_{b_j})]^2\} < \infty$$

for each $j = 1, \ldots, q$. Let g be a real-valued function defined on R^q which, in a neighborhood of $\boldsymbol{\eta} = (\eta^1, \ldots, \eta^q)$, has bounded second partial derivatives. Then, as $n \to \infty$ we have

$$n^{1/2}(\hat{\bar{\theta}} - \theta) \xrightarrow{d} N(0, \sigma_\theta^2),$$

where $g_j(\boldsymbol{\eta})$ denotes the first partial derivative of g with respect to its j-th argument evaluated at the mean $\boldsymbol{\eta}$,

$$f_1^j(\mathbf{Y}_1) = E\{f^j(\mathbf{y}_1, \mathbf{y}_2, \ldots, \mathbf{y}_{b_j}) | \mathbf{y}_1 = \mathbf{Y}_1\},$$

$$\phi_1^j(\mathbf{Y}_1) = f_1^j(\mathbf{Y}_1) - \eta^j,$$

$$\zeta^{jj'} = E\{\phi_1^j(\mathbf{y}_1)\phi_1^{j'}(\mathbf{y}_1)\},$$

and

$$\sigma_\theta^2 = \sum_{j=1}^{q} \sum_{j'=1}^{q} b_j b_{j'} g_j(\boldsymbol{\eta}) g_{j'}(\boldsymbol{\eta}) \zeta^{jj'}.$$

PROOF. See Arvesen (1969). ☐

The next theorem shows that the jackknife estimator of variance correctly estimates the asymptotic variance of $\hat{\hat{\theta}}$.

Theorem B.6. *Let g be a real-valued function defined on R^q which has continuous first partial derivatives in a neighborhood of $\boldsymbol{\eta}$. Let the remaining conditions of Theorem B.5 be given. Then, as $n \to \infty$ we have*

$$n v_J(\hat{\hat{\theta}}) \xrightarrow{p} \sigma_\theta^2.$$

PROOF. See Arvesen (1969). ☐

By Theorems B.5 and B.6 it follows that the pivotal statistic

$$t_J = (\hat{\hat{\theta}} - \theta)/\sqrt{v_J(\hat{\hat{\theta}})} \tag{B.3.1}$$

is asymptotically distributed as a standard normal random variable.

In presenting these results, it has been assumed that $n = mk \to \infty$ with $k = n$ and $m = 1$. The results may be repeated with only slight modification if $k \to \infty$ with $m > 1$. On the other hand, if k is fixed and $m \to \infty$, then the pivotal statistic (B.3.1) converges to a Student's t distribution with $(k - 1)$ degrees of freedom.

Arvesen's results extend directly to other with-replacement sampling schemes. In the most general case, consider a multistage sample where the primary sampling units (PSUs) are selected pps wr. Assume that subsampling is performed independently in the various PSUs, including duplicate PSUs. Theorems B.5 and B.6 apply to this situation provided that the values y_{ij} are replaced by

$$y_{ij} = \hat{y}_{ij}/p_i,$$

where $\hat{y}_{ij}$ is an estimator of the total of the j-th variable within the i-th PSU and p_i is the corresponding per-draw selection probability.

Krewski (1978b), following Nandi and Sen (1963), has extended Arvesen's results to the case of simple random sampling without replacement. In describing his results, we shall employ the concept of a sequence {Π} of finite populations and require that $n \to \infty$, $N \to \infty$, and $\lambda = n/N \to \lambda_0 < 1$. Once again we let $n = km$ with $k = n$ and $m = 1$.

Define

$$f_c^j(\mathbf{Y}_1, \ldots, \mathbf{Y}_c) = E\{f^j(\mathbf{y}_1, \ldots, \mathbf{y}_{b_j})|\mathbf{y}_1 = \mathbf{Y}_1, \ldots, \mathbf{y}_c = \mathbf{Y}_c\}$$

and

$$\phi_c^j(\mathbf{Y}_1, \ldots, \mathbf{Y}_c) = f_c^j(\mathbf{Y}_1, \ldots, \mathbf{Y}_c) - \eta^j$$

for $c = 1, \ldots, b_j$. As before, we let

$$\zeta^{jj'} = E\{\phi_1^j(\mathbf{y}_1)\phi_1^{j'}(\mathbf{y}_1)\}.$$

Define the $(q \times q)$ covariance matrix

$$\mathbf{Z} = (\zeta^{jj'}).$$

The asymptotic normality of the estimator $\hat{\bar{\theta}}$ and the consistency of the jackknife variance estimator are established in the following two theorems.

Theorem B.7. *Let g be a real-valued function defined on R^q with bounded second partial derivatives in a neighborhood of $\boldsymbol{\eta} = (\eta^1, \ldots, \eta^q)'$. Assume that $\mathbf{Z}$ converges to a positive definite matrix $\mathbf{Z}_0 = (\zeta_0^{jj'})$ as $n \to \infty$, $N \to \infty$, and $\lambda \to \lambda_0 < 1$, that $\sup E\{|\phi_{b_j}^j(\mathbf{y}_1, \ldots, \mathbf{y}_{b_j})|^{2+\delta}\} < \infty$ for some $\delta > 0$ and $j = 1, \ldots, q$, and that $\eta^j \to \eta_0^j$ for $j = 1, \ldots, q$. Then, we have*

$$\sqrt{n}(\hat{\bar{\theta}} - \theta) \to N(0, (1 - \lambda_0)\sigma_\theta^2),$$

where

$$\sigma_\theta^2 = \sum_{j=1}^q \sum_{j'=1}^q b_j b_{j'} g_j(\boldsymbol{\eta}_0) g_{j'}(\boldsymbol{\eta}_0) \zeta_0^{jj'}.$$

PROOF. See Krewski (1978b).   □

Theorem B.8. *Let g be a real-valued function defined on R^q with continuous first partial derivatives in a neighborhood of $\boldsymbol{\eta}$. Let the remaining conditions of Theorem B.7 be given. Then, as $n \to \infty$ and $\lambda \to \lambda_0$, we have*

$$nv_J(\hat{\bar{\theta}}) \overset{P}{\to} \sigma_\theta^2.$$

PROOF. See Krewski (1978b).

Theorems B.7 and B.8 show that the pivotal statistic

$$t_J = \frac{\hat{\bar{\theta}} - \theta}{\sqrt{(1 - \lambda_0)v_J(\theta)}}$$

is asymptotically a standard normal random variable.

Theorems B.7 and B.8 were stated in terms of the traditional jackknife estimator, with pseudovalue defined by

$$\hat{\theta}_i = n\hat{\theta} - (n - 1)\hat{\theta}_{(i)}.$$

These results extend simply to the generalized jackknife estimator (see Gray and Schucany (1972)) with

$$\hat{\theta}_i = (1 - R)^{-1}(\hat{\theta} - R\hat{\theta}_{(i)}).$$

The traditional jackknife is the special case of this more general formulation with $R = (n - 1)/n$. The Jones (1974) jackknife, introduced in Chapter 4, is the special case with $R = n^{-1}(N - n + 1)^{-1}(N - n)(n - 1)$. Thus, asymptotic normality and consistency of the variance estimator apply to the Jones jackknife as well as to the traditional jackknife. See Krewski (1978b) for a fuller discussion of these results.

Theorems B.7 and B.8 were also stated in terms of the Quenouille estimator the jackknife estimator of variance

$$v_J(\hat{\bar{\theta}}) = n^{-1}(n-1)^{-1} \sum_{i=1}^{n} (\hat{\theta}_i - \hat{\bar{\theta}})^2.$$

Similar results may be obtained for the parent sample estimator $\hat{\theta}$ and for the alternative jackknife estimator of variance

$$v_J(\hat{\theta}) = n^{-1}(n-1)^{-1} \sum_{i=1}^{n} (\hat{\theta}_i - \hat{\theta})^2.$$

Indeed, it follows from these various results that each of the four statistics

$$\frac{\hat{\bar{\theta}} - \theta}{\sqrt{v_J(\theta)}},$$

$$\frac{\hat{\theta} - \theta}{\sqrt{v_J(\theta)}},$$

$$\frac{\hat{\bar{\theta}} - \theta}{\sqrt{v_J(\hat{\theta})}},$$

$$\frac{\hat{\theta} - \theta}{\sqrt{v_J(\hat{\theta})}},$$

(B.3.2)

converges to a standard normal random variable $N(0, 1)$ as $n \to \infty$, $N \to \infty$, and $\lambda \to \lambda_0 < 1$, and thus each may be employed as a pivotal quantity for making an inference about θ.

All of the asymptotic results presented here for srs wor apply to arbitrary configurations of m and k, provided $n \to \infty$, $N \to \infty$, and $\lambda \to \lambda_0 < 1$, with m fixed. On the other hand, if we fix k and permit $m \to \infty$ with $\lambda \to \lambda_0 < 1$, then the four pivotal quantities in (B.3.2) converge to a Student's t random variable with $k - 1$ degrees of freedom. This result is identical with that stated earlier for srs wr.

Next, we turn attention to other variance estimation methods and look briefly at their asymptotic properties. We shall continue to assume srs wor.

Let $\hat{\mathbf{d}}$ denote the $(q \times 1)$ vector of first partial derivatives of $\hat{\theta}$ evaluated at $\mathbf{U} = (U^1, U^2, \ldots, U^q)'$. This vector is an estimator of

$$\mathbf{d} = (g_1(\boldsymbol{\eta}), g_2(\boldsymbol{\eta}), \ldots, g_q(\boldsymbol{\eta}))',$$

where the derivatives are evaluated at the mean $\boldsymbol{\eta} = (\eta^1, \eta^2, \ldots, \eta^q)$. Let $\hat{\boldsymbol{\Omega}}$ denote the $(q \times q)$ matrix with typical element

$$\hat{\Omega}_{jj'} = n^{-1}(n-1)^{-1} \sum_{i=1}^{n} (U_i^j - U^j)(U_i^{j'} - U^{j'}).$$

This is a jackknife estimator of the covariance matrix $\mathbf{\Omega} = (\Omega_{jj'})$ of U, where

$$\Omega_{jj'} = b_j b_{j'} \zeta^{jj'}.$$

Then a Taylor series estimator of the variance of $\hat{\theta}$ is given by

$$v_{\text{TS}}(\hat{\theta}) = \hat{\mathbf{d}}' \hat{\mathbf{\Omega}} \hat{\mathbf{d}}.$$

The following theorem establishes the probability limit of the Taylor series estimator.

Theorem B.9. *Let the conditions of Theorem B.7 hold. Then, as $n \to \infty$, $N \to \infty$, and $\lambda \to \lambda_0 < 1$ we have*

$$n v_{\text{TS}}(\hat{\theta}) \overset{p}{\to} \sigma_\theta^2.$$

PROOF. See Krewski (1978b).   □

Similar results may be obtained when the estimated covariance matrix $\hat{\mathbf{\Omega}}$ is based upon a random group estimator (with m fixed) or the jackknife applied to grouped data.

Combining Theorems B.7 and B.9, it follows that the pivotal statistic

$$t_{\text{TS}} = \frac{\hat{\theta} - \theta}{\sqrt{(1 - \lambda_0) v_{\text{TS}}(\hat{\theta})}} \qquad (\text{B.3.3})$$

is asymptotically a standard normal random variable. As a practical matter, the finite population correction, $1 - \lambda_0$, may be ignored whenever the sampling fraction is negligible.

Results analogous to Theorem B.9 and equation (B.3.3) may be obtained for the random group and balanced half-samples estimators of variance.

Finally, we turn briefly to the unequal probability without replacement sampling designs, where few asymptotic results are available. Exact large sample theory for certain specialized unequal probability without replacement designs is presented by Hájek (1964) and Rosén (1972). Also see Isaki and Fuller (1982). But none of these authors discuss the asymptotic properties of the variance estimators treated in this book. Campbell (1980) presents the beginnings of a general asymptotic theory for the without replacement designs, but more development is needed. Thus, at this point in time, the use of the various variance estimators in connection with such designs is justified mainly by the asymptotic theory for pps wr sampling.

Transformations

C.1. Introduction

Transformations find wide areas of application in the statistical sciences. It often seems advantageous to conduct an analysis on a transformed data set, rather than on the original data set. Transformations are most often motivated by the need or desire to

  (i) obtain a parsimonious model representation for the data set,
 (ii) obtain a homogeneous variance structure,
(iii) obtain normality for the distributions,
(iv) achieve some combination of the above.

Transformations are used widely in such areas as time series analysis, econometrics, biometrics, and the analysis of statistical experiments. But they have not received much attention in the survey literature. A possible explanation is that many survey organizations have emphasized the production of simple descriptive statistics, as opposed to analytical studies of the survey population.

In this appendix we show how transformations might usefully be applied to the problems studied in this book. We also present a simple empirical study of one specific transformation, Fisher's well-known z-transformation of the correlation coefficient. Our purpose here is mainly to draw attention to the possible utility of data transformations for survey sampling problems and to encourage further research in this area. Aside from the z-transformation, little is known about the behavior of transformations in finite population sampling, and so recommendations are withheld pending the outcome of future research.

C.2. How to Apply Transformations to Variance Estimation Problems

The methods for variance estimation discussed in Chapters 2, 3, and 4 (i.e., random group, balanced half sample, and jackknife) are closely related in that they each produce k estimators $\hat{\theta}_\alpha$ of the unknown parameter θ. The variance of the parent sample estimator, $\hat{\theta}$, is then estimated by $v(\hat{\theta})$, where $v(\hat{\theta})$ is proportional to the sum of squares

$$\sum_{\alpha=1}^{k} (\hat{\theta}_\alpha - \hat{\theta})^2.$$

When an interval estimate of θ is required, normal theory is usually invoked, resulting in the interval

$$(\hat{\theta} \pm c\sqrt{v(\hat{\theta})}), \qquad (C.2.1)$$

where c is the tabular value from either the normal or Student's t distributions. As an alternative to (C.2.1) we may consider $\hat{\theta}$ as a point estimator of θ, or an estimator of variance proportional to the sum of squares $\sum(\hat{\theta}_\alpha - \hat{\theta})^2$.

In Chapter 2 we assessed the quality of a variance estimator $v(\hat{\theta})$ by its variance $\text{Var}\{v(\hat{\theta})\}$ or by its relative variance $\text{RelVar}\{v(\hat{\theta})\} = \text{Var}\{v(\hat{\theta})\}/ E^2\{v(\hat{\theta})\}$. Another attractive criteria involves assessment of quality in terms of the interval estimates resulting from use of $v(\hat{\theta})$. We have found in our empirical work that these criteria are not necessarily in agreement with one another. Sometimes one variance estimator will produce "better" confidence intervals, while another will be "better" from the standpoint of minimum relative variance.

Specifically, the quality of the interval estimator given by (C.2.1), and thus also of the variance estimator $v(\hat{\theta})$, may be assessed in repeated sampling by the percentage of intervals that contain the true parameter θ. A given method may be said to be "good" if this percentage is roughly $100(1 - \alpha)\%$, not higher or lower, where $(1 - \alpha)$ is the nominal confidence level. Usually, good interval estimates are produced if and only if the subsample estimators $\hat{\theta}_\alpha(\alpha = 1, \ldots, k)$ behave as a random sample from a normal distribution with homogeneous variance. Often this is not the case because the distribution of $\hat{\theta}_\alpha$ is excessively skewed.

Normality can be achieved in some cases by use of a suitable transformation of the data, say ϕ. Given this circumstance, an interval estimate for θ is produced in two steps:

(1) an interval is produced for $\phi(\theta)$,
(2) an interval is produced for θ by transforming the $\phi(\theta)$-interval back to the original scale.

The first interval is

$$(\phi(\hat{\theta}) \pm c\sqrt{v(\phi(\hat{\theta}))}), \qquad (C.2.2)$$

where $v(\phi(\hat{\theta}))$ is proportional to the sum of squares

$$\sum_{\alpha=1}^{k} (\phi(\hat{\theta}_{\alpha}) - \phi(\hat{\theta}))^2.$$

Alternatively, we may work with

$$\widehat{\phi(\theta)} = \sum_{\alpha=1}^{k} \phi(\hat{\theta}_{\alpha})/k$$

as a point estimator of $\phi(\theta)$, or an estimator of variance proportional to the sum of squares

$$\sum_{\alpha=1}^{k} (\phi(\hat{\theta}_{\alpha}) - \widehat{\phi(\theta)})^2.$$

The second interval is

$$(\phi^{-1}(L), \phi^{-1}(U)), \tag{C.2.3}$$

where (L, U) denotes the first interval and ϕ^{-1} denotes the inverse transformation, i.e., $\phi^{-1}(\phi(x)) = x$. If the transformation ϕ is properly chosen, this two-step procedure can result in interval estimates that are superior to the direct interval (C.2.1).

We note that in some applications it may be sufficient to stop with the first interval (C.2.2), reporting the results on the ϕ-scale. In survey work, however, for the convenience of the survey sponsor and other users it is more common to report the results on the original scale. We also note that when the true variance $\text{Var}\{\hat{\theta}\}$ is "small", most reasonable transformations ϕ will produce approximately identical results, and approximately identical to the direct interval (C.2.1). This is because the $\hat{\theta}_{\alpha}$ will not vary greatly. If the transformation ϕ has a local linear quality (and most do), then it will approximate a linear transformation over the range of the $\hat{\theta}_{\alpha}$, and the two-step procedure will simply reproduce the direct interval (C.2.1). In this situation it makes little difference which transformation is used. For moderate to large true variance $\text{Var}\{\hat{\theta}\}$, however, nonidentical results will be obtained and it will be important to choose the transformation that conforms most closely with the conditions of normality and homogeneous variance.

C.3. Some Common Transformations

Bartlett (1947) describes several transformations that are used frequently in statistical analysis. See Table C.1. The main emphasis of these transformations is on obtaining a constant error variance in cases where the variance of the untransformed variate is a function of the mean. For example, a binomial proportion $\hat{\theta}$ with parameter θ has variance equal to $\theta(1 - \theta)/k$.

Table C.1. Some Common Transformations

Variance in Terms of Mean, θ	Transformation	Appropriate Variance on New Scale	Relevant Distribution
θ	$\sqrt{x}$, (or $\sqrt{x+\frac{1}{2}}$ for small integers)	0.25	Poisson
$\lambda^2\theta$		$0.25\lambda^2$	Empirical
$2\theta^2/(k-1)$	$\log_e x$	$2/(k-1)$	Sample variance
$\lambda^2\theta^2$	$\log_e x, \log_e (x+1)$ $\log_{10} x, \log_{10} (x+1)$	λ^2 $0.189\lambda^2$	Empirical
$\theta(1-\theta)/k$	$\mathrm{Sin}^{-1}\sqrt{x}$, (radians) $\mathrm{Sin}^{-1}\sqrt{x}$, (degrees)	$0.25/k$ $821/k$	Binomial
$\lambda^2\theta(1-\theta)$	$\mathrm{Sin}^{-1}\sqrt{x}$, (radians)	$0.25\lambda^2$	Empirical
$\lambda^2\theta^2(1-\theta)^2$	$\log_e [x/(1-x)]$	λ^2	Empirical
$(1-\theta^2)^2/(k-1)$	$\frac{1}{2}\log_e [(1+x)/(1-x)]$	$1/(k-3)$	Sample correlations
$\theta+\lambda^2\theta^2$	$\lambda^{-1}\,\mathrm{Sinh}^{-1}[\lambda\sqrt{x}]$, or $\lambda^{-1}\,\mathrm{Sinh}^{-1}[\lambda\sqrt{x+\frac{1}{2}}]$ for small integers	0.25	Negative binomial
$\mu^2(\theta+\lambda^2\theta^2)$		$0.25\mu^2$	Empirical

Source: Bartlett (1947).
Note: λ and μ are unknown parameters and k is the sample size.

The variance itself is a function of the mean. The transformation

$$\phi(\hat{\theta}) = \mathrm{Sin}^{-1}\sqrt{\hat{\theta}},$$

however, has variance proportional to k^{-1}, and the functional dependence between mean and variance is eliminated.

In general, if the variance of $\hat{\theta}$ is a known function of θ, say $\mathrm{Var}\{\hat{\theta}\} = \Psi(\theta)$, then a transformation of the data that makes the variance almost independent of θ is the indefinite integral

$$\phi(\theta) = \int d\theta/\sqrt{\Psi(\theta)}.$$

This formula is behind Bartlett's transformations cited in Table C.1. It is based on the linear term in the Taylor series expansion of $\phi(\hat{\theta})$ about the point θ.

Bartlett's transformations also tend to improve the closeness of the distribution to normality, which is our main concern here. On the original scale, the distribution of $\hat{\theta}$ may be subject to excessive skewness, which is eliminated after the transformation. Cressie (1981) has studied several of these transformations in connection with the jackknife method.

The Box–Cox (1964, 1982) family offers another potentially rich source of transformations that may be considered for survey data. Also see Bickel and Doksum (1981). This parametric family of transformations is defined by

$$\phi_1(\hat{\theta}) = \frac{\hat{\theta}^\lambda - 1}{\lambda}, \qquad \lambda \neq 0, \qquad \hat{\theta} > 0,$$

$$= \log \hat{\theta}, \qquad \lambda = 0, \qquad \hat{\theta} > 0,$$

or by

$$\phi_2(\hat{\theta}) = \frac{(\hat{\theta} + \lambda_2)^{\lambda_1} - 1}{\lambda_1}, \qquad \lambda_1 \neq 0, \qquad \hat{\theta} > -\lambda_2,$$

$$= \log(\hat{\theta} + \lambda_2), \qquad \lambda_1 = 0, \qquad \hat{\theta} > -\lambda_2,$$

where $\lambda, \lambda_1, \lambda_2$ are parameters. ϕ_1 is the one-parameter Box-Cox family of transformations; ϕ_2 is the two-parameter family.

The Box-Cox family was originally conceived as a data-dependent class of transformations (i.e., $\lambda, \lambda_1, \lambda_2$ determined from the data itself) in the context of linear statistical models. Parameter λ (or λ_1 and λ_2) was to be estimated by maximum likelihood methods or via Bayes' theorem. For the problem of variance estimation, the maximized log likelihood, except for a constant, is

$$\mathcal{L}_1(\lambda) = -k \log \hat{\sigma}_1(\lambda) + (\lambda - 1) \sum_{\alpha=1}^{k} \log \hat{\theta}_\alpha$$

$$\hat{\sigma}_1^2(\lambda) = k^{-1} \sum_{\alpha=1}^{k} (\phi_1(\hat{\theta}_\alpha) - \overline{\phi_1(\theta)})^2.$$

We may plot $\mathcal{L}_1(\lambda)$ versus λ and from this plot obtain the maximizing value of λ, say $\hat{\lambda}$. Then $\hat{\lambda}$ specifies the particular member of the Box-Cox family to be employed in subsequent analyses, such as in the preparation of a confidence interval for θ. Similar procedures are followed for the two-parameter family of transformations.

It may be unrealistic to allow the data itself to determine the values of the parameters in the context of variance estimation problems for complex sample surveys. Also, actual confidence levels for θ associated with a data-dependent λ (or λ_1 and λ_2) may not achieve the nominal levels specified by the survey statistician, although this is an issue in need of further study.

Much empirical research is needed concerning both the Box-Cox and the Bartlett transformations on a variety of data sets and on different survey parameters θ and estimators $\hat{\theta}$ of interest. Based on the empirical research, guidelines should be formulated concerning the applicability of the transformations to the various survey estimators and parameters. General principles should be established about which transformations work best for which survey problems. In future survey applications, then, the survey statistician would need only consult the general principles for a recommendation about which transformation (if any) is appropriate in the particular application. In this way, the dependence of the transformation on the data itself would be avoided, and a cumulative body of evidence about the appropriateness of the various transformations would build over time. One contribution to this cumulative process is described in the next section, where we report on an empirical study of Fisher's z-transformation.

C.4. An Empirical Study of Fisher's z-Transformation for the Correlation Coefficient

Fisher's z-transformation

$$z = \phi(\rho) = \frac{1}{2}\log\left(\frac{1+\rho}{1-\rho}\right)$$

is used widely in the analysis of the correlation, ρ, between two random variables, X and Y, particularly when (X, Y) is distributed as a bivariate normal random variable. The main emphasis of the transformation is on the elimination of the functional dependence between mean and variance. See Table C.1. This allows standard methods to be used in the construction of confidence intervals.

The asymptotic properties of z are presented in Anderson (1958). Briefly, if $\hat\rho$ is the sample correlation coefficient for a sample of size n from a bivariate normal distribution with true correlation ρ, then the statistic

$$\sqrt{n}(\hat\rho - \rho)/(1 - \rho^2)$$

is asymptotically distributed as a standard normal $N(0, 1)$ random variable. The asymptotic variance of $\hat\rho$, i.e., $(1 - \rho^2)^2/n$ is functionally dependent on ρ itself. On the other hand, the statistic

$$\sqrt{n}(\hat z - z)$$

is asymptotically distributed as a $N(0, 1)$ random variable, where $\hat z = \phi(\hat\rho)$. This shows that the z-transformation eliminates the functional relationship between mean and variance, i.e., the asymptotic variance n^{-1} is independent of ρ.

To illustrate the ideas in Sections C.2 and C.3, we present the results of a small empirical study of the effectiveness of z. Our results were originally reported in Mulry and Wolter (1981). Similar results were recently reported by Efron (1981), who worked with some small computer-generated populations. In our study we find that the z-transformation improves the performance of confidence intervals based on the random group, jackknife, and balanced half-samples estimators.

We assume that a simple random sample of size n is selected without replacement from a finite population of size N. The finite population correlation coefficient is

$$\rho = \frac{\sum\limits_{i}^{N} (X_i - \bar{X})(Y_i - \bar{Y})}{\left\{\sum\limits_{i}^{N} (X_i - \bar{X})^2\right\}^{1/2} \left\{\sum\limits_{i}^{N} (Y_i - \bar{Y})^2\right\}^{1/2}} .$$

The usual estimator of ρ and the random group, balanced half-sample, jackknife, Taylor series, and normal-theory estimators of Var$\{\hat\rho\}$ are given by

$$\hat\rho = \frac{\sum\limits_{i}^{n} (x_i - \bar{x})(y_i - \bar{y})}{\left\{\sum\limits_{i}^{n} (x_i - \bar{x})^2\right\}^{1/2} \left\{\sum\limits_{i}^{n} (y_i - \bar{y})^2\right\}^{1/2}} ,$$

$$v_{\text{RG}}(\hat\rho) = \frac{1}{k(k-1)} \sum_{\alpha}^{k} (\hat\rho_\alpha - \hat\rho)^2,$$

$$v_{\text{BHS}}^{\dagger}(\hat\rho) = \frac{1}{4k} \sum_{\alpha}^{k} (\hat\rho_\alpha - \hat\rho_\alpha^c)^2,$$

$$v_{\text{BHS}}(\hat\rho) = \frac{1}{k} \sum_{\alpha}^{k} (\hat\rho_\alpha - \hat\rho)^2,$$

$$v_{\text{J}}(\hat\rho) = \frac{1}{k(k-1)} \sum_{\alpha}^{k} (\hat\rho_\alpha - \hat\rho)^2,$$

$$v_{\text{TS}}(\hat\rho) = \frac{1}{n(n-1)} \sum_{i}^{n} \hat{r}_i^2 ,$$

and

$$v_{\text{NT}}(\hat\rho) = (1 - \hat\rho^2)^2/n,$$

respectively.

For the random group estimator, the sample is divided at random into k groups of size m (we assume $n = mk$), and $\hat\rho_\alpha$ is the estimator of ρ obtained from the α-th group. For the balanced half-sample estimator, $n/2$ pseudo-strata are formed by pairing the observations in the order in which they were selected. Then, v_{BHS} is based on k balanced half-samples, each containing one unit from each pseudostratum, and $\hat\rho_\alpha$ is the estimator based on the α-th half-sample. The estimator $v_{\text{BHS}}^{\dagger}$ is also based on the k balanced half-samples, where $\hat\rho_\alpha^c$ is based upon the half-sample which is complementary to the α-th half-sample. For the jackknife estimator, the sample is divided at random into k groups, and the pseudovalue $\hat\rho_\alpha$ is defined by

$$\hat\rho_\alpha = k\hat\rho - (k-1)\hat\rho_{(\alpha)},$$

where $\hat\rho_{(\alpha)}$ is the estimator of ρ obtained from the sample after deleting the α-th group.

For the Taylor series estimator, we express $\hat{\rho}$ as follows:

$$\hat{\rho}(\bar{u}, \bar{v}, \bar{w}, \bar{x}, \bar{y}) = \frac{\bar{w} - \bar{x}\bar{y}}{(\bar{u} - \bar{x}^2)^{1/2}(\bar{v} - \bar{y}^2)^{1/2}},$$

where $U_i = X_i^2$, $V_i = Y_i^2$, and $W_i = X_i Y_i$. Then,

$$\hat{r}_i = \hat{d}_1 u_i + \hat{d}_2 v_i + \hat{d}_3 w_i + \hat{d}_4 x_i + \hat{d}_5 y_i,$$

where $(\hat{d}_1, \hat{d}_2, \hat{d}_3, \hat{d}_4, \hat{d}_5)$ is the vector of partial derivatives of $\hat{\rho}$ with respect to its five arguments evaluated at the point $(\bar{u}, \bar{v}, \bar{w}, \bar{x}, \bar{y})$.

Alternative variance estimators may be obtained by using squared deviations from $\hat{\rho} = k^{-1} \sum_\alpha^k \hat{\rho}_\alpha$. An alternative Taylor series estimator may be obtained by grouping the $\hat{r}_i$ and then applying the random group, balanced half-samples, or jackknife estimator to the group means. None of these alternatives are addressed specifically in this study.

The data used in this study were collected in the 1972–73 Consumer Expenditure Survey, sponsored by the U.S. Bureau of Labor Statistics and conducted by the U.S. Bureau of the Census. The correlation between monthly grocery store purchases and annual income was investigated. The data refer to 1972 annual income and average monthly grocery purchases during the first quarter of 1973. An experimental file of 4532 consumer units who responded to all the grocery and income categories during the first quarter of 1973 was created and treated as the finite population of interest.

The population mean of the income variable for the 4532 consumer units is $14,006.60 and the standard deviation is $12,075.42. The mean and standard deviation of monthly grocery store purchases are $146.30 and $84.84 respectively. The true correlation between annual income and monthly grocery store purchases is $\rho = 0.3584$.[1] Figure C.1 presents a scatter plot of the data.

To investigate the properties of the variance estimators, 1000 samples (srs wor) of size $n = 60$, 120, and 480 were selected from the population of consumer units. These sample sizes correspond roughly to the sampling fractions 0.013, 0.026, and 0.106, respectively. For each sample size, the following were computed:

a. the mean and variance of $\hat{\rho}$
b. the mean and variance of $v_{RG}(\hat{\rho})$
c. the mean and variance of $v_J(\hat{\rho})$
d. the mean and variance of $v_{TS}(\hat{\rho})$
e. the mean and variance of $v_{NT}(\hat{\rho})$
f. the mean and variance of $v_{BHS}(\hat{\rho})$
g. the mean and variance of $v_{BHS}^\dagger(\hat{\rho})$

[1] The grocery store purchases include purchases made with food stamps. This probably tends to depress the correlation.

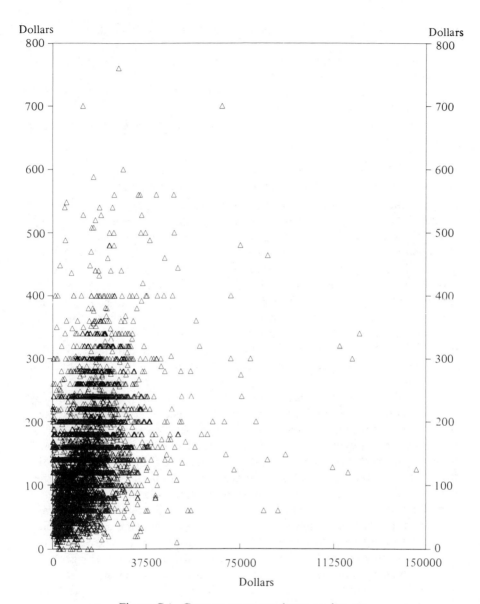

Figure C.1. Grocery store purchases vs. income.

h. proportion of confidence intervals formed using $v_{RG}(\hat\rho)$ that contain the true ρ
i. proportion of confidence intervals formed using $v_J(\hat\rho)$ that contain the true ρ
j. proportion of confidence intervals formed using $v_{TS}(\hat\rho)$ that contain the true ρ

k. proportion of confidence intervals formed using $v_{NT}(\hat{\rho})$ that contain the true ρ

l. proportion of confidence intervals formed using $v_{BHS}(\hat{\rho})$ that contain the true ρ

m. proportion of confidence intervals formed using $v_{BHS}^{\dagger}(\hat{\rho})$ that contain the true ρ

n. coverage rates in h, i, j, k, l, m for confidence intervals constructed using Fisher's z-transformation.

For all confidence intervals, the value of the constant c was taken as the tabular value from the standard normal $N(0, 1)$ distribution.

The Monte Carlo properties of the variance estimators are presented in Tables C.2 to C.6. Table C.2 gives the bias, variance, and mean square error (MSE) of the estimators. We observe that most of the estimators are downward biased, but that v_J is upward biased and v_{BHS} is nearly unbiased. The jackknife estimator v_J also tends to have the largest variance and MSE. Taylor series v_{TS} has reasonably good properties except in the case of the smallest sample size $n = 60$, where the bias is relatively large. The normal theory variance estimator has very small variance, but unacceptably large (in absolute value) bias. The variances of v_{RG} and v_J are inversely related to k, as might be expected from the theoretical developments in Section 2.6. Any one of v_{RG}, v_{TS}, v_{BHS}, or $v_{BHS}^{\dagger}$ might be recommended on the basis of these results.

Alternatively, we might judge the quality of the variance estimators by the difference between nominal and true confidence interval coverage rates. See Table C.3 for these results. Notice that for the Taylor series and normal theory estimators there are sharp differences between the Monte Carlo and nominal coverage rates. We conclude that neither estimator provides satisfactory confidence intervals for the sample sizes studied here. On the other hand, v_{RG}, v_J, v_{BHS}, and $v_{BHS}^{\dagger}$ all provide similar and relatively better confidence intervals. Even in these cases, however, the Monte Carlo coverage rates are too small. The confidence intervals tend to err on the side of being larger than the true ρ because the estimator $\hat{\rho}$ is upward biased and the variance estimators tend (except for v_J) to be downward biased. The problem is made worse by the fact that $\hat{\rho}$ and its variance estimators tend to be negatively correlated. Table C.5 gives the Monte Carlo correlations. Thus, the confidence intervals tend to be too narrow, particularly when $\hat{\rho}$ is too large. Finally, note that jackknife confidence intervals are competitive with confidence intervals formed using other variance estimators, whereas the jackknife could not be recommended on the basis of its own properties as given in Table C.2. The reverse is true of the Taylor series estimator.

Table C.4 shows the confidence interval coverage rates when the z-transformation is used. We observe substantial improvement in the confidence intervals associated with v_{RG}, v_J, v_{BHS}, and $v_{BHS}^{\dagger}$. Confidence intervals associated with v_{RG} and $v_{BHS}^{\dagger}$ are now particularly good, with very

Table C.2. Monte Carlo Properties of Estimators of Var$\{\hat{\rho}\}$

Estimator	Bias $\times 10^2$	Variance $\times 10^4$	MSE $\times 10^4$
Random Group			
$(n, k, m) = (60, 12, 5)$	−0.068	0.503	0.508
$(n, k, m) = (60, 6, 10)$	−0.194	0.966	1.004
$(n, k, m) = (120, 24, 5)$	−0.160	0.049	0.075
$(n, k, m) = (480, 32, 15)$	−0.193	0.002	0.039
Jackknife			
$(n, k, m) = (60, 12, 5)$	0.293	4.110	4.195
$(n, k, m) = (60, 60, 1)$	0.320	3.689	3.791
$(n, k, m) = (120, 24, 5)$	0.124	0.880	0.896
Taylor Series			
$n = 60$	−0.453	0.507	0.713
$n = 120$	−0.199	0.159	0.199
$n = 480$	−0.114	0.014	0.027
Normal Theory			
$n = 60$	−0.602	0.090	0.453
$n = 120$	−0.387	0.012	0.162
$n = 480$	−0.220	0.0003	0.049
Balanced Half-Samples			
$n = 60$	0.072	1.646	1.651
$n = 120$	0.020	0.386	0.386
Balanced Half-Samples[†]			
$n = 60$	−0.123	1.070	1.085
$n = 120$	−0.070	0.279	0.284

Note: The Monte Carlo expectation and variance of $\hat{\rho}$ are

n	$E\{\hat{\rho}\}$	Var$\{\hat{\rho}\} \times 10^2$
60	0.415	1.774
120	0.401	0.974
480	0.388	0.370

little discrepancy between the Monte Carlo and nominal coverage rates. The intervals still tend to miss ρ on the high side, but this effect is much diminished vis-à-vis the untransformed intervals. A partial explanation for the reduction in the asymmetry of the error is that $\hat{z} = \phi(\hat{\rho})$ and the estimators of its variance tend to be correlated to a lesser degree than the correlation between $\hat{\rho}$ and its variance estimators. See Table C.6.

Even on the transformed scale, however, confidence intervals associated with v_{TS} and v_{NT} perform badly. The transformation does not seem to

Table C.3. Monte Carlo Confidence Intervals for ρ

Estimator	90% Confidence Interval			95% Confidence Interval		
	% Contain ρ	% ρ ≤ Lower Bound	% ρ ≥ Upper Bound	% Contain ρ	% ρ ≤ Lower Bound	% ρ ≥ Upper Bound
Random Group						
$(n, k, m) = (60, 12, 5)$	82.9	15.3	1.8	88.1	11.3	0.6
$(n, k, m) = (60, 6, 10)$	78.3	19.1	2.6	84.0	14.9	1.7
$(n, k, m) = (120, 24, 5)$	80.9	17.3	1.8	87.1	11.9	1.0
$(n, k, m) = (480, 32, 15)$	65.7	28.4	5.9	74.5	22.5	3.0
Jackknife						
$(n, k, m) = (60, 12, 5)$	79.4	18.5	2.1	85.5	13.1	1.4
$(n, k, m) = (60, 60, 1)$	81.7	16.7	1.6	87.8	11.3	0.9
$(n, k, m) = (120, 24, 5)$	78.4	20.1	1.5	85.7	13.1	1.2
Taylor Series						
$n = 60$	74.6	21.3	4.1	82.8	14.9	2.3
$n = 120$	76.6	21.4	2.0	83.7	15.0	1.3
$n = 480$	70.6	21.6	7.8	77.6	15.8	6.6
Normal Theory						
$n = 60$	74.8	22.6	2.6	82.0	17.0	1.0
$n = 120$	73.0	24.8	2.2	79.9	18.7	1.4
$n = 480$	61.2	31.1	7.7	70.3	24.9	4.8
Balanced Half-Samples						
$n = 60$	81.6	17.1	1.3	87.2	12.2	0.6
$n = 120$	80.2	18.7	1.1	86.9	12.4	0.7
Balanced Half-Samples[†]						
$n = 60$	79.6	18.7	1.7	86.4	12.9	0.7
$n = 120$	78.7	19.8	1.5	85.4	13.5	1.1

Table C.4. Monte Carlo Confidence Intervals for $z = \phi(\rho)$

Estimator	90% Confidence Interval			95% Confidence Interval		
	% Contain ρ	% $\rho \leq$ Lower Bound	% $\rho \geq$ Upper Bound	% Contain ρ	% $\rho \leq$ Lower Bound	% $\rho \geq$ Upper Bound
Random Group						
$(n, k, m) = (60, 12, 5)$	91.4	8.1	0.5	96.2	3.7	0.1
$(n, k, m) = (60, 6, 10)$	84.2	14.1	1.7	89.1	9.9	1.0
$(n, k, m) = (120, 24, 5)$	91.6	7.8	0.6	95.2	4.5	0.3
$(n, k, m) = (480, 32, 15)$	74.6	22.1	3.3	82.7	16.2	1.1
Jackknife						
$(n, k, m) = (60, 12, 5)$	82.5	15.4	2.1	89.5	9.0	1.5
$(n, k, m) = (60, 60, 1)$	85.2	13.0	1.8	91.0	8.0	1.0
$(n, k, m) = (120, 24, 5)$	82.2	16.2	1.6	88.6	10.2	1.2
Taylor Series						
$n = 60$	66.7	28.4	4.9	74.6	22.8	2.6
$n = 120$	68.4	28.7	2.9	75.9	22.6	1.5
$n = 480$	62.7	28.7	8.6	71.0	21.5	7.5
Normal Theory						
$n = 60$	78.0	19.2	2.6	85.5	13.1	1.4
$n = 120$	74.5	22.7	2.8	83.3	15.2	1.5
$n = 480$	62.0	30.0	8.0	71.7	22.8	5.5
Balanced Half-Samples						
$n = 60$	85.6	13.1	1.3	91.0	8.3	0.7
$n = 120$	84.3	14.6	1.1	89.8	9.4	0.8
Balanced Half-Samples[†]						
$n = 60$	91.2	6.2	2.6	96.3	2.2	1.5
$n = 120$	90.8	7.8	1.4	95.8	3.2	1.0

Table C.5. Monte Carlo Correlation
Between $\hat{\rho}$ and $v(\hat{\rho})$

Variance Estimator	Correlation
Random Group	
$(n, k, m) = (60, 12, 5)$	-0.30
$(n, k, m) = (60, 6, 10)$	-0.28
$(n, k, m) = (120, 24, 5)$	-0.30
$(n, k, m) = (480, 32, 15)$	-0.46
Jackknife	
$(n, k, m) = (60, 12, 5)$	-0.32
$(n, k, m) = (60, 60, 1)$	-0.33
$(n, k, m) = (120, 24, 5)$	-0.31
Taylor Series	
$n = 60$	-0.28
$n = 120$	-0.23
$n = 480$	0.06
Normal Theory	
$n = 60$	-1.00
$n = 120$	-1.00
$n = 480$	-1.00
Balanced Half-Samples	
$n = 60$	-0.37
$n = 120$	-0.36
Balanced Half-Samples[†]	
$n = 60$	-0.37
$n = 120$	-0.31

improve these intervals, and in fact seems to make the Taylor series intervals worse.

Based on the results presented here, we recommend the z-transformation for making inferences about the finite population correlation coefficient, particularly when used with the random group, jackknife, or balanced half-sample variance estimators. The normal-theory estimator seems sensitive to the assumed distributional form and is not recommended for populations that depart from normality to the degree observed in the present population of consumer units. The Taylor series estimator is not recommended for inferential purposes either, although, this estimator does have reasonably good properties in its own right.

Table C.6. Monte Carlo Correlation
Between $\hat{z} = \phi(\hat{\rho})$ and $v(\hat{z})$

Variance Estimator	Correlation
Random Group	
$(n, k, m) = (60, 12, 5)$	-0.03
$(n, k, m) = (60, 6, 10)$	-0.01
$(n, k, m) = (120, 24, 5)$	-0.03
$(n, k, m) = (480, 32, 15)$	-0.22
Jackknife	
$(n, k, m) = (60, 12, 5)$	-0.06
$(n, k, m) = (60, 60, 1)$	-0.04
$(n, k, m) = (120, 24, 5)$	-0.08
Taylor Series	
$n = 60$	-0.28
$n = 120$	-0.23
$n = 480$	0.05
Normal Theory	
$n = 60$	0.00
$n = 120$	0.00
$n = 480$	0.00
Balanced Half-Samples	
$n = 60$	-0.02
$n = 120$	-0.06
Balanced Half-Samples[†]	
$n = 60$	0.42
$n = 120$	0.35

The Effect of Measurement Errors on Variance Estimation

We shall now introduce measurement (or response) errors and look briefly at the properties of variance estimators when the data are contaminated by such errors.

Throughout the book we have assumed that the response, say Y_i, for a given individual i is equal to that individual's "true value." Now we shall assume that the data may be adequately described by the additive error model

$$Y_i = \mu_i + e_i, \qquad (D.1)$$

$i = 1, \ldots, N$. The errors e_i are assumed to be $(0, \sigma_i^2)$ random variables, and the means μ_i are taken to be the "true values." Depending on the circumstances of a particular sample survey, the errors e_i may or may not be correlated with one another. In the sequel we shall make clear our assumptions about the correlation structure.

Model (D.1) is about the simplest model imaginable for representing measurement error. Many extensions of the model have been given in the literature. For general discussion of the basic model and extensions, see Hansen, Hurwitz, and Bershad (1961), Hansen, Hurwitz, and Pritzker (1964), Koch (1973), and Cochran (1977). The simple model (D.1) is adequate for our present purposes.

It should be observed that (D.1) is a conceptual model, where the Y_i and e_i are attached to the N units in the population prior to sampling. This situation differs from some of the previous literature on response errors, where it is assumed that the errors e_i are generated only for units selected into the sample. Our stronger assumption is necessary in order to interchange certain expectation operators. See, e.g., equation (D.7).

We shall assume that it is desired to estimate some parameter θ of the finite population. For the moment, we assume that the estimator of θ is of the form

$$\hat{\theta} = \sum_{i=1}^{N} w_i t_i Y_i, \qquad (D.2)$$

where the w_i are fixed weights attached to the units in the population, the t_i are indicator random variables

$$t_i = 1 \qquad \text{if } i \in s$$
$$= 0 \qquad \text{if } i \notin s,$$

and s denotes the sample. Form (D.2) includes many of the estimators found in survey sampling practice.

We are interested in estimators of the variance of $\hat{\theta}$ and in studying the properties of such estimators in the presence of model (D.1). Many authors, including those cited above, have studied the effects of measurement errors on the true variance of $\hat{\theta}$. We shall review this work and then go on to consider the problem of variance estimation, a problem where little is available in the published literature.

Before beginning, it is important to establish a clear notation for the different kinds of expectations that will be needed. There are two sources of randomness in this work. One concerns the sampling design, which is in the control of the survey statistician. All information about the design is encoded in the indicator variables t_i. We shall let E_d and Var_d denote the expectation and variance operators with respect to the sampling design. The other source of randomness concerns the distribution, say ξ, of the measurement (or response) errors e_i. We shall let $\mathscr{E}$ and $\mathscr{V}\!ar$ denote the expectation and variance operators with respect to the ξ-distribution. Finally, combining both sources of randomness, we shall let the unsubscripted symbols E and Var denote total expectation and total variance, respectively. The reader will note the following connections between the different operators:

(1) $E = E_d \mathscr{E} = \mathscr{E} E_d$

and

(2) $\text{Var} = E_d \mathscr{V}\!ar + \text{Var}_d \mathscr{E} = \mathscr{E} \text{Var}_d + \mathscr{V}\!ar E_d$.

Summarizing the notation, we have

Source	Operators
Sampling Design	E_d, Var_d
ξ	$\mathscr{E}$, $\mathscr{V}\!ar$
Total	E, Var.

The total variance of $\hat{\theta}$ may be written as

$$\text{Var}\{\hat{\theta}\} = \text{Var}_d\, \mathscr{E}\{\hat{\theta}\} + \text{E}_d\, \mathscr{V}\!ar\{\hat{\theta}\}.$$

Total	Sampling	Response
Variance	Variance	Variance

$$\text{(D.3)}$$

The *sampling variance* is the component of variability that arises because observations are made on a random sample, and not on the full population. This component is the total variance when measurement error is not present. The *response variance* is the component of variability that arises because of the errors of measurement e_i. This component is present even when the entire population is enumerated!

It is easily seen that the sampling variance is

$$\text{Var}_d\, \mathscr{E}\{\hat{\theta}\} = \text{Var}_d\left\{ \sum_{i=1}^{N} w_i t_i \mu_i \right\}$$

$$= \sum_{i=1}^{N} w_i^2 \mu_i^2 \pi_i(1 - \pi_i) + \sum_{i \neq j}^{N} w_i w_j \mu_i \mu_j (\pi_{ij} - \pi_i \pi_j), \quad \text{(D.4)}$$

where, as usual, π_i denotes the probability that the i-th unit is drawn into the sample and π_{ij} denotes the probability that both the i-th and j-th units are drawn into the sample. Equation (D.4) follows from the fact that

$$\mathscr{E}\{Y_i\} = \mu_i,$$

$$\text{Var}_d\{t_i\} = \pi_i(1 - \pi_i),$$

and

$$\text{Cov}_d\{t_i, t_j\} = \pi_{ij} - \pi_i \pi_j.$$

Now let $\sigma_{ij} = \mathscr{E}\{e_i e_j\}$ denote the ξ-covariance between the errors e_i and e_j, $i \neq j$. Then, the response variance is

$$\text{E}_d\, \mathscr{V}\!ar\{\hat{\theta}\} = \text{E}_d\left\{ \sum_{i=1}^{N} w_i^2 t_i^2 \sigma_i^2 + \sum_{i \neq j}^{N} w_i w_j t_i t_j \sigma_{ij} \right\}$$

$$= \sum_{i=1}^{N} w_i^2 \pi_i \sigma_i^2 + \sum_{i \neq j}^{N} w_i w_j \pi_{ij} \sigma_{ij}, \quad \text{(D.5)}$$

because $\text{E}_d\{t_i^2\} = \pi_i$, $\text{E}_d\{t_i t_j\} = \pi_{ij}$, and $\mathscr{V}\!ar\{Y_i\} = \sigma_i^2$. Combining (D.4) and (D.5) gives

Theorem D.1. *The total variance of an estimator of the form $\hat{\theta} = \sum w_i t_i Y_i$ is*

given by

$$\text{Var}\{\hat{\theta}\} = \sum_{i=1}^{N} w_i^2 \mu_i^2 \pi_i (1 - \pi_i) + \sum_{i \neq j} \sum w_i w_j \mu_i \mu_j (\pi_{ij} - \pi_i \pi_j)$$

$$+ \sum_{i=1}^{N} w_i^2 \pi_i \sigma_i^2 + \sum_{i \neq j} \sum w_i w_j \pi_{ij} \sigma_{ij}.^1$$

The last term on the right side is omitted when the measurement errors are uncorrelated.   □

EXAMPLE D.1. For srs wor and $\hat{\theta} = \bar{y}$, we have the familiar expression

$$\text{Var}\{\bar{y}\} = (1 - f)n^{-1}S_\mu^2 + n^{-1}\sigma^2\{1 + (n - 1)\rho\}, \quad (D.6)$$

where

$$S_\mu^2 = (N - 1)^{-1} \sum_{i=1}^{N} (\mu_i - \bar{M})^2,$$

$$\bar{M} = N^{-1} \sum_{i=1}^{N} \mu_i,$$

$$\sigma^2 = N^{-1} \sum_{i=1}^{N} \sigma_i^2,$$

$$\sigma^2\rho = N^{-1}(N - 1)^{-1} \sum_{i \neq j} \sum \sigma_{ij},$$

$$f = n/N.$$

This follows from Theorem D.1 with $w_i = 1/n$, $\pi_i = n/N$, and $\pi_{ij} = n(n - 1)/n(N - 1)$. The term involving $\sigma^2\rho$ is omitted whenever the errors are uncorrelated.   □

The expressions for total variance presented in Theorem D.1 and equation (D.6) have appeared previously in the literature. To investigate potential estimators of variance, however, it is useful to work with an alternative expression, obtained by interchanging the order of expectations. The alternative expression is

$$\text{Var}\{\hat{\theta}\} = \mathscr{E} \, \text{Var}_d\{\hat{\theta}\} + \mathscr{V}\!ar \, \text{E}_d\{\hat{\theta}\}. \quad (D.7)$$

---

[1] Some authors permit the distribution of Y_i to depend not only on unit i, but also on other units in the sample s. See Hansen, Hurwitz, and Bershad (1961) for an example involving a housing survey. Given this circumstance we may have $\mathscr{E}\{Y_i|s\} = \mu_{is} \neq \mu_i = \mathscr{E}\{Y_i|t_i = 1\}$ and $\mathscr{V}\!ar\{Y_i|s\} = \sigma_{is}^2 \neq \sigma_i^2 = \mathscr{V}\!ar\{Y_i|t_i = 1\}$. A nonzero covariance (or interaction) then exists between sampling error and measurement error. See, e.g., Koch (1973). In the simple additive model considered here, it is assumed that $\mu_{is} = \mu_i$ and $\sigma_{is}^2 = \sigma_i^2$, and thus the interaction component vanishes.

Neither of the components on the right side of (D.7) corresponds precisely to the components of (D.3).

Define

$$\tilde{\theta} = \sum_{i=1}^{N} w_i t_i \mu_i,$$

the estimator of the same functional form as $\hat{\theta}$ with the means μ_i replacing the response variables Y_i. The estimators $\hat{\theta}$ and $\tilde{\theta}$ are identical whenever measurement error is absent. We shall assume that there exists a design-unbiased estimator of the design-variance of $\tilde{\theta}$. That is, there exists an estimator $v(\tilde{\theta})$ such that

$$E_d\{v(\tilde{\theta})\} = \text{Var}_d\{\tilde{\theta}\}.$$

Such estimators have been discussed in this book and are discussed extensively in the traditional survey sampling texts.

Now define the "copy" of $v(\tilde{\theta})$, say $v_c(\hat{\theta})$, by replacing the μ_i by the responses Y_i. We shall view $v_c(\hat{\theta})$ as an estimator of the total variance of $\hat{\theta}$. The bias of this estimator is described in the following theorem.

Theorem D.2. *The bias of $v_c(\hat{\theta})$ as an estimator of the total variance of $\hat{\theta}$ is given by*

$$\text{Bias}\{v_c(\hat{\theta})\} = -\mathcal{V}a\imath\, E_d\{\hat{\theta}\} = -\sum_{i=1}^{N} w_i^2 \pi_i^2 \sigma_i^2 - \sum_{i \neq j}^{N} w_i w_j \pi_i \pi_j \sigma_{ij}.$$

PROOF. *By definition, $v(\tilde{\theta})$ is a design-unbiased estimator of $\text{Var}_d\{\tilde{\theta}\}$. Because this must be true for any characteristic of interest, we have*

$$E_d\{v_c(\hat{\theta})\} = \text{Var}_d\{\hat{\theta}\}.$$

Therefore,

$$E\{v_c(\hat{\theta})\} = \mathcal{E}\,\text{Var}_d\{\hat{\theta}\}$$

and the result follows by the decomposition (D.7).   □

The "copy" $v_c(\hat{\theta})$ may or may not be seriously biased, depending on the correlated component of the total variance. The following two examples illustrate these findings.

EXAMPLE D.2. We continue the first example, assuming srs wor and $\hat{\theta} = \bar{y}$. For this problem, the familiar variance estimators are

$$v(\tilde{\theta}) = (1 - f)s_\mu^2 / n$$

$$s_\mu^2 = (n - 1)^{-1} \sum_{i=1}^{n} (\mu_i - \bar{\mu})^2$$

$$\bar{\mu} = n^{-1} \sum_{i=1}^{n} \mu_i$$

and

$$v_c(\bar{y}) = (1 - f.)s_y^2/n$$

$$s_y^2 = (n - 1)^{-1} \sum_{i=1}^{n} (y_i - \bar{y})^2$$

$$\bar{y} = n^{-1} \sum_{i=1}^{n} y_i.$$

By Theorem D.2, the bias in the variance estimator is

$$\text{Bias}\{v_c(\bar{y})\} = -N^{-1}\sigma^2\{1 + (N - 1)\rho\}.$$

When measurement errors are uncorrelated, the bias reduces to

$$\text{Bias}\{v_c(\bar{y})\} = -N^{-1}\sigma^2,$$

and this will be unimportant whenever the sampling fraction f is negligible. If the fpc is omitted from the variance calculations, we note that

$$\text{Bias}\{s_y^2/n\} = N^{-1}S_\mu^2 - \sigma^2\rho,$$

reducing to

$$\text{Bias}\{s_y^2/n\} = N^{-1}S_\mu^2$$

for uncorrelated errors. Thus, even when measurement errors are uncorrelated, we are forced to accept a downward bias in the response variance (estimator with fpc) or upward bias in the sampling variance (estimator without fpc). $\qquad\square$

EXAMPLE D.3. We assume a πps sampling scheme with $\hat{\theta} = \hat{Y}$, the Horvitz-Thompson estimator of the population total, i.e., $w_i = \pi_i^{-1} = (np_i)^{-1}$. Assuming positive joint inclusion probabilities, $\pi_{ij} > 0$, the Yates and Grundy (1953) estimator is unbiased for the design variance of $\hat{\theta}$. See Section 1.4. The "copy" is then

$$v_c(\hat{Y}) = \sum_{i=1}^{n} \sum_{j>i}^{n} \{(\pi_i\pi_j - \pi_{ij})/\pi_{ij}\}(y_i/\pi_i - y_j/\pi_j)^2,$$

and by Theorem D.2 its bias must be

$$\text{Bias}\{v_c(\hat{Y})\} = -N\sigma^2\{1 + (N - 1)\rho\},$$

where

$$\sigma^2 = N^{-1} \sum_{i=1}^{N} \sigma_i^2$$

$$\sigma^2\rho = N^{-1}(N - 1)^{-1} \sum\sum_{i\neq j} \sigma_{ij}.$$

The bias reduces to $-N\sigma^2$ in the case of uncorrelated errors. On several occasions in this book we have also discussed the possibility of estimating the variance of $\hat{\theta} = \hat{Y}$ by the traditional formula for pps wr sampling

$$v_{\mathrm{wr}}(\tilde{\theta}) = n^{-1}(n-1)^{-1} \sum_{i=1}^{n} (\mu_i/p_i - \tilde{\theta})^2.$$

As was demonstrated in Section 2.4.5, this is a biased estimator of the design variance of $\tilde{\theta}$, with bias given by

$$\mathrm{Bias}_d\{v_{\mathrm{wr}}(\tilde{\theta})\} = \frac{n}{n-1} (\mathrm{Var}_d\{\tilde{\theta}_{\mathrm{wr}}\} - \mathrm{Var}_d\{\tilde{\theta}_{\pi\mathrm{ps}}\}),$$

where the first and second terms on the right side denote the variance of $\tilde{\theta}$ given with and without replacement sampling, respectively. Let $v_{\mathrm{wr,c}}(\hat{Y})$ denote the "copy" of $v_{\mathrm{wr}}(\tilde{\theta})$. Then, following the development of Theorem D.2, the bias of the "copy" as an estimator of $\mathrm{Var}\{\hat{Y}\}$ is

$$\mathrm{Bias}\{v_{\mathrm{wr,c}}(\hat{Y})\} = -N\sigma^2\{1+(N-1)\rho\} + \mathscr{E}\frac{n}{n-1}(\mathrm{Var}_d\{\hat{Y}_{\mathrm{wr}}\} - \mathrm{Var}_d\{\hat{Y}_{\pi\mathrm{ps}}\}).$$

The second term on the right side is the "price" to be paid for "copying" a biased estimator of the variance of $\tilde{\theta}$.   □

In most surveys of human populations, there tends to be a positive-valued correlated component of response variance $\sigma^2\rho$. This is particularly so when the enumeration is made via personal visit. See Bailar (1968, 1979) for some examples. Whenever such correlation occurs, there is a potential for both (1) an important increase in the total variance and (2) a serious bias in the variance estimator. The first point is illustrated in the first example, where we note (cf. (D.6)) that the total variance is order n^{-1}, except for an order 1 term in $\sigma^2\rho$. This latter term may result in an important increase in total variance relative to the situation where measurement errors are uncorrelated. The second point is illustrated in the second and third examples. We not only observe a bias in the variance estimator, but we see that the bias involves the order 1 term in $\sigma^2\rho$. Roughly speaking, this term is left out of the variance calculations, resulting in an order 1 downward bias!

Even when measurement errors are uncorrelated there is a bias in the variance estimators. This too is illustrated in the second and third examples. The bias is less harmful in this case, however, and is unimportant when the sampling fraction is negligible.

One might despair at this point, thinking that there is no hope for producing satisfactory variance estimates in the presence of correlated measurement error. Fortunately, some of the variance estimating methods discussed earlier in this book may provide a satisfactory solution.

To see this, let us assume that the correlated component arises strictly from the effects of interviewers. This assumption is fairly reasonable; most research on the correlated component points to the interviewer as the primary cause of the correlation. We note, however, that coders, supervisors, and the like may also contribute to this component.

We shall consider the random group estimator of variance. Similar results can be given for some of the other estimators studied in this book. We shall assume

(1) there are k random groups
(2) interviewers assignments are completely nested *within* random groups
(3) interviewers have a common effect on the ξ-distribution, i.e.,

$$\mathscr{E}\{e_i\} = 0$$

$$\mathscr{E}\{e_i^2\} = \sigma_i^2$$

$$\mathscr{E}\{e_i e_j\} = \sigma_{ij} \quad \text{if units } i \text{ and } j \text{ are enumerated by the same interviewer}$$

$$= 0 \quad \text{if units } i \text{ and } j \text{ are enumerated by different interviewers}$$

and these moments do not depend upon which interviewer enumerates the i-th and j-th units.

The parent sample estimator $\hat{\theta}$ is still as defined in (D.2). The estimator for the α-th random group is defined by

$$\hat{\theta}_\alpha = \sum_{i=1}^{N} w_{i(\alpha)} t_{i(\alpha)} Y_i,$$

where

$$t_{i(\alpha)} = 1 \quad \text{if the } i\text{-th unit is included in the } \alpha\text{-th random group } s_\alpha$$

$$= 0 \quad \text{otherwise}$$

and the $w_{i(\alpha)} = k w_i$ are the weights associated with the α-th random group. Because the estimators are linear we have

$$\hat{\theta} = k^{-1} \sum_{\alpha=1}^{k} \hat{\theta}_\alpha.$$

By our assumptions, the $\hat{\theta}_\alpha$ are ξ-uncorrelated, given the sample and its partition into random groups, and it follows that

$$\mathscr{V}a\imath\{\hat{\theta}\} = k^{-1} \mathscr{V}a\imath\{\hat{\theta}_\alpha\}. \tag{D.8}$$

The ξ-variance of $\hat{\theta}_\alpha$ is

$$\mathscr{V}a\imath\{\hat{\theta}_\alpha\} = \sum_{i=1}^{N} w_{i(\alpha)}^2 t_{i(\alpha)}^2 \sigma_i^2 + \sum_{i \neq j}^{N} w_{i(\alpha)} w_{j(\alpha)} t_{i(\alpha)} t_{j(\alpha)} \sigma_{ij}$$

and

$$E_d \mathscr{V}a\imath(\hat{\theta}_\alpha) = \sum_{i=1}^{N} w_{i(\alpha)}^2 (k^{-1} \pi_i) \sigma_i^2 + \sum_{i \neq j}^{N} w_{i(\alpha)} w_{j(\alpha)} (k^{-1} \phi_{j|i} \pi_{ij}) \sigma_{ij}$$

$$= k \sum_{i=1}^{N} w_i^2 \pi_i \sigma_i^2 + k \sum_{i \neq j}^{N} w_i w_j \phi_{j|i} \pi_{ij} \sigma_{ij}, \tag{D.9}$$

where $\phi_{j|i}$ is the conditional probability that unit j is included in the α-th random group, given that unit i is included in the α-th random group and that both units i and j are included in the parent sample. Combining (D.4), (D.8), and (D.9) gives the following result:

Theorem D.3. *Given assumptions* (1)–(3), *the total variance of* $\hat{\theta}$ *is*

$$\text{Var}\{\hat{\theta}\} = \sum_{i=1}^{N} w_i^2 \mu_i^2 \pi_i(1 - \pi_i) + \sum\sum_{i \neq j} w_i w_j \mu_i \mu_j (\pi_{ij} - \pi_i \pi_j)$$

$$+ \sum_{i=1}^{N} w_i^2 \pi_i \sigma_i^2 + \sum\sum_{i \neq j} w_i w_j \phi_{j|i} \pi_{ij} \sigma_{ij}. \qquad \square$$

The first and second terms on the right side of the above expression constitute the *sampling variance*, while the third and fourth terms constitute the *response variance*.

Comparing this expression with the corresponding expression in Theorem D.1 shows that the sampling variance is the same, but the correlated component of response variance is diminished by the factor $\phi_{j|i}$. The diminution in the correlated component arises because the measurement errors are assumed to be correlated within, and not between, interviewer assignments. This effect will be present whether or not the groups referenced in assumption (1) are formed at random. In fact, the correlated component will always be diminished roughly by a factor that is inversely proportional to the number of interviewers. By forming groups at random, we achieve both the reduction in the true variance and a rigorous estimator of variance, as we shall show in Theorem D.4. By forming groups in a nonrandom way, however, we achieve the reduction in the true variance but render that variance nonestimable.

EXAMPLE D.4. To illustrate the effect, note that $\phi_{j|i} = (m - 1)/(n - 1) \doteq k^{-1}$ for srs wor, where $m = n/k$. Thus, for large k, the correlated component is diminished very substantially when the errors e_i can be assumed to be uncorrelated between interviewer assignments. Specifically, for $\hat{\theta} = \bar{y}$, we now have

$$\text{Var}\{\bar{y}\} = (1 - f)n^{-1}S_\mu^2 + n^{-1}\sigma^2\{1 + (m - 1)\rho\}.$$

The term in $\sigma^2 \rho$ is now order k^{-1}, whereas in the earlier work this term was order 1. $\qquad \square$

By definition, the random group estimator of variance is

$$v_{\text{RG}}(\hat{\theta}) = k^{-1}(k - 1)^{-1} \sum_{\alpha=1}^{k} (\hat{\theta}_\alpha - \hat{\theta})^2 = 2^{-1}k^{-2}(k - 1)^{-1} \sum\sum_{\alpha \neq \beta}^{k} (\hat{\theta}_\alpha - \hat{\theta}_\beta)^2.$$

Assuming that the random group estimators are symmetrically defined,[2] we see that the total expectation of v_{RG} is given by

$$E\{v_{RG}(\hat{\theta})\} = (2k)^{-1}E\{(\hat{\theta}_\alpha - \hat{\theta}_\beta)^2\} = k^{-1}(\text{Var}\{\hat{\theta}_\alpha\} - \text{Cov}\{\hat{\theta}_\alpha, \hat{\theta}_\beta\}).$$
(D.10)

The following theorem establishes the bias of the random group estimator.

Theorem D.4. *Given assumptions* (1)–(3) *and that the $\hat{\theta}_\alpha$ are symmetrically defined, the total bias of the random group estimator of variance is*

$$\text{Bias}\{v_{RG}(\hat{\theta})\} = \sum_{i=1}^{N} w_i^2 \mu_i^2 \pi_i^2 - \sum_{i \neq j}^{N} w_i w_j \mu_i \mu_j (k\nu_{j|i}\pi_{ij} - \pi_i\pi_j),$$

where $\nu_{j|i}$ is the conditional probability that unit j is included in random group β, given that unit i is included in random group α ($\alpha \neq \beta$) and that both i and j are in the parent sample. In other words, the bias arises solely from the sampling distribution and not from the ξ-distribution. In particular, the bias is not an order 1 function of $\sigma^2\rho$.

PROOF. By (D.8) and (D.3),

$$\text{Var}\{\hat{\theta}\} = k^{-1} \text{Var}\{\hat{\theta}_\alpha\} + \text{Var}_d \, \mathscr{E}\{\hat{\theta}\} - k^{-1} \text{Var}_d \, \mathscr{E}\{\hat{\theta}_\alpha\}$$

$$= k^{-1} \text{Var}\{\hat{\theta}_\alpha\} + (1 - k^{-1}) \text{Cov}_d\{\mathscr{E}\hat{\theta}_\alpha, \mathscr{E}\hat{\theta}_\beta\}.$$

Combining this result with (D.10) and remembering that the $\hat{\theta}_\alpha$ are ξ-uncorrelated gives

$$\text{Bias}\{v_{RG}(\hat{\theta})\} = -\text{Cov}_d\{\mathscr{E}\hat{\theta}_\alpha, \mathscr{E}\hat{\theta}_\beta\}.$$

The theorem follows from

$$\text{Cov}_d\{\mathscr{E}\hat{\theta}_\alpha, \mathscr{E}\hat{\theta}_\beta\} = \sum_{i=1}^{N} \sum_{j=1}^{N} w_{i(\alpha)} w_{j(\beta)} \mu_i \mu_j \, \text{Cov}_d\{t_{i(\alpha)}, t_{j(\beta)}\}$$

$$= -\sum_{i=1}^{N} w_i^2 \mu_i^2 \pi_i^2 + \sum_{i \neq j}^{N} w_i w_j \mu_i \mu_j (k\nu_{j|i}\pi_{ij} - \pi_i\pi_j). \qquad \square$$

Some examples will illustrate the nature of the bias of v_{RG}.

EXAMPLE D.5. Again consider srs wor with $\hat{\theta} = \bar{y}$. From Theorem D.4 we have

$$\text{Bias}\{v_{RG}(\bar{y})\} = S_\mu^2 / N,$$

because $w_i = 1/n$, $\pi_i = n/N$, $\pi_{ij} = n(n-1)/N(N-1)$, $\nu_{j|i} = m/(n-1)$.

---

[2] By "symmetrically defined," we mean that the random groups are each of equal size and that the $\hat{\theta}_\alpha$ are defined by the same functional form. This assumption ensures that the $\hat{\theta}_\alpha$ are identically distributed.

This bias is unimportant in comparison with the bias displayed in Example D.2. The bias component in $\sigma^2\rho$ has now been eliminated. Moreover, the remaining bias will be unimportant whenever the sampling fraction $f = n/N$ is negligible. $\square$

EXAMPLE D.6. Let us assume πps sampling with $\hat\theta = \hat Y$, the Horvitz–Thompson estimator of the population total. In this case, $w_i = \pi_i^{-1}$ and $\nu_{j|i} = m/(n-1)$. Thus,

$$\text{Bias}\{v_{RG}(\hat Y)\} = \sum_{i=1}^{N} \mu_i^2 - \sum_{i\neq j}^{N} \mu_i\mu_j\left(\frac{n}{n-1}\frac{\pi_{ij}}{\pi_i\pi_j} - 1\right)$$

$$= \frac{n}{n-1}(\text{Var}\{\tilde\theta_{\text{wr}}\} - \text{Var}\{\tilde\theta_{\pi\text{ps}}\}),$$

where $\tilde\theta = \sum w_i t_i \mu_i$ and $\text{Var}\{\tilde\theta_{\text{wr}}\}$ and $\text{Var}\{\tilde\theta_{\pi\text{ps}}\}$ are variances assuming with and without replacement sampling, respectively. Compare this work with Example D.3. The bias component in $\sigma^2\rho$ has been eliminated. The residual bias is a function of the efficiency of πps sampling vis-à-vis pps wr sampling, and in the useful applications of πps sampling the bias will be positive.

EXAMPLE D.7. One of the most useful applications of Theorems D.3 and D.4 concerns cluster sampling. We shall assume a πps sample of n clusters, with possibly several stages of subsampling within the selected clusters. No restrictions are imposed on the subsampling design other than it be independent from cluster to cluster. For this problem, rule (iii), Section 2.4.1 is employed in the formation of random groups, and, to be consistent with assumptions (1)-(3), interviewer assignments are nested completely within clusters. Then, as we shall see, the bias in $v_{RG}(\hat\theta)$ arises solely in the between component of the sampling variance, and thus will be unimportant in many applications. Once again, the bias in $\sigma^2\rho$ is eliminated by use of the random group method. To show this effect, it will be convenient to adopt a double subscript notation. The estimator of θ is now

$$\hat\theta = \sum_{i=1}^{N}\sum_{j=1}^{M_i} w_{ij}t_{ij}Y_{ij},$$

where Y_{ij} denotes the j-th elementary unit in the i-th primary unit and the other symbols have a similar interpretation. We shall let $\hat\theta = \hat Y$, the Horvitz–Thompson estimator of the population total.

Let

$$\mu_i = \sum_{j=1}^{M_i} \mu_{ij}$$

denote the "true" total for the i-th primary unit. Then, by Theorems D.4,

2.4.5, and 2.4.6 it follows that

$$\text{Bias}\{v_{\text{RG}}(\hat{Y})\} = \frac{n}{n-1}\,(\text{Var}\{\tilde{\theta}_{\text{wr}}\} - \text{Var}\{\tilde{\theta}_{\pi\text{ps}}\}),$$

where

$$\tilde{\theta} = \sum w_i t_i \mu_i,$$

$$t_i = 1 \qquad \text{if the } i\text{-th primary is in sample}$$

$$\;\;= 0 \qquad \text{otherwise},$$

$$w_i = (np_i)^{-1},$$

p_i is the probability associated with the i-th primary unit, and $\text{Var}\{\tilde{\theta}_{\text{wr}}\}$ and $\text{Var}\{\tilde{\theta}_{\pi\text{ps}}\}$ denote the variances of $\tilde{\theta}$ assuming with and without replacement sampling, respectively. This expression confirms that the bias is in the between component of the sampling variance, and not in the within component nor in the response variance. In surveys where the between component of sampling variance is a negligible part of the total variance, the bias of v_{RG} will be unimportant. In any case the bias will tend to be positive in the useful applications of πps sampling. □

In summary, we have seen that the correlated component of response variance is eliminated entirely from the bias of the random group estimator of variance. The main requirements needed to achieve this result are that (1) k random groups be formed in accordance with the rules presented in Section 2.4.1, (2) interviewer assignments be nested completely within random groups, and (3) measurement errors be uncorrelated between interviewer assignments. Requirements (2) and (3) imply that the random group estimators $\hat{\theta}_\alpha$ are ξ-uncorrelated.

Our results also extend to more complicated situations where coders, supervisors and the like may potentially induce a correlation between the e_i. In this case, one needs to nest the coder and supervisor (etc.) assignments within random groups. This procedure ensures that the $\hat{\theta}_\alpha$ will be ξ-uncorrelated and that the results of our theorems will be valid.

The nesting techniques, intended to induce ξ-uncorrelated $\hat{\theta}_\alpha$, were studied by Mahalanobis (1939) as early as the 1930s under the name "interpenetrating subsamples." The terminology survives to the present day with authors concerned with components of response variability. Although many benefits accrue from the use of these techniques, one disadvantage is that the nesting of interviewer assignments may tend to slightly reduce flexibility and marginally increase costs. The extent of this problem will vary with each survey application.

The work done here may be extended in a number of directions. First, we have been working with estimators of the general form given in (D.2). Estimators which are nonlinear function of statistics of form (D.2) may be handled by using Taylor series approximations. In this way, our results extend to a very wide class of survey problems. Second, we have been working with the random group estimator of variance. Extensions of the results may be obtained for the jackknife and balanced half-sample estimators of variance. In the case of jackknife, for example, one begins by forming random groups; proceeds to nest interviewer (and possibly coder etc.) assignments within random groups; and then forms pseudovalues by discarding random groups from the parent sample. Third, the measurement error model (D.1) assumed here involved a simple additive structure. Extensions of the results could be given for more complicated models. Finally, we have been attempting to show in rather simple terms the impact of response errors upon the statistical properties of estimators of total variance. We have not discussed operational strategies for randomizing interviewers' assignments in actual field work. In actual practice, it is common to pair together two (or more) interviewers within a primary sampling unit (or within some latter stage sampling unit) and to randomly assign the corresponding elementary units to the interviewers. See, e.g., Bailar (1979). Depending upon how the randomization of assignments is actually accomplished, it will be possible to estimate various components of variance in addition to estimating the total variance.

Computer Software for Variance Estimation[1]

To implement the methods of variance estimation described in this book, one needs to have computer software of known quality and capability. One can write original software for this purpose or purchase a commercially available software package. The choice between writing and purchasing the software raises complex questions that we shall treat briefly in this appendix.

As this book goes to press, computer software for the analysis of complex sample survey data must be regarded as in its infancy. Developments are occurring rapidly, however, and progress is being made.

As of June 1984 there were no less than 14 different programs or program packages for variance estimation for complex sample surveys. A list of the programs and the affiliated institutions is presented in Table E.1. This list only includes programs or program packages that are portable to some degree and that were, are, or have some expectation of becoming commercially available. Even within this list, the programs vary considerably with respect to portability and availability. In addition, there are hundreds of special-purpose, nonportable, nonavailable variance programs maintained by the various survey organizations around the world. But no attempt is made here to catalogue these programs.

Before implementing any of these software packages, the potential user needs to have a fairly clear idea of the characteristics and features of "good" software. This information is needed in order to appraise the quality and capabilities of the alternative software packages so that an informed decision can be made about which package is best for a particular application. The

---

[1] This appendix is partially based upon a series of papers by Francis, Kaplan, and Sedransk and a recent unpublished paper by Smith. See Francis and Sedransk (1976, 1979); Kaplan, Francis, and Sedransk (1979a, 1979b); Kaplan (1979); and Smith (1984).

Table E.1. Variance Estimation Programs

Program Name	Affiliated Institution
A. BELLHOUSE	University of Western Ontario
B. CAUSEY	U.S. Bureau of the Census
C. CLUSTERS	World Fertility Survey
D. FINSYS-2	Colorado State University and U.S. Forest Service
E. HESBRR	U.S. National Center for Health Statistics
F. NASSTIM, NASSTVAR	Westat, Inc.
G. OSIRIS IV	University of Michigan
H. PASS	U.S. Social Security Administration
I. RGSP	Rothamsted Experimental Station
J. SPLITHALVES	Australian Bureau of Statistics
K. SUDAAN	Research Triangle Institute
L. SUPER CARP	Iowa State University
M. U–SP	University of Kent
N. VTAB and SMED83	Swedish National Central Bureau of Statistics

following characteristics and features are potentially important:

1. Input

a. Flexibility
b. Calculation of weights
c. Finite correction terms
d. Convenient to learn and use
e. Good recoding system
f. Missing value codes

2. Output

a. Echo all user commands
b. Clear labeling
c. Documentation of output clear, concise, self-explanatory
d. Options of providing estimates by stratum, cluster group, various stages of sampling

3. Accuracy

a. Computational
b. Appropriateness

4. Cost or efficiency

Here is what Francis and Sedransk (1979) say about these characteristics:

> Ideally it [the software package] should have great flexibility in dealing with various designs. The program should allow the user to describe his design exactly, accounting for strata, clusters, various stages of sampling, and various types of case weighting. The program should also be able to calculate weights from the data, if enough information is present. Finite population correction factors (f.p.c.'s) should be available if a user requests them. In particular, for "collapsed strata" methods, it would be desirable to have an option available for recalculation of new case weights and new f.p.c.'s derived from the original case weights and f.p.c.'s.
>
> If a program is to be of general use it must be reasonably convenient to learn and use. Such a program will not only be more effective, but will be easier to check and debug; and this, in turn, will improve accuracy. A good recoding system would allow for easy calculation of estimates for subpopulations. Missing value codes should exist and the program should be specific about its treatment of missing values, and small sample sizes (e.g., cluster sample sizes of zero or one).
>
> An essential feature is accuracy which depends on two things: the formula used and its computation by the program. First, *computational* accuracy should be required of every program. Second, the *formula* should be appropriate for the sample design employed. For example, in variance estimation an estimate of the variability in the lower stages of the sample should be given, and the effect of all f.p.c.'s should be considered.
>
> The output should echo all the user commands: all options which were specified should be clearly repeated, including a description of the design. The labelling should be clear, and allow the user flexibility in naming his variables. Additional useful output would include: (1) estimates for each stratum and any user-specified group of clusters; (2) design effects and (3) estimates of variability by stage of sampling.
>
> The documentation of the output should be clear, concise and self-explanatory. It should also provide references which clearly explain the statistical techniques programmed.
>
> Finally, since sample surveys frequently involve large amounts of data, the difficult question of efficiency, in terms of I/O and CPU time, must be addressed. Timing and execution costs can be compared. But one must be very careful: different manufacturers' machines will produce different costs; the same machine at different installations will have different charging algorithms; and even on a particular machine these measures are dependent on the machine load at the time of execution.

In addition to appraising a program's capabilities and features with respect to these criteria, one should also consider testing the program on some benchmark data sets where the true answers are known. Such investigation can test a program's computational accuracy and provide insight (at a level of detail not usually encountered in program manuals) into the methodology implemented in the program. To illustrate these ideas, Table E.2 presents six simple benchmark data sets. The design assumed here involves two stages of sampling within L strata. The number of PSUs in the h-th stratum is denoted by N_h and n_h denotes the sample size. M_{hi} denotes the size of the (h, i)-th PSU and m_{hi} denotes the subsampling size.

Table E.2. Six Benchmark Data Sets

I. Data set 1

Stratum		Cluster Number	M_{1i}	m_{1i}	Values of Observations
1	$N_1 = 15\ n_1 = 3$	1	10	5	1, 2, 3, 4, 5
		2	10	5	2, 3, 4, 5, 6
		3	10	5	3, 4, 5, 6, 7

II. Data set 2

Stratum		Cluster Number	M_{hi}	m_{hi}	Values of Observations
1	$N_1 = 15\ n_1 = 3$	1	10	5	1, 2, 3, 4, 5
		2	10	5	2, 3, 4, 5, 6
		3	10	5	3, 4, 5, 6, 7
2	$N_2 = 15\ n_2 = 3$	1	10	5	1, 2, 3, 4, 5
		2	10	5	2, 3, 4, 5, 6
		3	10	5	3, 4, 5, 6, 7
3	$N_3 = 15\ n_3 = 3$	1	10	5	1, 2, 3, 4, 5
		2	10	5	2, 3, 4, 5, 6
		3	10	5	3, 4, 5, 6, 7

III. Modified[a] data set 1

Stratum		Cluster Number	M_{41}	m_{41}	Values of Observations
4	$N_4 = 15\ n_4 = 1$	1	10	5	1, 2, 3, 4, 5

IV. Modified[a] data set 2

Stratum		Cluster Number	M_{41}	m_{41}	Values of Observations
4	$N_4 = 1 \; n_4 = 1$	1	10	5	1, 2, 3, 4, 5

V. Modified[a] data set 3

Stratum		Cluster Number	M_{4i}	m_{4i}	Values of Observations
4	$N_4 = 14 \; n_4 = 3$	1	10	5	1, 2, 3, 4, 5
		2	10	5	2, 3, 4, 5, 6
		3	10	1	3

VI. Modified[a] data set 4

Stratum		Cluster Number	M_{4i}	m_{4i}	Values of Observations
4	$N_4 = 15 \; n_4 = 3$	1	10	5	1, 2, 3, 4, 5
		2	10	5	2, 3, 4, 5, 6
		3	1	1	3

Source: Francis and Sedransk (1979).

[a] The first three strata of data sets III, IV, V and VI are identical to data set 2.

For these small data sets, one is able to compute true answers by hand and compare them to answers produced by a software package.

Benchmark data sets should be chosen so as to test as many features of the software as possible. Our data sets I and II are rather straightforward and should produce few surprises. Data set III may be revealing because only one primary unit is selected in the fourth stratum. A program would need to do some collapsing of strata in order to produce a variance estimate. Data set IV contains a self-representing PSU and its treatment should be checked. Data sets V and VI also contain samples of size one, but at the second stage of sampling instead of at the first stage.

In addition to assessing the capabilities of the software, the potential user, purchaser, or developer needs to assess carefully the needs and requirements of their particular applications. Some key issues are

(a) Are computations needed for one or many kinds of survey designs?
(b) Will the surveys be one-time or recurring?
(c) Are computations to be limited to simple tabulations and associated variance estimates or will further statistical analysis of the data be undertaken?
(d) What kind of user is expected?
(e) What kind of hardware environment is anticipated?
(f) Is the software maintained by a reliable organization?
(g) What kind of internal support can be provided for the software?
(h) What are the costs of the software? Initial costs? Maintenance costs?

Issues (a)–(h) define the importance to the potential user of the assessment criteria and benchmark tests. Two examples will clarify this situation. First, if only one kind of survey design is anticipated then the importance of software *flexibility* is relatively diminished, whereas if many designs are anticipated then software *flexibility* assumes relatively greater importance. Second, if the main users are skilled mathematical statisticians or experts in statistical computing, then *convenience* (to learn and use) is relatively less important than if the main users are analysts in some other scientific field.

We suggest that the survey statistician assess needs and requirements for software first, and then evaluate the various options of developing or purchasing software in light of the requirements. The characteristics, features, and benchmark data sets cited earlier will be useful in conducting this evaluation.

We close this appendix by giving a brief description of the 14 variance programs cited earlier. We do not attempt comprehensive descriptions, nor do we undertake any comparisons or recommendations. The software markets are undergoing rapid change and enhancements to several of the existing programs are in progress; any detailed discussion or comparison here would surely be out of date before our own publication date. We note, however, that the 14 programs fall into two basic groups. Group I consists of OSIRIS IV, SUDAAN, and SUPER CARP, which are state-of-the-art, well-supported, portable programs. The remaining programs, Group II, are either ill-supported; undergoing preliminary development; or are not portable. Most potential users would be advised to concentrate their attentions on the Group I programs. Meanwhile, one eye should be kept on new developments in Group II, including BELLHOUSE, NASSTIM, and U-SP.

Each of the 14 program descriptions is followed by a table giving essential information on cost, documentation, source languages, hardware systems, operating systems, and name and address of the program's developer and distributor.

A. BELLHOUSE

This program is being developed at the University of Western Ontario by David Bellhouse. A prototype of the program may be available by fall 1985. At the present time only a few general details are available. The variance estimating methodology involves Theorems 11.1 and 11.2 of Cochran (1977), wherein sums of squares are calculated both within and between primary sampling units and are combined in a linear combination using a tree transversal algorithm to provide an estimate of the total variance. This methodology is closely related to the random group method, except the random group method does not typically involve the within sum of squares. For nonlinear estimators, the above methodology is used in combination with the appropriate Taylor series formula.

Table E.3. BELLHOUSE

Developed by:	Distributed by:
David Bellhouse	Same
Department of Statistical and Actuarial Science	
University of Western Ontario	
London	
Ontario N6A 5B9	
Canada	
Compatible with following computer systems:	Operating systems:
PRIME	
CDC CYBER	
Source Languages:	
FORTRAN	
Cost:	
Documentation:	
Program and documentation are currently under development.	

B. CAUSEY

This program was developed by Beverley Causey and Ralph Woodruff at the U.S. Bureau of the Census, but it is no longer supported nor available for distribution. This program is extremely flexible in the kinds of sampling designs and estimators that it can handle. A stratified design is assumed, and the variance within stratum is calculated according to simple random sampling, or from a user supplied subroutine for any other sampling design.

Table E.4. CAUSEY

Developed by:	Distributed by:
Beverley Causey	Can no longer be obtained,
Statistical Research Division	nor is it supported by
Census Bureau	its U.S. developer.
Washington, DC 20233	
Compatible with following computer systems:	Operating system:
UNIVAC	EXEC 8
Source languages:	
FORTRAN	
Cost:	
Not applicable	
Documentation:	

Woodruff, R. S. and Causey, B. D. (1976). Computerized Method for Approximating the Variance of a Complicated Estimate. *Journal of the American Statistical Association* **71** (354), 315–321.

Taylor series methods are used to estimate the variance of *any* nonlinear statistic. The program evaluates derivatives numerically, and the user must supply the main program and the function to be differentiated.

C. CLUSTERS

CLUSTERS (Computation and Listing of Useful Statistics on Errors of Sampling). was developed for the World Fertility Survey, International Statistical Institute, by V. Verma and M. Pearce. This program will estimate standard errors for clustered sampling designs. It applies to a variety of basic descriptive statistics, including proportions, means, ratios, and differences of proportions, means and ratios. Estimates and corresponding variance estimates can be made for the full sample or for subclasses of the sample. Random group variance estimators seem to be utilized in CLUSTERS, and Taylor series methods are used with the random group estimators to obtain variance estimates for ratio statistics. CLUSTERS allows for exclusion of exceptional values. The user can edit data by specifying values of a variable for which the observations with that value should not be included in the calculations. CLUSTERS also has recoding facilities that are useful for forming subclass estimates. The program requires a sort routine, such as the IBM SORT-MERGE procedure.

Table E.5. CLUSTERS

Developed by:	Distributed by:
V. Verma	Same
World Fertility Survey	
International Statistical Institute	
Voorburg,	
The Netherlands	

Compatible with following   Operating system:
  computer systems:

IBM 360/370		OS/DOS
HP 3000	MPE 11/111	
ICL 1900	GEORGE 2/3	
ILL 2900	SCOPE/KRONOS	
CDC 6000		

Source languages:
FORTRAN IV

Cost:

There is a one-time charge of $50.00 for universities, governments and nonprofit
  organizations, and commercial use.

Documentation:

International Statistical Institute (1978). *Users Manual for CLUSTERS.* WFS/Tech.
  770.

D. FINSYS-2

This system was developed by W. E. Frayer at the College of Forestry and
Natural Resources, Colorado State University, for use by the U.S. Forest
Service. It is restricted to certain sampling designs, including simple random,
stratified, double sampling, and sampling with partial replacement. Three
programs are included: EDIT-2, TABLE-2, and OUTPUT-2. EDIT-2 is an
independent editing and file updating system which is designed to apply a
specific set of data checks and cross-checks, record-by-record. The TABLE-2
program forms tables of means, variances, and covariances. And OUTPUT-2
prints tables of data with title, row, and column headings. The FINSYS
system was developed primarily for applications to forestry problems.

Table E.6. FINSYS-2

Developed by:	Distributed by:
W. E. Frayer	J. Barnard
School of Forestry	N.E. Forest Experiment Station
Colorado State University	U.S. Forest Service
Fort Collins, Colorado	Broomall, PA 19008
Compatible with following computer systems:	Operating systems:

Any computer with compiler for FORTRAN
  IV or higher order FORTRAN.
Minimum of 6 input/output units
  required.

Source languages:
FORTRAN

Cost:
Available at cost (tape cost) from Barnard.

Documentation:
User manuals published by the U.S. Forest Service and available from Barnard.

E. HESBRR

This program was developed by Gretchen Jones at the U.S. National Center
for Health Statistics (NCHS). It was developed originally to display and
analyze data from the Health Examination Survey. HESBRR is restricted
to sampling designs with n units (usually 2) selected per stratum. The
program can accommodate estimators such as means, proportions, and
ratios. The variance estimating methodology is the balanced half-sample
method. This methodology is used for both linear and nonlinear estimators.
There are extensive table generation facilities, and the program allows for
recoding, editing, and labeling.

Table E.7. HESBRR

Developed by:	Distributed by:
Gretchen K. Jones	Same
NCHS, Rm. 2-12	
3700 East-West Highway	
Hyattsville, MD 20782	

Table E.7. continued

Compatible with following computer system:	Operating systems:
IBM/370-168	VS

Source languages:
PL/1 version 5.4
Programs are executed using JCL catalogued procedure.

Cost:
No charge

Documentation:
HESBRR
(HES Variance and Crosstabulation Program)
Version 3 (Issued 01/31/83)

F. NASSTIM and NASSVAR

This package was developed at Westat, Inc. by D. Morgenstein. It operates
in a SAS (Statistical Analysis System) environment and uses the balanced
half-sample method for variance estimation. Any two-per-stratum sampling
design can be handled by this program, and by appropriate grouping or
collapsing other designs can be handled as well. Self-representing primary
sampling units are handled by designating pairs of half-samples within each
such unit. NASSTIM and NASSVAR grammar is similar to that used in
standard SAS procedures. The full range of SAS arithmetic operators and

Table E.8. NASSTIM and NASSVAR

Developed by:	Distributed by:
David Morgenstein	Same
Vice President	
Westat, Inc.	
1650 Research Boulevard	
Rockville, MD 20850	
Compatible with following computer systems:	Operating systems:
IBM and	
IBM look alikes that support SAS	

Source languages:

Cost:

Documentation:

functions may be used with this system. Estimates and variance estimates may be calculated for such survey statistics as proportions, means, ratios, and other arithmetic functions available in SAS. The system default is to produce estimates of total variance. There is an optional procedure known as "WITHIN" that provides a capability for producing estimates of components of variance.

G. OSIRIS IV

This system was developed by L. Kish, M. Frankel, and N. Van Eck at the Survey Research Center, University of Michigan. It consists of two main programs for estimating variances for complex survey designs. The &PSALMS program produces variance estimates for estimated totals, ratios, and differences of ratios. It uses the Taylor series methodology together with simple sums of squares and cross products. The &REPERR program produces variance estimates for means, correlation coefficients, and regression coefficients, and utilizes replication methods of estimation. Three alternative forms of replication are available, including random groups, balanced half-samples, and jackknife. The OSIRIS IV system is capable of handling large stratified, multistage designs. Variance estimation requires at least two primary selections per stratum, and when a sampling design does not conform to this restriction the user must first invoke the collapsed stratum technique. The system has recoding facilities that interface with &PSALMS and &REPERR. There are facilities for subclass designation and in a single run one can produce a range of estimates for both the total sample and an unlimited number of subclasses. The balanced half-sample procedure inherent in &REPERR is limited to between 4 and 88 strata. Some form of collapsing or partial balancing is needed to accommodate large numbers of strata. Missing data are deleted from the computations separately for each statistic under study.

Table E.9. OSIRIS IV

Developed by:	Distributed by:
Leslie Kish	Same
The Institute for Social Research	
Survey Research Center	
University of Michigan	
Ann Arbor, MI 48106	
Compatible with following computer systems:	Operating systems:
IBM 360/370 or	MTS
IBM compatible such as	OS/360
AMDAHL 470 V/6	MVS
	or equivalent

Table E.9. (continued)

Source languages:
FORTRAN IV, IBM ASSEMBLER

Cost according to the following schedule:

first year	annual renewal	
$2400	$1800	basic fee
$1600	$1200	government agencies & nonprofit institutions
$1200	$900	for institutions granting degrees
$900	$675	for Inter-university Consortium for Political and Social Research (ICPSR) members

Documentation:

Computer Support Group (1982). *OSIRIS VI: Statistical Analysis and Data Management Software System.* Survey Research Center, Institute for Social Research, University of Michigan.

Computer Support Group (1980). *Sampling Error Analysis in OSIRIS IV.* Survey Research Center, Institute for Social Research, University of Michigan.

H. PASS

PASS (Processor for Analysis of Statistical Surveys) was developed by R. Finch and D. Thompson of the U.S. Social Security Administration. It is capable of handling such survey design features as clustering and stratification and parameters such as means, totals, and ratios. Several variance estimating methodologies are available at the users option, including random groups, the collapsed stratum method, and the Taylor series method. PASS employs some UNIVAC dependent features. It is not portable to other computer systems at this time, although there is a plan to make it portable in the near future. The program documentation is unclear as to the precise range of designs and estimators that PASS is capable of handling.

Table E.10. PASS

Developed by:	Distributed by:
Don Thompson	Same
Division of Specialized Software Services	
Room 1120 Woodlawn Drive Building	
Social Security Administration	
6401 Security Boulevard	
Baltimore, MD 21235	

Table E.10. (continued)

Compatible with following computer systems:	Operating systems:
UNIVAC 1108	EXEC 8, Level 37R2

Source languages:
FORTRAN V (field data)
ASM Level 15RI
UNIVAC FURPUR Processor, Level 27R3A

Cost:

Documentation:

Division of Specialized Software Systems (1983). *PASS User Guide.* Office of Data Services, Social Security Administration, Baltimore, MD.

I. RGSP

The Rothamsted General Survey Program (RGSP) is a versatile package for tabulating survey and other data, including hierarchical data, and manipulating and printing the resulting tables. Composite tables, made up of the whole or parts of other tables, can be constructed; the printing facilities are very flexible. The system of table files enables tables to be condensed and rearranged for final presentation, after examination of preliminary prints, and permits surveys covering several regions or repeated at intervals to be tabulated batch by batch as the data become available,

Table E.11. RGSP

Developed by:	Distributed by:
F. Yates	RGSP Secretariate
Statistics Department	Computing Unit
Rothamsted Experimental Station	Rothamsted Experimental Station
Harpenden	Harpenden
Hertfordshire AL5 2JQ	Hertfordshire AL5 2JQ
England	England
Compatible with following computer systems:	Operating systems:
ICL 4-70	MULTIJOB
ICL 1900	GEORGE II
ICL 2900	VME
ICL ME29	VME
IBM 360/370	- - -
CDC 6400	- - -
DEC 10	- - -

Table E.11. (continued)

DEC VAX	VMS VSN 3
IBM 4331	- - -
PRIME	PRIMOS

Source languages:
FORTRAN 66

Cost:

Fee	Concern
£50 media charge (includes 1 doc. set)	British Universities
£500 + VAT	British Government Ministries and agencies
£500	Universities foreign to Great Britain
£1000 + VAT	Other bodies

Documentation:

Documentation is available from Rothamsted at the following prices:

Reference Manual, Part 1	£10.00
Reference Manual, Part 2	£5.00

A general description is included in the 4th edition of *Sampling Methods for Censuses and Surveys* by F. Yates (Griffin, 1981). This does not cover some recent improvements to Part 1, of which the standard execution program is the most important.

---

with later summarization. An updating facility enables late returns and corrections to be incorporated without program alteration or retabulation of the earlier data. Standard errors are not provided automatically, but can be calculated for virtually any type of design, including hierarchical data. Interfaces with GLIM and GENSTAT are provided.

RGSP was developed by F. Yates, formerly Head of the Rothamsted Statistics Department, and has been extended in partnership with J. D. Beasley and B. M. Church. It has been in use at Rothamsted and other agricultural institutes since 1972, and is available commercially. Though no longer formally supported, advice may be sought and RES will respond when possible. There are no current plans for further development. If conversion is desired to a computer for which no version is currently available, terms must be negotiated with Rothamsted. Full conversion instructions and test programs are available.

J. SPLITHALVES

At the Australian Bureau of Statistics (ABS), variance estimation is handled by the SPLITHALVES programs, called E28, E31, E32, E33, and E34; by an ultimate cluster program, E35; and by the program E37, QUANTILES

VARIANCE ESTIMATION. These program modules estimate variances of levels, ratios, differences, and differences of ratios. The programs are a part of ABS's Survey Facilities system, which provides a wide range of capabilities for the design and analysis of survey data. The system is integrated with the Table Producing Language (TPL), developed at the U.S. Bureau of Labor Statistics, and with SAS, so that variances may be computed and reports prepared in a single run. The documentation is readable and gives sufficient detail in a clear and straightforward manner so as to make it understood. The Survey Facilities system utilizes the ABS data dictionary and JOL (Job Organization Language) for job control and job submission user interface, and as a result, is not easily portable. A potential user would need to provide surrogates for the dictionary and for JOL.

Table E.12. SPLITHALVES

Developed by:	Distributed by:
Australian Bureau of Statistics	J. R. Pryor
P.O. Box 10	Australian Bureau of Statistics
Belconnen A.C.T. 2616	
Australia	

Compatible with following computer systems: Operating systems:

FACOM M382
IBM compatible systems

Source languages:
PL1, SAS

Cost:
Users are charged for providing them with a magnetic tape.

Documentation:
User manuals available from the Australian Bureau of Statistics.

K. SUDAAN

The SUDAAN system was developed at the Research Triangle Institute by B. V. Shah. It is adequate for handling most survey designs with stratification or clustering. At the present time SUDAAN consists of four general procedures, and these can be used only in conjunction with the SAS system. The Institute has long range plans to develop SUDAAN further for the analysis of survey data. The procedures can handle estimation and variance estimation for means, proportions, ratios and regression coefficients. The variance methodology involves Taylor series approximations in conjunction with textbook-type variance formulas (see Chapter 1).

The four procedures are:

1. SESUDAAN—for computing standard errors for totals, means, and proportions of the type $\sum wy/\sum w$, or their differences by various subgroups of the population.
2. RATIOEST—for computing standard errors for ratio estimators of the type $\sum wy/\sum wx$;
3. SURREGR—for computing standard errors for regression coefficients;
4. RTIFREQS—for computing standard errors for weighted frequencies and percentages.

Table E.13. SUDAAN

Developed by:	Distributed by:
B. V. Shah	Same
Research Triangle Institute	
P.O. Box 12194	
Research Triangle Park	
North Carolina 27709	
Compatible with following computer systems:	Operating systems:
Compatible with computers which can accommodate SAS	OS/MVS 360/370

Source languages:

FORTRAN, ASSEMBLER
SUDAAN consists of several SAS procedures

Cost:

$500 per year

Documentation:

Shah, B. V. (1981). *SESUDAAN: Standard Errors Program for Computing of Standardized Rates from Sample Survey Data.* Research Triangle Institute, NC.
Shah, B. V. (1982). *RTIFREQS: Program to Compute Weighted Frequencies, Percentages, and their Standard Errors.* Research Triangle Institute, NC.
Shah, B. V. (1981). *RATIOEST: Standard Errors Program for Computing of Ratio Estimates from Sample Survey Data.* Research Triangle Institute, NC.
Holt, M. M. (1979). *SURREGR: Standard Errors of Regression Coefficients from Sample Survey Data.* Research Triangle Institute, NC.

L. SUPER CARP

SUPER CARP (Cluster Analysis and Regression Program) was developed at Iowa State University by W. Fuller, M. Hidiroglou, and R. Hickman. This program computes estimates and variance estimates for multistage, stratified sampling designs with arbitrary probabilities of selection. Most sampling designs used in practice will fall within this general framework. It can handle estimated totals, means, ratios, and differences of means or ratios. Ratio and regression estimators are included. For variance estimation, SUPER CARP mainly relies upon the unbiased estimators (as found in Chapter 1). For nonlinear statistics, the Taylor series method is used for variance estimation. Collapsed stratum options are available in case of small sample sizes within some of the strata. The program can be adopted to compute the Yates–Grundy formula for unequal probability sampling, two-per-stratum, without replacement. SUPER CARP computes estimates and variance estimates both for the total sample and for any user specified subpopulations. There are data editing or screening facilities built into SUPER CARP. There are also extensive facilities for regression analysis, with various assumed error structures. There are algorithms for the solution of errors-in-variables problems and for the analysis of categorical data.

Table E.14. SUPER CARP

Developed by:	Distributed by:
Wayne Fuller	Same
Department of Statistics	
Iowa State University	
Ames, Iowa 50010	
Compatible with following computer systems:	Operating systems:
IBM 360 and 370	OS, TSO,
UNIVAC 1100	and others
Any computer with a FORTRAN compiler.	
Source languages:	
FORTRAN G	
Cost:	
$150.00	
Documentation:	
Hidiroglou, M. A., Fuller, W. A., and Hickman, R. D. (1980). *SUPER CARP.* Iowa State University, Ames, Iowa.	

M. U–SP

The U–SP program is a microcomputer-based package for the analysis of survey data. It was developed at the University of Kent at Canterbury by a project team consisting of G. B. Wetherill, P. M. North, C. Daffin, P. Duncombe, and M. G. Hills. The main emphasis in developing U–SP was placed on user friendliness. The package works interactively through a series of menus and prompts presented to the user. Little technical expertise, either statistical or computer, is required to use this system. U–SP mainly handles general analysis of survey data. Its capabilities for variance estimation have not yet been fully developed, and we have no information at this time as to the methodologies that are or will be employed. As the variance estimating capabilities are completed, however, this should be a very attractive package. U–SP has been in use by several island nations in the Pacific.

Table E.15. U–SP (A User-Friendly Survey Analysis Package)

Developed by:	Distributed by:
Applied Statistics Research Unit	Applied Statistics Research
Mathematical Institute	Unit
University of Kent	Mathematical Institute
Canterbury, Kent, CT2 7NF	University of Kent
England	Canterbury, Kent, CT2 7NF
	England
Compatible with following computer systems:	Operating systems:
Any machine that has APL68000	MIRAGE, or any system with
(for the Motorola 68000 chip).	which APL68000 is
Minimum RAM requirement is 256K	compatible.
Source languages:	
APL	

Cost: (Approximate) Multi-user systems: main frames, minis £ Sterling 2500
                                              Multi-user micros £ Sterling 1500
                                              Single-user systems £   750

Documentation:

Not yet finalized.

N. VTAB and SMED 83

The VTAB program was developed by the Swedish National Central Bureau of Statistics. It is part of a package for table production and it is highly flexible for subgroups. Variance estimates are calculated for all types of table estimates such as totals, percentages and means. Summation of sub-units is possible up to the sampling unit level before variance calculations. For nonlinear statistics, variance estimates are obtained by the Taylor series formula. Four options are available for different estimators and sampling plans: (1) simple or stratified random sampling, (2) stratified random sampling with subsampling among nonrespondents, (3) grouped stratified sampling—variance estimator is obtained by the collapsed stratum technique, (4) the simple variance among the units belonging to the same cell of a table.

SMED83 is based on the TAB68-system for table production developed by Statistics Sweden. This makes it possible to specify the table in a very free manner. In the same table it is possible to have estimates of totals, means and ratios and their estimated standard errors or coefficient of variation. The sample designs considered are stratified random sampling without replacement. The possibility of calculating estimates and their approximate standard errors when subsampling among nonrespondents will be implemented during 1985.

Table E.16A. VTAB

Developed by:	Distributed by:
Statistics Sweden	No longer supported or distributed.
Compatible with following computer systems:	Operating systems:
IBM 360/370	OS/VS
Source languages:	
FORTRAN	
Cost:	
Documentation:	
Only in Swedish.	

Table E.16B. SMED83

Developed by:	Distributed by:
Statistics Sweden	Statistics Sweden
Compatible with following computer systems:	Operating systems:
IBM 370 or compatible	MVS
Source languages:	
PL/1	
Cost:	
After agreement	
Documentation:	
Handbook in English for TAB68. The extensions for SMED83 is described in an appendix in Swedish but can be translated to English if needed.	

References

Anderson, T. W. (1958). *An Introduction to Multivariate Statistical Analysis*. New York: John Wiley and Sons.

Arvesen, J. N. (1969). Jackknifing U-Statistics. *Annals of Mathematical Statistics* **40**, 2076-2100.

Arvesen, J. N. and Schmitz, T. (1970). Robust Procedures for Variance Component Problem Using the Jackknife. *Biometrics* **26**, 677-686.

Bailar, B. (1968). *Effects of Interviewers and Crew Leaders*. U.S. Bureau of the Census, Series ER 60, No. 7, Washington, D.C. 20233.

——— (1979). *Enumerator Variance in the 1970 Census*. U.S. Bureau of the Census, PHC(E)-13, Washington, D.C. 20233.

Barnett, F. C., Mullen, K., and Saw, J. G. (1967). Linear Estimates of a Population Scale Parameter. *Biometrika* **54**, 551-514.

Bartlett, M. S. (1947). The Use of Transformations. *Biometrics* **3**, 39-52.

Baumert, L, Golomb, S. W., and Hall, Jr., M. (1962). Discovery of a Hadamard Matrix of Order 92. *Bulletin of American Mathematical Society* **68**, 237-238.

Bean, J. A. (1975). Distribution and Properties of Variance Estimators for Complex Multistate Probability Sample. *Vital and Health Statistics*, Series 2, No. 65, National Center for Health Statistics, Public Health Service, Washington, D.C.

Bickel, P. J. and Doksum, K.A. (1981). An Analysis of Transformation Revisited. *Journal of the American Statistical Association* **76**, 296-311.

Bishop, Y., Fienberg, S., and Holland, P. (1975). *Discrete Multivariate Analysis*. MIT Press, Cambridge, Mass.

Bissell, A.F. and Ferguson, R. A. (1975). The Jackknife—Toy, Tool, or Two-edged Weapon?. *The Statistician* **24**, 79-100.

Borack, J. (1971), A General Theory of Balanced 1/N Sampling. Unpublished Ph.D. dissertation, Cornell University, Ithaca, NY.

Box, G. E. P. and Cox, D. R. (1964). An Analysis of Transformations. *Journal of the Royal Statistical Society, B* **43**, 177-182.

——— (1982). An Analysis of Transformation Revisited, Rebutted. *Journal of the American Statistical Association* **77**, 177-182.

Brillinger, D. R. (1964). The Asymptotic Behavior of Tukey's General Method of Setting Approximate Confidence Intervals (the Jackknife) When Applied to Maximum Likelihood Estimates. *Review of the International Statistical Institute* **32**, 202–206.

——— (1966). The Application of the Jackknife to the Analysis of Sample Surveys. *Commentary* **8**, 74–80.

Brooks, C. (1977). The Effect of Controlled Selection on the Between PSU Variance in the CPS. Unpublished manuscript, U.S. Bureau of the Census, Washington, D.C. 20233.

Bryant, E. E., Baird, J. T., and Miller, H. W. (1973). Sample Design and Estimation Procedures for a National Health Examination Survey of Children. *Vital and Health Statistics*, Series 2, No. 43, National Center for Health Statistics, Public Health Service, Washington, D.C.

Cahoon, L. (1977). CPS Variances—Parameters for Variances of Estimates of Level for Employment and Earnings. Unpublished Memorandum, U.S. Bureau of the Census, Washington, D.C. 20233.

Campbell, C. (1980). A Different View of Finite Population Estimation. *Proceedings of the Section on Survey Research Methods*, American Statistical Association, 319–324.

Carlson, M. D. (December 1974). The 1972–73 Consumer Expenditure Survey. *Monthly Labor Review*, **97**, 16–34.

Cassel, C. M., Sarndal, C. E., and Wretman, J. H. (1977). *Foundations of Inference in Survey Sampling*. New York: John Wiley and Sons.

Causey, B. D. (1982). One versus two per stratum. Unpublished manuscript, U.S. Bureau of the Census, Washington, D.C. 20233.

Chakrabarty, R. P. and Rao, J. N. K. (1968). The Bias and Stability of the Jackknife Variance Estimator in Ratio Estimation. *Journal of the American Statistical Association* **63**, 748.

Chapman, D. W. (1966). An Approximate Test for Independence Based on Replications of a Complex Sample Survey Design. Unpublished M.S. thesis, Cornell University.

Chapman, D. W. and Hansen, M. H. (1972). Estimating the Variance of Estimates Computed in the School Staffing Survey. A project prepared for the School Staffing Survey project, carried out under contract with Westat Research, Inc. for The U.S. Office of Education, Washington, D.C.

Cochran, W.G. (1946). Relative Accuracy of Systematic and Stratified Random Samples for a Certain Class of Populations. *Annals of Mathematical Statistics* **17**, 164–177.

——— (1977). *Sampling Techniques*, 3rd ed. New York: John Wiley and Sons.

Cox, D. R. (1954). The Mean and Coefficient of Variation of Range in Small Samples from Non-Normal Populations. *Biometrika* **41**, 469–481. Correction **42**, 277.

Cressie, N. (1981). Transformations and the Jackknife. *Journal of the Royal Statistical Society, B* **43**, 177–182.

Dalenius, T. (1957). *Sampling in Sweden*. Stockholm: Almquist and Wicksell.

Dalenius, T. and Hodges, Jr., J. L. (1959). Minimum Variance Stratification. *Journal of the American Statistical Association* **54**, 88–101.

David, H. A. (1970). *Order Statistics*. New York: John Wiley and Sons.

Deming, W. E. (1950). *Some Theory of Sampling*. New York Dover Publications.

——— (1956). On Simplification of Sampling Design through Replication with Equal Probabilities and Without Stages. *Journal of the American Statistical Association* **51**, 24–53.

——— (1960). *Sample Design in Business Research*. New York: John Wiley and Sons.

——— (1963). On the Correction of Mathematical Bias by Use of Replicated Design. *Metrika* **6**, 37–42.

Deng, L. Y and Wu, C. F. (1984). Estimation of Variance of the Regression Estimator. MRC Technical Summary Report # 2758, University of Wisconsin, Madison.

Dippo, C. S. (1981). Variance Estimation for Nonlinear Estimators Based Upon Stratified Samples from Finite Populations. Unpublished Ph.D. dissertation, The George Washington University, Washington, D.C. 20052.

Dippo, C. S. and Wolter, K. M. (1984). A Comparison of Variance Estimators Using the Taylor Series Approximation. *Proceedings of the Survey Research Section, American Statistical Association.*

Dippo, C. S., Fay, R. E., and Morganstein, D. H. (1984). Computing Variances from Complex Samples with Replicate Weights. *Proceedings of the Section on Survey Research Methods,* American Statistical Association.

Durbin, J. (1953). Some Results in Sampling Theory When the Units Are Selected with Unequal Probabilities. *Journal of the Royal Statistical Society, B* **15**, 262–269.

——— (1959). A Note on the Application of Quenouille's Method of Bias Reduction to the Estimation of Ratios. *Biometrika* **46**, 477–480.

——— (1967). Design of Multi-Stage Surveys for the Estimation of Sampling Errors. *Applied Statistics* **16**, 152–164.

Edelman, M. W. (1967). Curve Fitting of Keyfitz Variances. Unpublished memorandum, U.S. Bureau of the Census, Washington, D.C. 20233.

Efron, B. (1979). Bootstrap Methods: Another Look at the Jackknife. *Annals of Statistics* **7**, 1–26.

Efron, B. and Stein, C. (1981). The Jackknife Estimate of Variance. *Annals of Statistics* **9**, 586–596.

——— (1981). Nonparametric Estimates of Standard Error: The Jackknife, the Bootstrap and Other Methods. *Biometrika* **68**, 589–599.

——— (1982). *The Jackknife, the Bootstrap, and Other Resampling Plans.* Philadelphia: Society for Industrial and Applied Mathematics.

Erdos, P. and Renyi, A. (1959). On a Central Limit Theorem for Samples from a Finite Population. *Publ. Math. Inst. Hung. Acad. Sci.* **4**, 49–61.

Ernst, L. (1979). Increasing the Precision of the Partially Balanced Half-Sample Method of Variance Estimation. Unpublished manuscript, U.S. Bureau of the Census, Washington, D.C. 20233.

Ernst, L. (1981). A Constructive Solution for Two-Dimensional Controlled Selection Problems. *Proceedings of the Survey Research Methodology Section,* American Statistical Association.

Folsom, R. E., Bayless, D. L., and Shah, B. V. (1971). Jackknifing for Variance Components in Complex Sample Survey Design. *Proceedings of the Social Statistics Section,* American Statistical Association.

Frankel, M. R. (1971a). Inference From Clustered Samples. *Proceeding of the Social Statistics Section,* American Statistical Association.

——— (1971b). *Inference From Survey Samples.* Ann Arbor: Institute of Social Research, University of Michigan.

Francis, I. and Sedransk, J. (1976). Software Requirements for the Analysis of Surveys. *Proceedings 9th International Biometric Conference,* 228–253.

——— (1979). Comparing Software for Processing and Analyzing Survey Data. *Bulletin of the International Statistical Institute.*

Freeman, D. H. (1975). The Regression Analysis of Data from Complex Sample Surveys: An Empirical Investigation of Covariance Matrix Estimation. Unpublished Ph.D. dissertation, University of North Carolina.

Fuller, W. A., (1970). Sampling with Random Stratum Boundaries. *Journal of the Royal Statistical Society, B* **32**, 209–226.

——— (1975). Regression Analysis for Sample Surveys. *Sankhya, C* **37**, 117–132.

—— (1976). *Introduction to Statistical Time Series.* New York: John Wiley and Sons.

Fuller, W. A. and Isaki, C. T. (1981). Survey Design Under Superpopulation Models. In *Current Topics in Survey Sampling*, D. Krewski, J. N. K. Rao, and R. Platek (eds.). New York: Academic Press.

Gautschi, W. (1957). Some Remarks on Systematic Sampling. *Annals of Mathematical Statistics* **28**, 385-394.

Godambe, V. P. (1955). A Unified Theory of Sampling from Finite Populations. *Journal of the Royal Statistical Society, B* **17**, 269-278.

Gonzalez, M., Ogus, J., Shapiro, G., and Tepping, B. (1975). Standards for Discussion and Presentation of Errors in Survey and Census Data. *Journal of the American Statistical Association, Supplement* **70**, 5-23.

Goodman, R. and Kish, L. (1950). Controlled Selection—A Technique in Probability Sampling. *Journal of the American Statistical Association* **45**, 350-372.

Gray, H. L. and Schucany, W. R. (1972). *The Generalized Jackknife Statistics.* New York: Marcel Dekcer.

Gurney, M. (1969). Random Group Method for Estimating Variances. Unpublished manuscript, U.S. Bureau of the Census, Washington, D.C. 20233.

—— (1970a). McCarthy's Orthogonal Replication for Estimating Variances with Grounded Strata. *Technical Notes No. 3*, 13-16, U.S. Bureau of the Census, Washington, D.C. 20233.

—— (1970b). The Variance of the Replication Method for Estimating Variances for CPS Sample Design. *Technical Notes No. 3*, 7-12, U.S. Bureau of the Census, Washington, D.C. 20233.

Gurney, M. and Jewett, R. S. (1975), Constructing Orthogonal Replications for Variance Estimation. *Journal of the American Statistical Association* **70**, 819-821.

Hajek, J. (1960). Limiting Distributions in Simple Random Sampling From A Finite Population. *Publ. Math. Inst. Hung. Acad. Science* **5**, 361-374.

—— (1964). Asymptotic Theory of Rejective Sampling with Varying Probabilities From A Finite Population. *Annals of Mathematical Statistics* **35**, 1491-1523.

Hall, Jr., M. (1967). *Combinatorial Theory.* Waltham, Mass.: Blaisdell.

Hansen, M. H., Hurwitz, W. N, and Madow, W. G. (1953). *Sample Survey Methods and Theory*, 2 Volumes. New York: John Wiley and Sons.

Hansen, M. H., Hurwitz, W. N., and Bershad, M. A. (1961). Measurement Errors in Censuses and Surveys, *Bulletin of the International Statistical Institute* **38**, Part II, 359-374.

Hansen, M. H., Hurwitz, W. N., and Pritzker, L. (1964). The Estimation and Interpretation of Gross Differences and the Simple Response Variance. In C. R. Rao, ed., *Contributions to Statistics Presented to Professor P. C. Mahalanobis on the Occasion of his 70th Birthday*, Calcutta, Pergamon Press, Ltd., 111-136.

Hanson, R. H. (1978). *The Current Population Survey—Design and Methodology*, Technical Paper 40, U.S. Bureau of the Census, Washington, D.C. 20233.

Harter, H. L. (1959). The Use of Sample Quasi-Ranges in Estimating Population Standard Deviation. *Annals of Mathematical Statistics* **30**, 980-999. Correction **31**, 228.

Hartley, H. O. and Rao. J. N. K. (1962). Sampling With Unequal Probabilities and Without Replacement. *Annals of Mathematical Statistics* **33**, 350-374.

—— (1966). Systematic Samping With Unequal Probability and Without Replacement. *Journal of the American Statistical Association* **61**, 739-748.

Hartley, H. O., Rao, J. N. K., and Kiefer, G. (1969). Variance Estimation with One Unit Per Stratum. *Journal of the American Statistical Association* **64**, 841-851.

Hartley, H. O. and Sielken, R. L. (1975). A Super-Population Viewpoint for Finite Population Sampling. *Biometrics* **31**, 411-422.

Hanurav, T. V. (1966). Some Aspects of Unified Sampling Theory. *Sankhya, Ser. A* **28**, 175-204.

Heilbron, D. C. (1978). Comparison of the Estimators of the Variance of Systematic Sampling. *Biometrika* **65**, 429-433.

Hidiroglou, M. A. (1974). Estimation of Regression Parameters for Finite Populations. Unpublished Ph.D. dissertation, Iowa State University, Ames, Iowa.

Hoeffding, W. (1948). A Class of Statistics with Asymptotically Normal Distributions. *Annals of Mathematical Statistics* **19**, 293-325.

Holst, L. (1973). Some Limit Theorems with Applications in Sampling Theory. *Annals of Statistics* **1**, 644-658.

Horvitz, D. G. and Thompson, D. J. (1952). A Generalization of Sampling Without Replacement from a Finite Population. *Journal of the American Statistical Association* **47**, 663-685.

Huber, P. J. (1972). The Wald 1972 Lecture. Robust Statistics: A Review. *Annals of Mathematical Statistics* **43**, 1041-1067.

Isaki, C. T. and Pinciaro, S. J. (1977). The Random Group Estimator of Variance with Application to PPS Systematic Sampling. Unpublished manuscript, U.S. Bureau of the Census, Washington, D.C. 20233.

Isaki, C. T. and Fuller, W. A. (1982). Survey Design Under the Regression Superpopulation Model. *Journal of the American Statistical Association* **77**, 89-96.

Johnson, N. L. and Kotz, S. (1969). *Discrete Distributions.* Boston: Houghton Mifflin.

——— (1970a). *Continuous Univariate Distributions-1.* New York: John Wiley and Sons.

——— (1970b). *Continuous Univariate Distributions-2.* Boston: Houghton Mifflin.

Jones, H. L. (1974). Jackknife Estimation of Functions of Strata Means. *Biometrika* **61**, 343-348.

Kaplan, B. (1979). A Comparison of Methods and Programs for Computing Variances of Estimators from Complex Sample Surveys. Unpublished master's thesis, Cornell University.

Kaplan, Francis I. and Sedransk, J. (1979a). A Comparison of Methods and Programs for Computing Variances of Estimators from Complex Sample Surveys. *Proceedings of the Section on Survey Research Methods,* American Statistical Association, 97-100.

——— (1979b). Criteria for Comparing Programs for Computing Variances of Estimators from Complex Sample Surveys. *Proceedings of the 12th Annual Symposium on the Interface of Computer Science and Statistics,* 390-395.

Keyfitz, N. (1957). Estimates of Sampling Variance Where Two Units Are Selected from Each Stratum. *Journal of the American Statistical Association* **52**, 503-510.

Kish, L. (1965). *Survey Sampling.* New York: John Wiley and Sons.

Kish, L. and Frankel, M. R. (1968). Balanced Repeated Replication for Analytical Statistics. *Proceedings of the Social Statistics Section,* American Statistical Association.

——— (1970). Balanced Repeated Replication for Standard Errors. *Journal of the American Statistical Association* **65**, 1071-1094.

Koch, G. G. and Lemeshow, S. (1972). An Application of Multivariate Analysis to Complex Sample Survey Data. *Journal of the American Statistical Association* **67**, 780-782.

Koch, G. G. (1973). An Alternative Approach to Multivariate Response Error Models for Sample Survey Data with Applications to Estimators Involving Subclass Means. *Journal of the American Statistical Association* **63**, 906-913.

Koch, G. G., Freeman, D. H., and Freeman, J. L. (1975). Strategies in the Multivariate Analysis of Data from Complex Surveys. *International Statistical Review* **43**, 55-74.

Koop, J. C. (1960). On Theoretical Questions Underlying the Technique of Replicated or Interpenetrating Samples. *Proceedings of the Social Statistics Section, American Statistical Association.*

—— (1968). An Exercise in Ratio Estimation. *The American Statistician* **22**, 29–30.

—— (1971). On Splitting a Systematic Sample for Variance Estimation. *Annals of Mathematical Statistics* **42**, 1084–1087.

—— (1972). On the Derivation of Expected Value and Variance of Ratios Without the Use of Infinite Series Expansions. *Metrika* **19**, 156–170.

Krewski, D. (1978a), On the Stability of Some Replication Variance Estimators in the Linear Case. *Journal of Statistical Planning and Inference* **2**, 45–51.

—— (1978b). Jackknifing U-Statistics in Finite Populations. *Commun. Statisti.-Theory Meth.* **A7**(1), 1–12.

Krewski, D. and Rao, J. N. K. (1981). Inference from Stratified Samples: Properties of the Linearization, Jackknife and Balanced Repeated Replication Methods. *Annals of Statistics* **9**, 1010–1019.

Krewski, D. and Chakrabarty, R. P. (1981). On the Stability of the Jackknife Variance Estimator in Ratio Estimation. *Journal of Statistical Planning and Inference* **5**, 71–78.

Lee, K. H. (1972). The Use of Partially Balanced Designs for Half-Sample Replication Method of Variance Estimation. *Journal of the American Statistical Association* **67**, 324–334.

—— (1973a). Using Partially Balanced Designs for the Half-Sample Method of Variance Estimation. *Journal of the American Statistical Association* **68**, 612–614.

—— (1973b). Variance Estimation in Stratified Sampling. *Journal of the American Statistical Association* **66**, 336–342.

Lemeshow, S. (1976). The Use of Unique Statistical Weights for Estimating Variances with the Balanced Half-Sample Technique. *Proceedings of the Social Statistics Section,* American Statistical Association.

—— (1979). The Use of Unique Statistical Weights for Estimating Variances With the Balanced Half-Sample Technique. *Journal of Planning and Inference* **3**, 315–323.

Levy, P. S. (1971). A Comparison of Balanced Half-Sample Replication and Jackknife Estimators of the Variances of Ratio Estimates in Complex Sampling with Two Primary Sampling Units Per Stratum. Unpublished manuscript, National Center for Health Statistics, Rockville, MD.

Madow, W. G. (1948). On the Limiting Distributions of Estimates Based on Samples from Finite Universes. *Annals of Mathematical Statistics* **19**, 535–545.

Madow, W. G. and Madow, L. G. (1949). On the Theory of Systematic Sampling, I. *Annals of Mathematical Statistics* **15**, 1–24.

Mahalanobis, P. C. (1939). A Sample Survey of the Acreage Under Jute in Bengal. *Sankhya* **4**, 511–531.

—— (1946). Recent Experiments in Statistical Sampling in the Indian Statistical Institute. *Journal of the Royal Statistical Society* **109**, 325–378.

Majumdor, H. and Sen, P. K. (1978). Invariance Principles for Jackknifing U-Statistics for Finite Population Sampling and Some Applications. *Comm. Statist.-Theory Meth.* **A 7**, 1007–1025.

Mann, H. B. and Wald, A. (1943). On Stochastic Limit and Order Relationships. *Annals of Mathematical Statistics* **14**, 217–226.

Marks, E. S., Seltzer, W. and Krotki, K. (1974). *Population Growth Estimation: A Handbook of Vital Statistics Measurement.* Population Council, New York.

Matern, B. (1947). Methods of Estimating the Accuracy of Line and Sample Plot Surveys. *Medd. fr. Statens Skogsforskningsinstitut* **36**, 1–138.

McCarthy, P. J. (1966). Replication: An Approach to the Analysis of Data from Complex Surveys. *Vital and Health Statistics*, Series 2, No. 14, National Center for Health Statistics, Public Health Service, Washington, D.C.

McCarthy, P. J. (1969a). Pseudoreplication: Further Evaluation and Application of the Balanced Half-Sample Technique. *Vital and Health Statistics*, Series 2, No. 31, National Center for Health Statistics, Public Health Service, Washington, D.C.

―――― (1969b). Pseudoreplication: Half-Samples. *Review of the International Statistical Institute* **37**, 239-264.

Mellor, R. W. (1972). Subsample Replication Variance Estimators. Unpublished Ph.D. dissertation, Harvard University.

Miller, R. G., Jr. (1964). A Trustworthy Jackknife. *Annals of Mathematical Statistics* **35**, 1594-1605.

―――― (1968). Jackknife Variances. *Annals of Mathematical Statistics* **39**, 567-582.

―――― (1974a). The Jackknife—A Review. *Biometrika* **61**, 1-15.

―――― (1974b). An Unbalanced Jackknife. *Annals of Mathematical Statistics* **2**, 880-891.

Mosteller, F. (1971). The Jackknife. *Review of the International Statistical Institute* **39**, 363-368.

Mulry, M. H. and Wolter, K. M. (1981). The Effect of Fisher's Z-Transformation on Confidence Intervals for the Correlation Coefficient. *Proceedings of the Survey Research Section*, American Statistical Association.

Nandi, H. K. and Sen, P. K. (1963). Unbiased Estimation of the Parameters of a Finite Population. *Calcutta Statistical Association Bulletin* **12**, 124-147.

Nathan, G. (1973). Approximate Tests of Independence in Contingency Tables from Complex Stratified Samples. *Vital and Health Statistics*, Series 2, No. 53, National Center for Health Statistics, Public Health Service, Washington, D.C.

Osborne, J. G. (1942). Sampling Errors of Systematic and Random Surveys of Cover-Type Areas. *Journal of the American Statistical Association* **37**, 256-264.

Plackett, R. L. and Burman, J. P. (1946). The Design of Optimum Multifactorial Experiments. *Biometrika* **33**, 305-325.

Quenouille, M. H. (1949). Approximate Tests of Correlation in Time Series. *Journal of the Royal Statistical Society*, B **11**, 68-84.

―――― (1956). Notes on Bias in Estimation. *Biometrika* **43**, 353-360.

Raj, D. (1965). Variance Estimation in Randomized Systematic Sampling With Probability Proportional to Size. *Journal of the American Statistical Association* **60**, 278-284.

Raj, D. (1966). Some Remarks on a Simple Procedure of Sampling Without Replacement. *Journal of the American Statistical Association* **61**, 391-396.

―――― (1968). *Sampling Theory*. New York: McGraw-Hill.

Rao, J. N. K. (1963). On Three Procedures of Unequal Probability Sampling Without Replacement. *Journal of the American Statistical Association* **58**, 202-215.

―――― (1965). A Note on the Estimation of Ratios by Quenouille's Method. *Biometrika* **52**, 647-649.

Rao, J. N. K. and Webster, K. (1966). On Two Methods of Bias Reduction in the Estimation of Ratios. *Biometrika* **53**, 571-577.

―――― (1975). Unbiased Variance Estimation for Multistage Designs. *Sankhya*, C, **37**, 133-139.

Rao, P. S. R. S. and Rao, J. N. K. (1971). Small Sample Results for Ratio Estimators. *Biometrika* **58**, 625-630.

Rosen, B. (1972). Asymptotic Theory for Successive Sampling with Varying Probabilities without Replacement, I. *Annals of Mathematical Statistics* **43**, 373-397.

Royall, R. M. and Cumberland, W. G. (1978). Variance Estimation in Finite Population Sampling. *Journal of the American Statistical Association* **73**, 351–358.
———— (1981a). An Empirical Study of the Ratio Estimator and Estimators of Its
Variance. *Journal of the American Statistical Association* **76**, 66–88.
———— (1981b). The Finite-Population Linear Regression Estimator and Estimators
of Its Variance—An Empirical Study. *Journal of the American Statistical Associ-
ation* **76**, 924–930.
Schucany, W. R., Gray, H. L., and Owen, D. B. (1971). On Bias Reduction in
Estimation. *Journal of the American Statistical Association* **66**, 524–533.
Seth, G. R. (1966). On Collapsing Strata. *Journal of the Indian Society of Agricultural
Statistics* **18**, 1–3.
Shah, B. V. (1978). Variance Estimates for Complex Statistics from Multistage
Sample Surveys. *Survey Sampling and Measurement*, N. Krishnan Namboodiri
(ed.). Academic Press: New York.
Shapiro, G. M. and Bateman, D. V. (1978). A Better Alternative to the Collapsed
Stratum Variance Estimate. *Proceedings of the Section on Survey Research
Methods*, American Statistical Association.
Simmons, W. R. and Baird, J. T. (1968). Pseudo-replication in the NCHS Health
Examination Survey. *Proceedings of the Social Statistics Sections*, American
Statistical Association.
Smith, Phillip J. (1984). Variance Estimation Software Packages. Unpublished report,
U.S. Bureau of the Census, Washington, D.C. 20233.
Sukhatme, P. V. and Sukhatme, B. V. (1970). *Sampling Theory of Surveys with
Applications*. Ames: Iowa State University Press.
Tepping, B. J. (1968). Variance Estimation in Complex Surveys. *Proceedings of the
Social Statistics Section*, American Statistical Association.
———— (October 5, 1976), Exhibit No. 40. Before the Interstate Commerce Com-
mission, Southern Railway Company versus Seaboard Coast Line Railroad
Company.
Tomlin, P. (1974). Justification of the Functional Form of the GATT Curve and
Uniqueness of Parameters for the Numerator and Denominator of Proportions.
Unpublished memorandum, U.S. Bureau of the Census, Washington, D.C.
20233.
Tomlin, P., Cahoon, L., and Makens, P. (1977). Current Issues in Generalized
Variance Estimation and Some Documents which Address Those Issues.
Unpublished memorandum, U.S. Bureau of the Census, Washington, D.C.
20233.
Tukey, J. W. (1958). Bias and Confidence in Not Quite Large Samples. *Annals of
Mathematical Statistics* **29**, 614.
United Nations (1949). The Preparation of Sampling Survey Reports. Statistical
Papers, Series C, No. 1, Statistical Office of the United Nations, New York.
U.S. Bureau of the Census (1976). Census of Transportation, 1972. *Volume III
Commodity Transportation Survey, Part 3. Area Statistics, South and West
Regions and U.S. Summary*. U.S. Government Printing Office, Washington, D.C.
20402.
U.S. Bureau of the Census (1976). Census of Wholesale Trade, 1972. *Volume I,
Summary and Subject Statistics*. U.S. Government Printing Office, Washington,
D.C. 20402.
U.S. Department of Labor (September 1976). *Employment and Earnings*, U.S.
Government Printing Office, Washington, D.C. 20402.
U.S. Department of Labor (1978). Consumer Expenditure Survey: Integrated Diary
and Interview Survey Data, 1972–73. *Bulletin* 1992, Bureau of Labor Statistics,
Washington, D.C. 20212.

Wolter, K. M. *et al.* (1976). Sample Selection and Estimation Aspects of the Census Bureau's Monthly Business Surveys. *Proceedings of the Business and Economic Statistics Section,* American Statistical Association.

——— (1984). An Investigation of Some Estimators of Variance for Systematic Sampling. *Journal of the American Statistical Association* **79**, 781–790.

Woodruff, R. S. (1971). A Simple Method for Approximating the Variance of a Complicated Estimate. *Journal of the American Statistical Association* **66**, 411–414.

Wright, D. G. (1973). Description of Sample Design for the 1972 Commodity Transportation Survey. Unpublished Report, U.S. Bureau of the Census, Washington, D.C. 20233.

Wu, C. F. (1981). Estimation in Systematic Sampling with Supplementary Observations. Technical Report No. 644, University of Wisconsin, Madison.

Yates, F. (1949). *Sampling Methods for Censuses and Surveys.* London: Griffin.

Yates, F. and Grundy, P. M. (1953). Selection Without Replacement from Within Strata with Probability Proportional to Size. *Journal of the Royal Statistical Society, B* **15**, 253–261.

Zinger, A. (1980). Variance Estimation in Partially Systematic Sampling. *Journal of the American Statistical Association* **75**, 206–211.

Index